AF212632

METHODS IN MOLECULAR BIOLOGY

Series Editor
John M. Walker
School of Life and Medical Sciences
University of Hertfordshire
Hatfield, Hertfordshire, UK

For further volumes:
http://www.springer.com/series/7651

For over 35 years, biological scientists have come to rely on the research protocols and methodologies in the critically acclaimed *Methods in Molecular Biology* series. The series was the first to introduce the step-by-step protocols approach that has become the standard in all biomedical protocol publishing. Each protocol is provided in readily-reproducible step-by-step fashion, opening with an introductory overview, a list of the materials and reagents needed to complete the experiment, and followed by a detailed procedure that is supported with a helpful notes section offering tips and tricks of the trade as well as troubleshooting advice. These hallmark features were introduced by series editor Dr. John Walker and constitute the key ingredient in each and every volume of the *Methods in Molecular Biology* series. Tested and trusted, comprehensive and reliable, all protocols from the series are indexed in PubMed.

Models in Inflammation and Immunopharmacology

Edited by

Lucas Secchim Ribeiro

German Center for Neurodegenerative Diseases (DZNE), Bonn, Germany

Humana Press

Editor
Lucas Secchim Ribeiro
German Center for Neurodegenerative
Diseases (DZNE)
Bonn, Germany

ISSN 1064-3745 ISSN 1940-6029 (electronic)
Methods in Molecular Biology
ISBN 978-1-0716-5018-9 ISBN 978-1-0716-5019-6 (eBook)
https://doi.org/10.1007/978-1-0716-5019-6

© The Editor(s) (if applicable) and The Author(s), under exclusive license to Springer Science+Business Media, LLC, part of Springer Nature 2026

This work is subject to copyright. All rights are solely and exclusively licensed by the Publisher, whether the whole or part of the material is concerned, specifically the rights of translation, reprinting, reuse of illustrations, recitation, broadcasting, reproduction on microfilms or in any other physical way, and transmission or information storage and retrieval, electronic adaptation, computer software, or by similar or dissimilar methodology now known or hereafter developed.
The use of general descriptive names, registered names, trademarks, service marks, etc. in this publication does not imply, even in the absence of a specific statement, that such names are exempt from the relevant protective laws and regulations and therefore free for general use.
The publisher, the authors and the editors are safe to assume that the advice and information in this book are believed to be true and accurate at the date of publication. Neither the publisher nor the authors or the editors give a warranty, expressed or implied, with respect to the material contained herein or for any errors or omissions that may have been made. The publisher remains neutral with regard to jurisdictional claims in published maps and institutional affiliations.

This Humana imprint is published by the registered company Springer Science+Business Media, LLC, part of Springer Nature.
The registered company address is: 1 New York Plaza, New York, NY 10004, U.S.A.

If disposing of this product, please recycle the paper.

Preface

Scientific knowledge is the result of a long and intricate dialogue between empirical observation and theoretical reflection. In the biomedical sciences, this dialogue often takes the form of carefully designed experimental models, as controlled systems in which complexity is made approachable, and where the mysteries of disease and immunity are rendered into questions that can be tested, refined, and eventually answered. This volume emerges from that tradition, offering a detailed collection of in vitro and in vivo protocols developed to study inflammation, the interactions between host and microorganisms and methods to pharmacologically modulate these systems.

Inflammation is a fundamental biological process, that reflects a coordinated response to danger aimed at restoring homeostasis or, in some cases, exacerbate pathology. Understanding this phenomenon is not only central to our ability to interpret basic physiological mechanisms but is also crucial for the development of new diagnostic tools, therapeutic strategies, and preventive interventions such as vaccines. For that purpose, experimental models serve as bridges between the unseen biological complexity of the living world and the rational constructs of scientific thought. They allow us to operate variables with precision, to observe outcomes in a reproducible manner, and to explore the causal architecture of health and disease. The protocols collected in this book reflect the expertise and refinement of their authors, grounded in biological plausibility and empirical rigor. They are written with technical detail, so that researchers at all levels may reproduce them faithfully and contribute to the accumulation of knowledge through sound experimental practice.

Yet, the use of experimental models, particularly those related to animal research, brings with it not only scientific opportunity but also moral responsibility. Biomedical science has long depended on the use of animals to model the physiological and pathological processes of human biology. From the first immunization experiments conducted by Edward Jenner using cowpox to the testing of vaccines for polio, HIV, and most recently COVID-19, animal experimentation has provided irreplaceable insights. These models have underpinned our understanding of immunological memory, tolerance, cytokine signaling, and the dynamics of pathogen invasion, to name a few. They continue to be indispensable for studying systemic inflammatory responses, tissue-specific immune reactions, and the complex feedback loops that characterize host-pathogen interactions. Nonetheless, the legitimacy of this enterprise rests on the principles of ethical conduct. The scientific community has a profound duty to minimize suffering, ensure humane treatment, and uphold public trust. This ethical reflection is not in conflict with scientific ambition; rather, it must ground it. In an age of accelerating discovery, artificial intelligence, and personalized medicine, the rigor and reproducibility of preclinical models remain essential. They provide the scaffold upon which clinical translation can occur, but they must be designed and interpreted with care, humility, and an awareness of their limitations. No model perfectly recapitulates human disease, as all models are simplifications, shaped by practical constraints and conceptual frameworks. It is the responsibility of the scientist to understand where models illuminate truth and where they may obscure it.

The authors of the chapters in this volume have approached their task with this responsibility in mind. Their contributions span a range of biological systems, methodological approaches, and experimental goals. Some protocols describe long-established techniques, honed by years of application; others offer innovative adaptations or novel platforms. All are characterized by a commitment to precision, reproducibility, and ethical integrity. By providing these protocols in a clear, detailed, and accessible manner, this volume aims to serve as both a practical guide and a conceptual resource. It is intended not only for researchers actively engaged in the study of inflammation, immunopharmacology and host-microorganism interactions but also for educators, clinicians, and policymakers who seek to understand the foundations of experimental design in biomedicine. It is a contribution to the collective enterprise of science, rooted in the belief that knowledge, to be meaningful, must be both rigorously obtained and responsibly applied. Ultimately, the study of disease is the study of life under stress, when biology is confronted with disruption, adaptation, and recovery.

I am very grateful for the opportunity extended by Dr. John Walker and Anna Rakovsky to coordinate this significant compilation for the immunology research community. Their unwavering support, insightful suggestions, and steady guidance throughout the editorial journey have been truly invaluable. Most importantly, this endeavor owes its realization to the outstanding dedication and expertise of the contributing authors, whose hard work and commitment were essential to bringing this project to fruition. May this volume serve as a useful tool, a thoughtful companion, and a reminder of the responsibility we bear as scientists: to seek truth with integrity, to pursue progress with compassion, and to conduct our work with a reverence for both knowledge and life.

Bonn, Germany ***Lucas Secchim Ribeiro***

Contents

Contributors

BRUNO SILVA AMARAL • *Laboratório de Interação Microrganismo-Hospedeiro, Departamento de Microbiologia, Instituto de Ciências Biológicas, Universidade Federal de Minas Gerais, Belo Horizonte-MG, Brazil*

THALES AUGUSTO ANESTINO • *Departamento de Microbiologia, Instituto de Ciências Biológicas, Universidade Federal de Minas Gerais, Belo Horizonte, MG, Brazil*

RAFAELA RIBEIRO ALVARES BATISTA • *Laboratório de Interação Microrganismo-Hospedeiro (LIMHO), Departamento de Microbiologia, Instituto de Ciências Biológicas, Universidade Federal de Minas Gerais, Belo Horizonte, MG, Brazil; Centro de Pesquisa e Desenvolvimento de Fármacos (CPDF), Instituto de Ciências Biológicas, Universidade Federal de Minas Gerais, Belo Horizonte, MG, Brazil*

VIVIANE L. BATISTA • *Departamento de Microbiologia, Instituto de Ciências Biológicas, Universidade Federal de Minas Gerais, Belo Horizonte, MG, Brazil; Grupo de Pesquisa em Arboviroses, Instituto de Ciências Biológicas, Universidade Federal de Minas Gerais, Belo Horizonte, MG, Brazil*

VINÍCIUS AMORIM BELTRAMI • *Departamento de Morfologia, Instituto de Ciências Biológicas, Universidade Federal de Minas Gerais, Belo Horizonte, MG, Brazil; Grupo de Pesquisa em Arboviroses, Instituto de Ciências Biológicas, Universidade Federal de Minas Gerais, Belo Horizonte, MG, Brazil*

ANTÔNIO FELIPE SILVA CARVALHO • *Programa de Pós-Graduação em Análises Clínicas e Toxicológicas, Universidade Federal de Minas Gerais, Belo Horizonte-MG, Brazil*

JORDANA G. A. COELHO-DOS-REIS • *Departamento de Microbiologia, Instituto de Ciências Biológicas, Universidade Federal de Minas Gerais, Belo Horizonte-MG, Brazil*

VICTOR R. M. COSTA • *Institute of Biomedicine and Glycomics, Griffith University, South East Queensland, Australia; Departamento de Microbiologia, Instituto de Ciências Biológicas, Universidade Federal de Minas Gerais, Belo Horizonte, MG, Brazil; Grupo de Pesquisa em Arboviroses, Instituto de Ciências Biológicas, Universidade Federal de Minas Gerais, Belo Horizonte, MG, Brazil*

VIVIAN V. COSTA • *Departamento de Morfologia, Instituto de Ciências Biológicas, Universidade Federal de Minas Gerais, Belo Horizonte, MG, Brazil; Grupo de Pesquisa em Arboviroses, Instituto de Ciências Biológicas, Universidade Federal de Minas Gerais, Belo Horizonte, MG, Brazil*

MANOELA COUTO • *Departamento de Morfologia, Instituto de Ciências Biológicas, Universidade Federal de Minas Gerais, Belo Horizonte, MG, Brazil; Grupo de Pesquisa em Arboviroses, Instituto de Ciências Biológicas, Universidade Federal de Minas Gerais, Belo Horizonte, MG, Brazil*

FELIPE ROCHA DA SILVA SANTOS • *Grupo de Pesquisa em Arboviroses, Instituto de Ciências Biológicas, Universidade Federal de Minas Gerais, Belo Horizonte, MG, Brazil; Departamento de Morfologia, Instituto de Ciências Biológicas, Universidade Federal de Minas Gerais, Belo Horizonte, MG, Brazil*

Renan Vasconcelos da Graça-Filho • *Laboratório de Interação Microrganismo-Hospedeiro (LIMHO), Departamento de Microbiologia, Instituto de Ciências Biológicas, Universidade Federal de Minas Gerais, Belo Horizonte, MG, Brazil; Centro de Pesquisa e Desenvolvimento de Fármacos (CPDF), Instituto de Ciências Biológicas, Universidade Federal de Minas Gerais, Belo Horizonte, MG, Brazil*

Sabrina Berger da Silva • *Departamento de Patologia Geral, Programa de Pós-Graduação em Patologia, Instituto de Ciências Biológicas, Universidade Federal de Minas Gerais, Belo Horizonte, MG, Brazil*

Rafaela das Dores Pereira • *Departamento de Bioquímica e Imunologia, Instituto de Ciências Biológicas, Universidade Federal de Minas Gerais, Belo Horizonte, MG, Brazil*

Larisse de Souza Barbosa Lacerda • *Departamento de Morfologia, Instituto de Ciências Biológicas, Universidade Federal de Minas Gerais, Belo Horizonte, MG, Brazil*

Bárbara Maria de Amorim-Santos • *Laboratório de Interação Microrganismo-Hospedeiro, Departamento de Microbiologia, Instituto de Ciências Biológicas, Universidade Federal de Minas Gerais, Belo Horizonte, MG, Brazil*

Simone de Araújo • *Departamento de Fisiologia e Farmacologia, Instituto de Ciências Biológicas, Universidade Federal de Minas Gerais, Belo Horizonte, MG, Brazil*

Camila Bernardo de Brito • *Laboratório de Interação Microrganismo-Hospedeiro, Departamento de Microbiologia, Instituto de Ciências Biológicas, Universidade Federal de Minas Gerais, Belo Horizonte, MG, Brazil; Laboratório de Imunologia de Doenças Infecciosas, Departamento de Bioquímica e Imunologia, Instituto de Ciências Biológicas, Universidade Federal de Minas Gerais, Belo Horizonte, MG, Brazil*

Ana Caroline de Lima • *Laboratório de Interação Microrganismo-Hospedeiro (LIMHO), Departamento de Microbiologia, Instituto de Ciências Biológicas, Universidade Federal de Minas Gerais, Belo Horizonte, MG, Brazil; Centro de Pesquisa e Desenvolvimento de Fármacos (CPDF), Instituto de Ciências Biológicas, Universidade Federal de Minas Gerais, Belo Horizonte, MG, Brazil*

Aline Silva de Miranda • *Departamento de Morfologia, Instituto de Ciências Biológicas, Universidade Federal de Minas Gerais, Belo Horizonte, MG, Brazil*

Pedro Augusto de Oliveira • *Departamento de Microbiologia, Instituto de Ciências Biológicas, Universidade Federal de Minas Gerais, Belo Horizonte, MG, Brazil*

Carolina Braga de Resende • *Unidade de Pesquisa Clínica, Centro de Pesquisa e Desenvolvimento de Fármacos, Instituto de Ciências Biológicas, Universidade Federal de Minas Gerais, Belo Horizonte-MG, Brazil*

Carla Daiane Ferreira de Sousa • *Laboratório de Interação Microrganismo-Hospedeiro, Departamento de Microbiologia, Instituto de Ciências Biológicas, Universidade Federal de Minas Gerais, Belo Horizonte-MG, Brazil; Department of Neurology with Institute of Translational Neurology, University Hospital Münster, University of Münster, Münster, Germany*

Lirlândia Pires de Sousa • *Departamento de Análises Clínicas e Toxicológicas, Faculdade de Farmácia, Universidade Federal de Minas Gerais, Belo Horizonte-MG, Brazil*

Clara Soares de Souza • *Laboratório de Interação Microrganismo-Hospedeiro, Departamento de Microbiologia, Instituto de Ciências Biológicas, Universidade Federal de Minas Gerais, Belo Horizonte-MG, Brazil*

Luiz Pedro de Souza-Costa • *Departamento de Bioquímica e Imunologia, Instituto de Ciências Biológicas, Universidade Federal de Minas Gerais, Belo Horizonte, MG, Brazil*

Angélica Samer Lallo Dias • *Grupo de Pesquisa em Arboviroses, Instituto de Ciências Biológicas, Universidade Federal de Minas Gerais, Belo Horizonte, MG, Brazil; Departamento de Bioquímica e Imunologia, Instituto de Ciências Biológicas, Universidade Federal de Minas Gerais, Belo Horizonte, MG, Brazil*

Manoela Gonzaga Gontijo do Couto • *Departamento de Morfologia, Instituto de Ciências Biológicas, Universidade Federal de Minas Gerais, Belo Horizonte, MG, Brazil*

Anna Clara Paiva Menezes dos Santos • *Departamento de Microbiologia, Instituto de Ciências Biológicas, Universidade Federal de Minas Gerais, Belo Horizonte, MG, Brazil*

Caio Tavares Fagundes • *Laboratório de Interação Microrganismo-Hospedeiro (LIMHO), Departamento de Microbiologia, Instituto de Ciências Biológicas, Universidade Federal de Minas Gerais, Belo Horizonte, MG, Brazil; Centro de Pesquisa e Desenvolvimento de Fármacos (CPDF), Instituto de Ciências Biológicas, Universidade Federal de Minas Gerais, Belo Horizonte, MG, Brazil*

Franciel Batista Felix • *Departamento de Morfologia, Instituto de Ciências Biológicas, Universidade Federal de Minas Gerais, Belo Horizonte-MG, Brazil*

Lívia Maria Alves Ferreira • *Departamento de Microbiologia, Instituto de Ciências Biológicas, Universidade Federal de Minas Gerais, Belo Horizonte, MG, Brazil*

Luiza de Almeida Queiroz Ferreira • *Departamento de Odontologia Restauradora, Faculdade de Odontologia, Universidade Federal de Minas Gerais, Belo Horizonte, MG, Brazil*

Matheus R. Gonçalves • *Departamento de Microbiologia, Instituto de Ciências Biológicas, Universidade Federal de Minas Gerais, Belo Horizonte, MG, Brazil*

William Antonio Gonçalves • *Departamento de Morfologia, Instituto de Ciências Biológicas, Universidade Federal de Minas Gerais, Belo Horizonte, MG, Brazil; Faculdade de Medicina de Ribeirão Preto (FMRP-USP), Universidade de São Paulo, Ribeirão Preto, SP, Brazil*

Isadora Oliveira Gondim • *Departamento de Morfologia, Instituto de Ciências Biológicas, Universidade Federal de Minas Gerais, Belo Horizonte, MG, Brazil*

Pedro P. G. Guimarães • *Departamento de Fisiologia e Biofísica, Instituto de Ciências Biológicas, Universidade Federal de Minas Gerais, Belo Horizonte-MG, Brazil*

Eduardo Gutseit • *Departamento de Morfologia, Instituto de Ciências Biológicas, Universidade Federal de Minas Gerais, Belo Horizonte-MG, Brazil*

Fabiana Simão Machado • *Departamento de Bioquímica e Imunologia, Instituto de Ciências Biológicas, Universidade Federal de Minas Gerais, Belo Horizonte-MG, Brazil*

Mila Fernandes Moreira Madeira • *Departamento de Microbiologia, Instituto de Ciências Biológicas, Universidade Federal de Minas Gerais, Belo Horizonte, MG, Brazil*

Débora Gonzaga Martins • *Departamento de Morfologia, Instituto de Ciências Biológicas, Universidade Federal de Minas Gerais, Belo Horizonte-MG, Brazil*

Jennifer R. Martins • *Departamento de Bioquímica e Imunologia, Instituto de Ciências Biológicas, Universidade Federal de Minas Gerais, Belo Horizonte-MG, Brazil; Grupo de Pesquisa em Arboviroses, Instituto de Ciências Biológicas, Universidade Federal de Minas Gerais, Belo Horizonte, MG, Brazil; Departamento de Bioquímica e Imunologia, Instituto de Ciências Biológicas, Universidade Federal de Minas Gerais, Belo Horizonte, MG, Brazil*

Gabrielly Mattos • *Departamento de Bioquímica e Imunologia, Instituto de Ciências Biológicas, Universidade Federal de Minas Gerais, Belo Horizonte, MG, Brazil*

Luan Lopes Menezes • *Departamento de Patologia Geral, Programa de Pós-Graduação em Patologia, Instituto de Ciências Biológicas, Universidade Federal de Minas Gerais, Belo Horizonte, MG, Brazil*

Natália Nóbrega • *Departamento de Morfologia, Instituto de Ciências Biológicas, Universidade Federal de Minas Gerais, Belo Horizonte, MG, Brazil*

Albená Nunes-Silva • *Laboratório de Inflamação e Imunologia do Exercício (LABIIEX), Departamento de Educação Física e Esportes, Escola de Educação Física da Universidade Federal de Ouro Preto, Ouro Preto-MG, Brazil*

Giuliana do Nascimento Ozores • *Laboratório de Interação Microrganismo-Hospedeiro (LIMHO), Departamento de Microbiologia, Instituto de Ciências Biológicas, Universidade Federal de Minas Gerais, Belo Horizonte, MG, Brazil; Centro de Pesquisa e Desenvolvimento de Fármacos (CPDF), Instituto de Ciências Biológicas, Universidade Federal de Minas Gerais, Belo Horizonte, MG, Brazil*

Lara Penna • *Departamento de Patologia Geral, Programa de Pós-Graduação em Patologia, Instituto de Ciências Biológicas, Universidade Federal de Minas Gerais, Belo Horizonte, MG, Brazil*

Jéssica A. B. Pereira • *Departamento de Fisiologia e Biofísica, Instituto de Ciências Biológicas, Universidade Federal de Minas Gerais, Belo Horizonte, MG, Brazil; Grupo de Pesquisa em Arboviroses, Instituto de Ciências Biológicas, Universidade Federal de Minas Gerais, Belo Horizonte, MG, Brazil*

Jordane C. Pimenta • *Departamento de Morfologia, Instituto de Ciências Biológicas, Universidade Federal de Minas Gerais, Belo Horizonte, MG, Brazil; Grupo de Pesquisa em Arboviroses, Instituto de Ciências Biológicas, Universidade Federal de Minas Gerais, Belo Horizonte, MG, Brazil*

Vanessa Pinho • *Departamento de Morfologia, Instituto de Ciências Biológicas, Universidade Federal de Minas Gerais, Belo Horizonte-MG, Brazil*

Vinicius Leal Pitcella • *Departamento de Patologia Geral, Programa de Pós-Graduação em Patologia, Instituto de Ciências Biológicas, Universidade Federal de Minas Gerais, Belo Horizonte, MG, Brazil*

Celso Martins Queiroz-Junior • *Departamento de Morfologia, Grupo de Pesquisa em Arboviroses, Instituto de Ciências Biológicas, Universidade Federal de Minas Gerais, Belo Horizonte, MG, Brazil*

Filipe Resende • *Departamento de Morfologia, Instituto de Ciências Biológicas, Universidade Federal de Minas Gerais, Belo Horizonte, MG, Brazil*

Barbara Maximino Rezende • *Departamento de Patologia Geral, Programa de Pós-Graduação em Patologia, Instituto de Ciências Biológicas, Universidade Federal de Minas Gerais, Belo Horizonte, MG, Brazil*

Lucas Secchim Ribeiro • *Systems Medicine, German Center for Neurodegenerative Diseases (DZNE), Bonn, Germany*

Mayra Fernanda Ricci • *Departamento de Bioquímica e Imunologia, Instituto de Ciências Biológicas, Universidade Federal de Minas Gerais, Belo Horizonte-MG, Brazil*

Mariana Rates Gonzaga Santos • *Municipal Dentist, Belo Horizonte, MG, Brazil*

Patrícia Campi Santos • *Laboratório de Interação Microrganismo-Hospedeiro, Departamento de Microbiologia, Instituto de Ciências Biológicas, Universidade Federal de Minas Gerais, Belo Horizonte, MG, Brazil; German Center for Neurodegenerative Diseases (DZNE), Bonn, Germany*

Victor Rodrigues Santos • *Departamento de Morfologia, Instituto de Ciências Biológicas, Universidade Federal de Minas Gerais, Belo Horizonte, MG, Brazil*

Walison Nunes Silva • *Departamento de Fisiologia e Biofísica, Instituto de Ciências Biológicas, Universidade Federal de Minas Gerais, Belo Horizonte-MG, Brazil*

DANIELLE G. SOUZA • *Laboratório de Interação Microrganismo-Hospedeiro, Departamento de Microbiologia, Instituto de Ciências Biológicas, Universidade Federal de Minas Gerais, Belo Horizonte, MG, Brazil; Departamento de Microbiologia, Instituto de Ciências Biológicas, Universidade Federal de Minas Gerais, Belo Horizonte, MG, Brazil*

LUIZ PEDRO SOUZA-COSTA • *Grupo de Pesquisa em Arboviroses, Instituto de Ciências Biológicas, Universidade Federal de Minas Gerais, Belo Horizonte, MG, Brazil; Departamento de Bioquímica e Imunologia, Instituto de Ciências Biológicas, Universidade Federal de Minas Gerais, Belo Horizonte, MG, Brazil*

ANTONIO FELIPE SOUZA-GOMES • *Laboratório de Inflamação e Imunologia do Exercício (LABIIEX), Departamento de Educação Física e Esportes, Escola de Educação Física da Universidade Federal de Ouro Preto, Ouro Preto-MG, Brazil*

FERNANDA L. TANA • *Departamento de Bioquímica e Imunologia, Instituto de Ciências Biológicas, Universidade Federal de Minas Gerais, Belo Horizonte-MG, Brazil*

ANTÔNIO LÚCIO TEIXEIRA • *Neuropsychiatry Program, McGovern Medical School, University of Texas Health Science Center at Houston, Houston, TX, USA*

DANIELLE CUNHA TEIXEIRA • *Departamento de Morfologia, Instituto de Ciências Biológicas, Universidade Federal de Minas Gerais, Belo Horizonte, MG, Brazil; Grupo de Pesquisa em Arboviroses, Instituto de Ciências Biológicas, Universidade Federal de Minas Gerais, Belo Horizonte, MG, Brazil*

MAURO MARTINS TEIXEIRA • *Departamento de Bioquímica e Imunologia, Instituto de Ciências Biológicas, Universidade Federal de Minas Gerais, Belo Horizonte-MG, Brazil*

JULIANA PRISCILA VAGO • *Departamento de Morfologia, Instituto de Ciências Biológicas, Universidade Federal de Minas Gerais, Belo Horizonte-MG, Brazil*

Chapter 1

Handling and Experimentation with Germ-Free Mice

Camila Bernardo de Brito, Bárbara Maria de Amorim-Santos, and Danielle G. Souza

Abstract

Immediately after birth, mammals are largely colonized by microorganisms, with the gastrointestinal tract being the most commonly colonized organ. Over the years, several studies have shown that the intestinal microbiota is important for various physiological functions of the host. Gnotobiotic animal models are frequently used to better understand how the microbiota influences health and disease scenarios. Among gnotobiotic models, germ-free (GF) animals were first used in 1895, but it was not until 60 years later that germ-free colonies were suitable for large-scale experiments. The use of GF mice is an interesting and rich tool for studying the microbiome. However, their maintenance is a complex process that needs to be done carefully. In this chapter, we describe step by step how to manage and manipulate the gut microbiota of GF mice.

Key words Gnotobiotic mice, Microbiota, Monocolonization, Conventionalization, Germ-free mice

1 Introduction

The gastrointestinal (GI) tract contains commensal bacteria and a variety of other microorganisms such as viruses, fungi, protozoa, and archaea. Among these microorganisms, commensal bacteria constitute the vast majority of the gut microbiota population [1–3]. Several studies have demonstrated the relationship and importance of the microbiota in the development and maintenance of overall host physiological homeostasis [1, 4–7]. However, the study of human microbial communities is challenging because of the difficulty in targeting the human microbiota and separating microbial effects from host genetic and environmental factors [8]. Thus, gnotobiotic animal models are often used to better understand how microbiota influence not only GI function, but also various physiological processes in different systems [3, 8–10].

Animal models can be used to study a wide range of interactions between cells, tissues, and systems in health and disease, as

Lucas Secchim Ribeiro (ed.), *Models in Inflammation and Immunopharmacology*, Methods in Molecular Biology, vol. 2993, https://doi.org/10.1007/978-1-0716-5019-6_1,
© The Author(s), under exclusive license to Springer Science+Business Media, LLC, part of Springer Nature 2026

well as microorganism-host interactions [11]. Among these models, one offers a better perspective for microbiota studies: the gnotobiological animal model. Pasteur was the first to hypothesize that the microbiota might be essential or beneficial to host life [5, 12]. Germ-free is a biological state characterized by the complete absence of microorganisms [13], although virus-like particles have already been found in some tissues of mice, possibly transmitted vertically via the embryo or via breast milk [14]. Despite some physiological differences, such as enlargement of the cecum and changes in the digestive system, germ-free animals are very similar to their conventional counterparts [8, 11, 13]. Notwithstanding the complexity of colony establishment and maintenance, germ-free mice enabled understanding of the role of the microbiota in host homeostasis, e.g., protection against pathogens [2, 16], maintenance of the intestinal barrier [17], establishment and maintenance of inflammatory responses [2, 18–21], behavioral disorders, and neurological diseases [20, 22–24].

The establishment of a colony of GF mice is a complex process involving several steps (Fig. 1). Based on the theory that the fetus is sterile in utero [12], a cesarean section is performed on pregnant dams 24 hours before the expected delivery [1]. In a sterile environment, the uterus must be clamped [2], and once it is collected, it must be placed in a container with a sterile buffered solution with a temperature between 36 °C and 37 °C to protect it from possible contamination and thermal shock [3]. The uterus is placed in an isolator and rinsed with sterile distilled water [4]. The pups are removed, and the surviving animals are transferred to a germ-free foster mother or artificially fed [5]. The animals must be kept for at least eight generations to avoid contamination of the colony [6], and after all these generations the colony is ready for experiments [7] [9, 12, 13, 25]. In this chapter, guidelines for handling, keeping, and performing experiments with germ-free mice are presented.

2 Materials

1. Surgical procedure gloves, sterile and non-sterile.
2. 70% (v/v) ethanol.
3. Laminar flow cabinet.
4. Sterile microisolators for GF mice (e.g., Microisolator cage type II) (*see* **Note 1**).
5. Thioglycolate broth medium: 0.5 g/L of L-cystine, 2.5 g/L of sodium chloride, 5.5 g/L of glucose, 5 g/L of yeast extract, 15 g/L of pancreatic digest of casein and 0.5 g/L of Sodium thioglycolate.

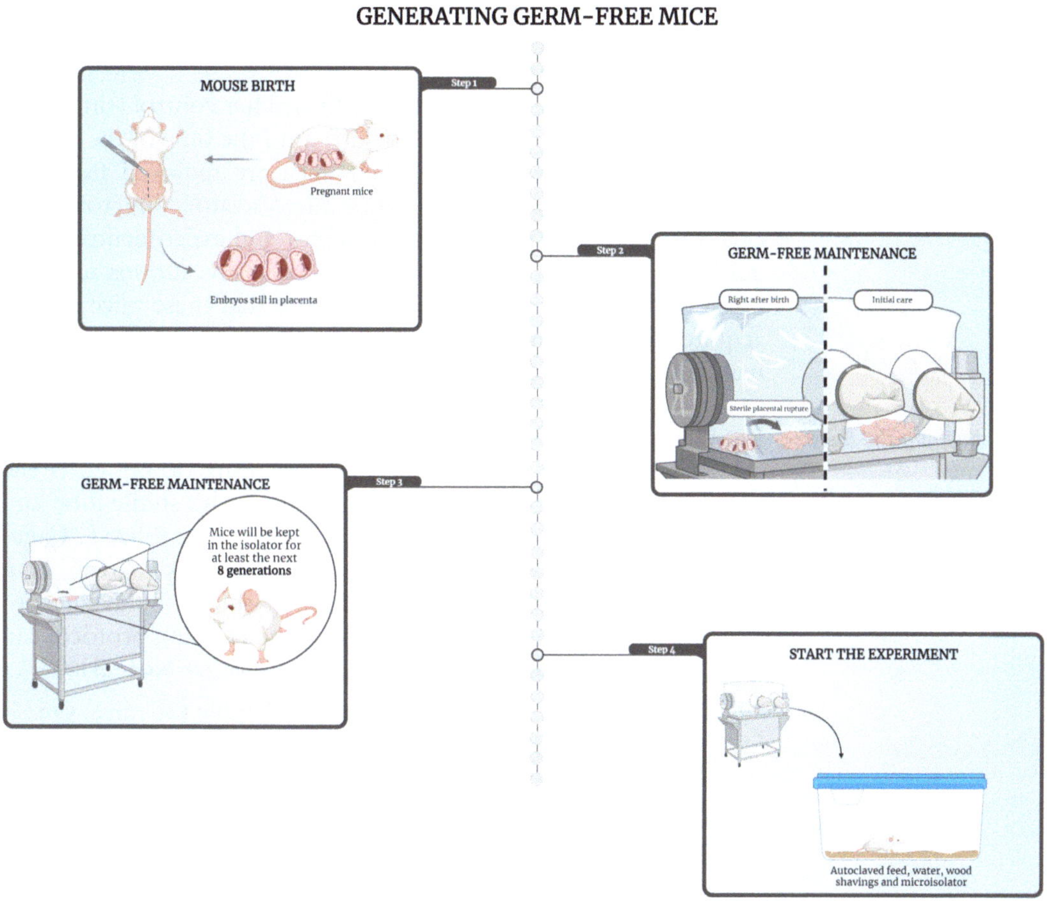

Fig. 1 Generation of germ-free mice. The uterus containing the embryos is removed from the pregnant woman by caesarean section. This procedure is performed in a sterile environment. Once the uterus is removed, it is placed in an isolator and the embryos are released. They are raised or artificially fed by a foster mother and kept there for at least eiggt generations. After it is determined that the animals are germ-free, they can be transferred to properly autoclaved microisolators for experimental purposes. (Created with BioRender.com)

6. Sabouraud broth medium: 40 g/L of dextrose and 10 g/L Peptone.
7. Curved mice gavage needle (18 G).
8. Sterile microisolators for GF mice containing wood shavings.
9. Sterile water flasks.
10. Sterile water and mouse chow.
11. Bacteriological loop.
12. Sterile 1.5 mL microcentrifuge tube.
13. Sterile 0.9% sodium chloride solution.
14. Sterile gavage needle.
15. Bacterium inoculum at 10^9 CFU/mL (e.g., *Escherichia coli* or *Bacteroides thetaiotaomicron*) or another pre-established inoculum.

3 Methods

GF mice must be kept in isolators with air filter control (single-tier isolators or isolators with flexible film) until the time of the experiment (*see* **Notes 1** and **2**). When mice are removed from the isolator, it must be done in sterile microisolators (microisolator cage type II, with filter). When caring for and experimenting with germ-free mice outside of isolators, several precautions must be taken to avoid colonization and to ensure that these mice do not acquire contaminating microorganisms. This protocol describes all materials and procedures necessary for handling GF mice.

3.1 Keeping the Mice Free from Microorganisms

1. GF mice must be housed in microisolators capable of filtering the supply air to ensure that microorganisms do not enter the cage. GF mice should always be provided with sterile food and water, and microisolators should always be sterilized. Cages should be changed two to three times per week to keep them clean. However, unnecessary handling should be avoided to prevent contamination. In addition, the cleaning protocol can be adapted to the experimental conditions (*see* **Note 3**).
2. Two investigators are often needed to handle GF mice. One is responsible for handling the mice and sterile materials, and the other is responsible for handling non-sterile materials, such as the outer part of the microisolators.
3. First, turn on the laminar flow cabinet and sterilize it with UV light (at least 20 min). Apply 70% alcohol to all surfaces (Fig. 2a);
4. Investigator 1 must put on sterile gloves. From this moment on, this investigator will only touch sterile material (Fig. 2b).
5. Investigator 2 must place all necessary materials (pre-sterilized with 70% alcohol) in the laminar flow cabinet to perform the protocol. He must also apply 70% alcohol to the outer surface of the microisolator containing the mice to be placed in the laminar flow cabinet (Fig. 2c).
6. Investigator 2 must open the microisolator with the mice and the clean microisolator. Then investigator 1 transfers all the animals to the new microisolator, adding new vials of sterile water and food pellets (Fig. 2d–f).
7. Investigator 2 closes the cage and ends the procedure (Fig. 2g).

3.2 Preparing Water and Mouse Feed

GF mice should receive autoclaved water. For this purpose, an autoclavable bottle containing filtered drinking water is used and autoclaved at 121 °C for 15 min. After that, the water should be brought to room temperature before being offered to the mice. A container that allows the passage and filtration of air should be used

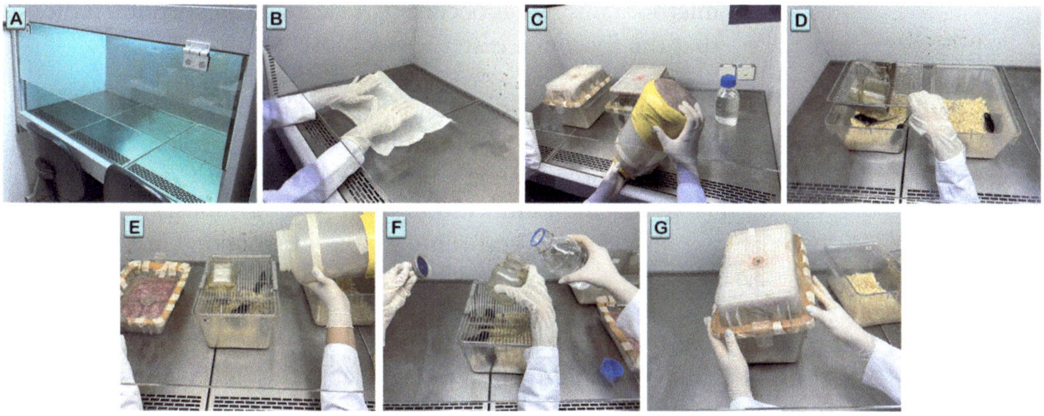

Fig. 2 Keeping germ-free mice. To change animal cages, follow seven main steps: (**a**) turn on the UV light in the laminar flow cabinet; (**b**) the first investigator puts on sterile gloves; (**c**) the second investigator puts in all materials; (**d**) investigator 1 takes the animals out of the cages; (**e**, **f**) investigator 1 adds water and food in the new cage; (**g**) investigator 2 finishes the procedure

to prepare the mouse food (Fig. 2c). However, before autoclaving the mouse food, it must be prepared by warming it up in an autoclave. This step is important to ensure that all spores of the microorganisms have germinated before autoclaving.

1. Place the food in an air-permeable container with a filter (to ensure sterility) (Fig. 2c).
2. Place the container in an autoclave and wait until the temperature reaches 121 °C.
3. Then, immediately turn off the autoclave and wait for the temperature and pressure to drop.
4. Store the food at room temperature for 12 to 24 h and then autoclave at 121 °C for 15 min. Wait until it reaches room temperature before offering it to the mice.

3.3 Confirmation of the Absence of Microorganisms in Mouse Feces

To verify the sterility of GF mice, proceed as follows:

1. Investigator 1 handles and immobilizes GF mice according to the procedures described in Subheading 3.1.
2. Investigator 2 removes two fecal pellets directly from the anal cavity of the mice and stores them in a sterile 1.5 mL tube, without touching the animal.
3. Transfer the fecal pellets to the thioglycolate medium using a bacteriological loop.
4. Incubate the culture media in a bacteriological incubator at 37°.
5. After 24 and 48 h of incubation, check the turbidity of the culture medium to determine if visible growth of

microorganisms has occurred. A clear culture medium indicates that the specimen is sterile (*see* **Note 4**).

3.4 Experimentation with Germ-Free Mice

In this section, we describe how conventionalization of GF mice (with the complete microbiota transfer) or monocolonization (with specific microorganisms) is performed (Fig. 3). Conventionalization (Fig. 3a) and monocolonization (Fig. 3b) must be performed by two investigators according to the instructions in Subheading 3.1.

3.4.1 Conventionalization

1. Collect feces from conventional mice in laminar flow cabinets (under sterile conditions). Place the feces in a 1.5 mL microcentrifuge tube.
2. Homogenize and dilute the feces in saline (10% w/v). Then wait a few minutes for sedimentation.
3. Using a sterile oral gavage needle, orally administer 100 μL of the supernatant to each mouse.
4. Twenty-one days after administration, verify that the microbiota has established as described in Subheading 3.3 (*see* **Note 4**).

3.4.2 Monocolonization

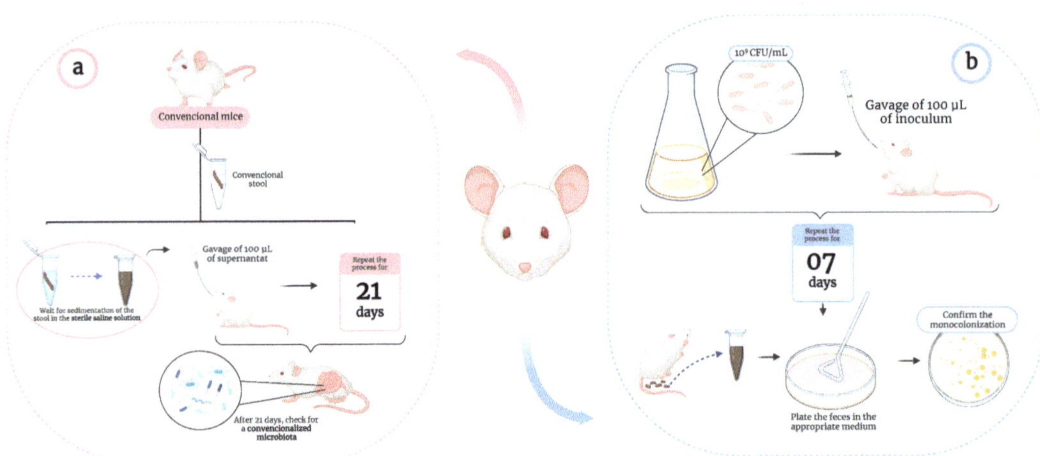

Fig. 3 Experiments with germ-free mice. Protocols for experiments with these animals include (**a**) Conventionalization: it consists of using the feces of conventional mice. This feces is collected in a sterile environment with a laminar flow cabinet and stored in sterile 1.5 mL tubes. After collection, sterile saline is added to the tube. After sedimentation of the contents, 100 μL of the supernatant is collected using a sterile syringe. Germ-free mice are gavaged once and after 21 days the stool is collected and a culture is established to verify effective conventionalization. (**b**) A monocolonization may also be performed: First, an inoculum of the microorganism of interest is prepared and adjusted to a ratio of 109 CFU/mL. From this inoculum, 100 μL is taken and administered orally (via gavage) to the animals once. After 7 days, the feces from mice are collected, a solution is prepared with sterile saline, and the supernatant is plated onto medium specific for the microorganism of interest to verify monocolonization. (Prepared using BioRender.com)

1. Prepare bacterial inoculum at 10^9 CFU/mL or a pre-established inoculum, according to bacteria sample standardization. To prepare the inoculum, first thaw the bacterial sample in a suitable culture medium. Then, centrifuge the inoculum at 5000 RPM for 10 min and resuspend the pellet in saline (NaCl 0.9%). The optical density (600 nm) is then measured using a suitable spectrophotometer and the number of bacteria (*see* **Note 5**) is adjusted to 10^8 CFU/100uL.
2. Using a sterile oral gavage needle, administer 100uL (10^8 CFU) of the bacterial inoculum to each mouse.
3. Seven days after administration, check that colonization has occurred by plating the GF feces of the mice in appropriate medium and examining the microscopic features (*see* **Note 6**).

4 Notes

1. When handling microisolators, it is important that a filter (often made of glass wool) is placed in areas where air entry is possible.
2. It is recommended that chronic experiments be conducted in isolating bubbles instead of cages (e.g., dietary obesity model and neurobehavioral disorders).
3. It is important to emphasize that due to the size of the microisolator, it is advisable to keep a maximum of four mice per microisolator.
4. If the medium is not turbid, you can use a Gram stain to confirm the absence of microorganisms.
5. The equivalence of OD 600 nm and CFU is different for different bacteria. Therefore, the density of each bacteria must be standardized before performing the experiment.
6. All procedures must be performed under sterile conditions.

Funding

This study was supported by Instituto Nacional de Ciência e Tecnologia (INCT) em dengue e interação microrganismo hospedeiro (Grant 465425/2014-3), Coordenação de Aperfeiçoamento de Pessoal de Nível Superior (CAPES), Conselho Nacional de Desenvolvimento Científico e Tecnológico (CNPq), and FAPEMIG. The authors also indicate that they have no potential conflicts of interest.

References

1. Adak A, Khan MR (2019) An insight into gut microbiota and its functionalities. Cell Mol Life Sci 76:473–493
2. Fagundes CT, Amaral FA, Vieira AT et al (2012) Transient TLR activation restores inflammatory response and ability to control pulmonary bacterial infection in germfree mice. J Immunol 188:1411–1420. https://doi.org/10.4049/jimmunol.1101682
3. Turner PV (2018) The role of the gut microbiota on animal model reproducibility. Anim Model Exp Med 1:109–115
4. Fagundes CT, Antô FAA, Teixeira L et al (2012) Adapting to environmental stresses: the role of the microbiota in controlling innate immunity and behavioral responses. Immunol Rev 245(1):250–264
5. Kubelkova K, Benuchova M, Kozakova H et al (2016) Gnotobiotic mouse model's contribution to understanding host–pathogen interactions. Cell Mol Life Sci 73:3961–3969. https://doi.org/10.1007/s00018-016-2341-8
6. Liu X, Chen Y, Zhang S, Dong L (2021) Gut microbiota-mediated immunomodulation in tumor. J Exp Clin Cancer Res 40:221
7. Schoeler M, Caesar R (2019) Dietary lipids, gut microbiota and lipid metabolism. Rev Endocr Metab Disord 20:461–472
8. Bhattarai Y, Kashyap PC (2016) Germ-free mice model for studying host–microbial interactions. Methods Mol Biol 1438:123–135. https://doi.org/10.1007/978-1-4939-3661-8_8
9. Arvidsson C, Hallén A, Bäckhed F (2012) Generating and Analyzing germ-free mice. In: Current protocols in mouse biology. Wiley, Hoboken
10. Kamareddine L, Najjar H, Sohail MU et al (2020) The microbiota and gut-related disorders: insights from animal models. Cells 9: 2401
11. Grover M, Kashyap PC (2014) Germ-free mice as a model to study effect of gut microbiota on host physiology. Neurogastroenterol Motil 26: 745–748
12. Qv L, Yang Z, Yao M et al (2020) Methods for establishment and maintenance of germ-free rat models. Front Microbiol 11. https://doi.org/10.3389/fmicb.2020.01148
13. Uzbay T (2019) Germ-free animal experiments in the gut microbiota studies. Curr Opin Pharmacol 49:6–10
14. Kajima M, Pollard M (1965) Detection of viruslike particles in germ-free mice. J Bacteriol 90(5):1448–1454
15. Nicholson JK, Holmes E, Kinross J et al (1979) (2012) host-gut microbiota metabolic interactions. Science 336:1262–1267
16. Costa MC, Santos JRA, Ribeiro MJA et al (2016) The absence of microbiota delays the inflammatory response to Cryptococcus gattii. Int J Med Microbiol 306:187–195. https://doi.org/10.1016/j.ijmm.2016.03.010
17. Schuijt TJ, van der Poll T, de Vos WM, Wiersinga WJ (2013) The intestinal microbiota and host immune interactions in the critically ill. Trends Microbiol 21:221–229
18. Amaral FA, Sachs D, Costa VV et al (2008) Commensal microbiota is fundamental for the development of inflammatory pain. Proc Natl Acad Sci U S A 105(6):2193–2197
19. Cisalpino D, Fagundes CT, Brito CB et al (2017) Microbiota-induced antibodies are essential for host inflammatory responsiveness to sterile and infectious stimuli. J Immunol 198:4096–4106. https://doi.org/10.4049/jimmunol.1600852
20. Fagundes CT, Amaral FA, Teixeira AL et al (2012) Adapting to environmental stresses: the role of the microbiota in controlling innate immunity and behavioral responses. Immunol Rev 245:250–264. https://doi.org/10.1111/j.1600-065X.2011.01077.x
21. Souza DG, Vieira AT, Soares AC et al (2004) The essential role of the intestinal microbiota in facilitating acute inflammatory responses. J Immunol 173:4137–4146. https://doi.org/10.4049/jimmunol.173.6.4137
22. Borre YE, Moloney RD, Clarke G et al (2014) The impact of microbiota on brain and behavior: mechanisms & therapeutic potential. Adv Exp Med Biol 817:373–403. https://doi.org/10.1007/978-1-4939-0897-4_17
23. Horne R, Foster JA (2018) Metabolic and microbiota measures as peripheral biomarkers in major depressive disorder. Front. Psychiatry 9:513
24. Zheng P, Zeng B, Liu M et al (2019) The gut microbiome from patients with schizophrenia modulates the glutamate-glutamine-GABA cycle and schizophrenia-relevant behaviors in mice. Sci Adv 5(2):eaau8317
25. Lavin R, DiBenedetto N, Yeliseyev V et al (2018) Gnotobiotic and conventional mouse systems to support microbiota based studies. Curr Protoc Immunol 121. https://doi.org/10.1002/cpim.48

Chapter 2

Modeling Gut Barrier Dysfunction and Subsequent Opportunistic Infection by *Candida albicans* During Chemotherapy-Induced Intestinal Mucositis in Mice

Rafaela Ribeiro Alvares Batista, Renan Vasconcelos da Graça-Filho, Giuliana do Nascimento Ozores, Ana Caroline de Lima, and Caio Tavares Fagundes

Abstract

Chemotherapeutic agents exert antineoplastic effects by damaging the DNA of cancer cells, leading to cell death. However, this effect is not specific, and may affect healthy cells and cause several adverse events such as myelosuppression and gastrointestinal mucositis. The term gastrointestinal mucositis refers to lesions in the mucosa, which include the oral cavity, pharynx, larynx, stomach, and intestine. The focus of this chapter will be the chemotherapy-induced intestinal damage, which can lead to weight loss, diarrhea, constipation, and increased predisposition to infections. Despite its severity, there is no treatment available for mucositis, and the pathophysiology of the disease has not been fully elucidated so far. This highlights the relevance of studying the mechanism of the disease and possible intervention strategies in intestinal mucositis. Here, we provide a detailed description of the main methodologies involved in the evaluation of changes induced by chemotherapy in a murine model, namely, 5-Fluorouracil (5-FU)-induced mucositis. These parameters include (i) assessment of body weight, (ii) stool consistency, (iii) fecal occult blood, and (iv) intestinal permeability. In addition, we describe a protocol to model opportunistic infection after chemotherapy, by exposing mice treated with 5-FU to *Candida albicans* in the drinking water.

Key words Cancer chemotherapy, Intestinal permeability, Opportunistic pathogens, Disease activity index, Occult blood test, Evans' blue dye, Colony-forming units

1 Introduction

Intestinal mucositis is a common adverse effect of chemotherapeutics, affecting about 40–100% of patients [1]. The main symptoms include diarrhea or constipation, nausea, weight loss, and infections, which can lead to sepsis and death. In these settings,

Authors Rafaela Ribeiro Alvares Batista and Renan Vasconcelos da Graça-Filho have equally contributed to this chapter.

Lucas Secchim Ribeiro (ed.), *Models in Inflammation and Immunopharmacology*, Methods in Molecular Biology, vol. 2993, https://doi.org/10.1007/978-1-0716-5019-6_2,
© The Author(s), under exclusive license to Springer Science+Business Media, LLC, part of Springer Nature 2026

chemotherapy might be interrupted to avoid severe complications. Nowadays, there is no treatment available for mucositis [2]. Intestinal damage involves chemotherapy-induced apoptosis of cells in the epithelial crypts by direct DNA damage, which leads to enhanced release of pro-inflammatory mediators, resulting in epithelial disruption [3]. In addition to inflammation, compromised integrity of the intestinal epithelial barrier may be caused by reduced expression of occludin (a protein relevant for maintaining tight junctions and intestinal barrier integrity) and nuclear factor erythroid 2-related factor 2 (Nrf2) (a transcription factor associated with the regulation of antioxidants and apoptotic proteins) in the small intestine [4]. Secondary to loss of intestinal barrier, dysbiosis and subsequent microbial translocation may occur [5]. Dysbiosis refers to microbiota alterations in composition and metabolic functions, which contribute to disease development such as inflammatory bowel disease, metabolic syndrome, multiple sclerosis, and depression [6]. Dysbiosis also predisposes to enhanced severity of intestinal mucositis, due to compromised modulation of the inflammatory response, immunoregulation, intestinal permeability, and oxidative stress [7]. Besides, the chosen chemotherapeutic regime can determine mucositis severity [8, 9].

In experimental animals, especially rodents, mucositis is often modeled by systemic injection of chemotherapeutic drugs, and several chemotherapeutic regimes have been previously published [10]. Assessment of disease activity index is obtained from stool analysis (consistency and presence of occult blood). Diarrhea is an outcome of inflammatory bowel diseases (IBD), involving impaired intestinal barrier integrity, dysbiosis, and hydroelectrolytic imbalance in the intestinal lumen [11]. Among the several methods available to assess stool bleeding, the method using Meyer's reagent has been extensively used. It is based on the reduction of phenolphthalein by the zinc present in the phthalic anhydride reagent. In the presence of blood in stool, erythrocyte lysis occurs and peroxidase is released, which reacts with hydrogen peroxide, added later. The oxygen released from this reaction deoxidizes phthalic anhydride to phenolphthalein, resulting in a reddish color at alkaline pH [12].

Intestinal permeability may be analyzed as an indicator of epithelial barrier function. A simple and straightforward method is the Evans' blue assay, which is based on the binding of Evans' blue dye to proteins, such as serum albumin, and it is commonly performed to assess vascular permeability [13]. Under pathological conditions with increased vascular permeability, albumin bound to this dye can leak out of the bloodstream, turning tissues into a bluish color. Similarly, Evans' blue, administered by oral gavage, binds to proteins in the gut lumen, which leak out through the epithelial mucosa upon an increase in intestinal permeability. Then, plasma and adjacent tissues can acquire the blue coloration from the dye.

This dye is extracted from tissues by incubation with formamide and quantified in plasma and tissue extracts directly by spectrophotometry.

Some studies report that intestinal dysbiosis caused by cytotoxic chemotherapy may favor opportunistic infection with *Candida albicans* [14, 15]. *C. albicans* is a common commensal yeast of the microbiota of humans and some mammals. This species colonizes the mucous surfaces of healthy individuals and is considered a regular component of the microbiota of the gastrointestinal and genitourinary tract [16]. In healthy hosts, the resident eubiotic bacterial microbiota plays an important role in controlling *C. albicans* outgrowth and colonization of the mucous membranes [17]. However, when the bacterial community is disturbed by immunosuppressive agents, some bacteria may overgrow and form mutualistic associations with *C. albicans* [14, 18]. This process can lead to a continuous dysbiosis that amplifies the damage to the mucosa. Recent studies have demonstrated synergistic associations between *C. albicans* and intestinal bacterial microbiota, such as *Enterococcus faecalis*, in models of mucositis induced by 5-Fluorouracil (5-FU) [14, 18]. Hence, exposure of 5-FU-treated mice to *C. albicans* yeasts in the drinking water leads to sustained colonization by the fungus and subsequent systemic infection, as demonstrated in this chapter.

In this chapter, we will describe the detailed methodology to (1) model and to assess disease activity indexes during 5-FU-induced mucositis in mice; (2) to assess occult blood in feces samples obtained from 5-FU-treated mice using Meyer's reagent; (3) to quantify intestinal permeability using Evans' blue dye in samples from 5-FU-injected mice; (4) to model systemic infection by *C. albicans* through the gastrointestinal route in mice that received 5-FU chemotherapy.

2 Materials

2.1 Mucositis Induction After Chemotherapy with 5-FU

2.1.1 Animals, Randomization, and Experimental Design

1. Wild-type (WT) female or male C57/BL6 mice (8–9 weeks old) entrained to a 12 h light/dark cycle.
2. Individually ventilated cages, shaving bed, filtered water, and food.
3. Ventilated shelves for cages.

2.1.2 Chemotherapy

1. 50 mg/mL of 5-fluorouracil (5-FU).
2. Phosphate-buffered saline (PBS).
3. 1 mL Syringes.

4. 30G needles.
5. Adjustable volume pipettes and pipette tips.
6. 2 mL microtubes.
7. 15 mL tubes.
8. Weight scale.

2.1.3 Clinical Score Assessment

Body Weight and Morbidity Assessment

1. Laboratory notebook.
2. Mice.
3. Weight scale.
4. Cage wire lid.

Examining Fecal Consistency

1. Empty cages.
2. Previously weighted 1.5-mL Microtubes.
3. Wooden rod.
4. Forceps.

Detecting Occult Blood in Feces Using Meyer's Reagent

1. Microtubes with mice's feces.
2. 0.9% w/v NaCl in distilled water.
3. Adjustable volume pipettes and pipette tips.
4. Microtube centrifuge.
5. 96-well flat-bottomed microplate.
6. Meyer's reagent (20% w/v sodium hydroxide, 2% w/v phenolphthalein, 10% w/v zinc powder in distilled water).
7. 3% v/v H_2O_2 in distilled water.
8. Plate-adapted spectrophotometer.

2.2 Determination of Intestinal Permeability Using Evans' Blue Dye

2.2.1 Evans' Blue Dye Preparation and Administration

1. 125 mg/mL Evans blue, diluted in filtered water.
2. 0.45 μm syringe filter.
3. 15 mL tubes.
4. 1 mL Syringes.
5. Gavage needle.
6. Mice.

2.2.2 Euthanasia and Collection of Blood and Small Intestine Samples

1. Mice.
2. Surgical material (scissors, forceps).
3. Anesthetic (a mixture of 1 mL of Ketamine at 100 mg/mL, 1 mL of Xylazine at 20 mg/mL, and 3 mL of PBS).
4. 1 mL Syringes.
5. 30G needles.

6. Adjustable volume pipettes and pipette tips.
7. Heparinized pipette tips and 1.5 mL microtubes (*see* **step 3** from Subheading 3.2.2).
8. Heparin solution (15 IU/mL in PBS).
9. Petri dishes (previously weighed).

2.2.3 Obtaining Plasma and Extracting Evans' Blue Dye from the Small Intestine

1. Microtube centrifuge.
2. Centrifuged blood from mice.
3. Petri dishes with small intestine samples.
4. 37 °C incubator.
5. Formamide (P.A, 99.6%).
6. Weight scale.
7. Adjustable volume pipettes and pipette tips.
8. 0.5 mL microtubes.

2.2.4 Standard Curve Preparation and Dye Quantification in Microplates

1. 125 mg/mL Evans blue, diluted in filtered water.
2. 1.5 mL microtubes.
3. Adjustable volume pipettes and pipette tips.
4. 8 points of Evans blue serial dilution curve in PBS from 20 μg/mL (plasma samples) or 200 μg/mL (tissue samples).
5. 100 μL of plasma samples diluted (1:2) in PBS.
6. 400 μL of Evans' blue dye extracts obtained from tissue samples.
7. Water (blank for standard curve).
8. Flat-bottomed 96-well microplate.
9. Plate-adapted spectrophotometer.

***2.3 Infection of 5-FU-Treated Mice with* C. albicans**

2.3.1 C. albicans *Inoculum Preparation*

1. Sterile Erlenmeyer flask (250 mL).
2. Autoclaved 2% w/v Sabouraud Dextrose Broth.
3. *Candida albicans* BSC1 strain stock cryotube.
4. Temperature-controlled shaking incubator.
5. Sterile Phosphate-Buffered Saline (PBS).
6. Adjustable volume pipettes and pipette tips.
7. Neubauer Chamber.
8. Centrifuge tube (50 mL).
9. Refrigerated microcentrifuge compatible with 50 mL tubes.
10. Optical microscope with 400× magnification.
11. Sterile drinking water.

2.3.2 Animal Exposure to C. albicans *Inoculum*

1. Female 8–9 weeks old C57BL/6 mice subjected to 5-FU-induced mucositis protocol (*see* Subheading 3.1.2).
2. *C. albicans* inoculum (*see* Subheading 3.3.1).
3. Centrifuge tube (50 mL).
4. Stainless steel caps for drinking bottles.

2.3.3 Euthanasia and Fungal Burden Evaluation (Stool, Small Intestine, Cecum, and Kidneys)

1. 1–3 stool pellets from each mouse.
2. Microtubes (1.5 and 2.0 mL).
3. 0.9% w/v NaCl solution in distilled water.
4. Vortex.
5. Bucket of ice.
6. Analytical weight scale.
7. Cotton stick.
8. 70% v/v ethanol.
9. Surgical instruments (forceps and scissors).
10. Sabouraud Dextrose Agar with Gentamicin 0.1 mg/L and Vancomycin 0.01 mg/mL in polystyrene (90 × 15 mm) Petri dishes.
11. Marker pen.
12. Adjustable volume pipettes and pipette tips.
13. Mortar and pestle.
14. Sterile Drigalski handle.
15. Anesthetic solution (ketamine and xylazine mixture in PBS).
16. PBS.
17. Biochemical Oxygen Demand (B.O.D.) incubator.

3 Methods

3.1 Mucositis Induction After Chemotherapy with 5-FU

The experimental design of the present model of 5-FU-induced intestinal mucositis and the evaluated parameters are shown in Fig. 1.

3.1.1 Animals, Randomization, and Experimental Design

1. Identify each of the animals with a number.
2. Raffle the numbers to compose the groups.
3. Place the animals in individually ventilated cages, with free access to food and water in a 12-hour light/dark cycle (*see* **Note 1**).
4. Keep notes of the animals' identifications in each group.

3.1.2 Chemotherapy

1. Weigh mice from the same group using the weight scale.
2. Calculate the arithmetic mean of the values obtained from the weight of the animals.
3. Calculate the chemotherapy volume for 75 mg/Kg dose of 5-FU (500 mg/10 mL) (*see* **Note 2**).
4. To obtain the total volume of chemotherapy for the group, multiply the value obtained above by the number of animals in the group.
5. Dilute the chemotherapy drug in PBS using the pipettes, tips, and tubes.
6. Inject 100 μL of 5-FU intraperitoneally or of PBS in mice from the corresponding groups using the syringe and 30G needles.
7. Repeat the procedures for each of the 3 days of the protocol.

3.1.3 Clinical Score Assessment

Parameters evaluated to obtain the clinical score are described in Table 1 and include body weight loss, piloerection, posture, and activity patterns, as well as fecal consistency and blood content. For each sign, a score value is attributed, as shown in Table 2. Figure 2a shows a PBS-treated mouse, and Fig. 2b shows a 5-FU-treated mouse at the 5th day of the experimental protocol depicted in Fig. 1. Parameters should be assessed in each mouse at 1, 3, 5, 8, and 10 days of the experimental protocol. Analyze the individual

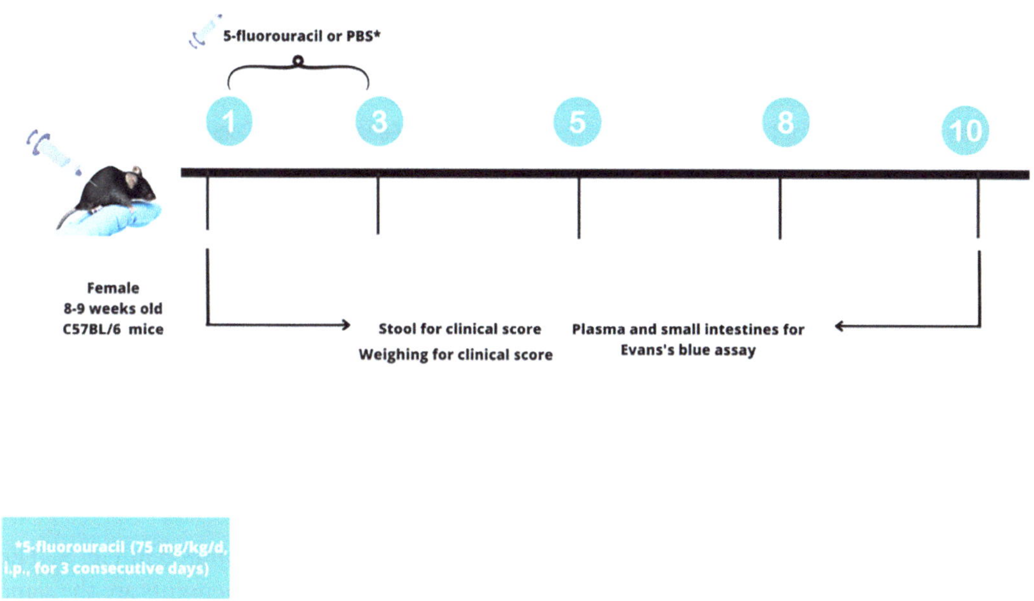

Fig. 1 Experimental design for the 5-FU-induced intestinal mucositis model. C57BL/6 female mice were submitted to the intestinal mucositis induction protocol after injection with 5-FU (75 mg/Kg/day; 100 μL i.p.; 3 days) or PBS (100 μL i.p.; 3 days), and then evaluated for the clinical score and intestinal permeability at the indicated timepoints

Table 1
Clinical score components used to characterize disease activity

Clinical score				
Body weight loss (% of initial weight)	**Morbidity signs**	**Fecal consistency**	**Occult blood in feces**	**Final score**
Up to 5%	Absent	Normal	Absent	0
Up to 10%	Piloerection	Pasty	Present	1
Up to 15%	Piloerection and hunched posture	Liquid		2
Up to 20%	Piloerection, hunched posture, and inactivity			3
Greater than 20%				4
Final clinical score The sum of the score for each of the four parameters evaluated				**(0–10)**

Clinical score evaluates disease activity and corresponds to the sum of the score values assigned to each of the following parameters: percentage of body weight loss, morbidity signs, fecal consistency, and occult blood in feces, ranging from 0 to 4, as shown in the table. The percentage of body weight loss is determined relative to the initial weight, and for each range of this percentage (up to 5%, 10%, 15%, 20%, or greater than 20%), a score value is assigned. Signs of morbidity include piloerection, hunched posture, and inactivity. Stool consistency is analyzed by smashing the stool sample using a wooden rod, and fecal occult blood is assessed using Meyer's reagent. The maximum score, therefore, is 10.

Table 2
Clinical score of PBS or 5-FU-treated animals on the 5th day of the experimental protocol

Clinical score					
Mouse (Fig. 2)	**Body weight loss**	**Morbidity signs**	**Fecal consistency**	**Occult blood in feces**	**Final score**
A (PBS)	0 (gained 2%)	0 (absent)	0 (normal)	0 (absent)	0
B (5-FU)	3 (lost 17%)	3 (piloerection, hunched posture and inactivity)	0 (normal)	1 (present)	7

Clinical scores of two mice were determined on the 5th day of the experimental protocol to assess the disease activity index. The mouse treated with PBS gained 2% of body weight (score 0) and showed no changes in morbidity (score 0) and fecal parameters (score 0). The sum of these scores is 0. The mouse that received 5-FU injections presented body weight loss of up to 20% of the initial weight, corresponding to a score of 3. The mouse also showed piloerection, hunched posture, and inactivity (score 3); stool consistency was normal (score 0), with the presence of occult blood in the stool (score 1). The sum of these score values corresponds to the final clinical score (score 7).

parameters in each mouse at a time, as demonstrated in Fig. 3. The procedures for the analyses in each of those timepoints are the same and are described below:

Body Weight and Morbidity Assessment

These signals compound the morbidity parameter and indicate the severity of the disease.

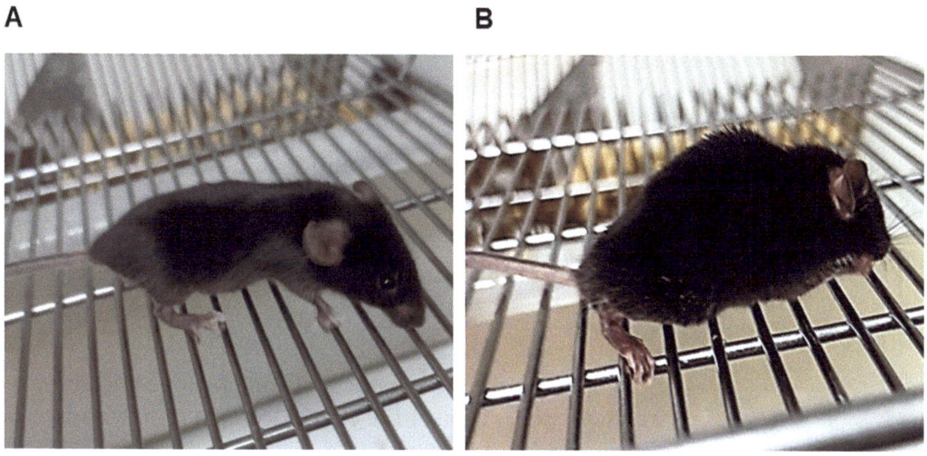

Fig. 2 Signs of morbidity in mice after treatment with 5-FU. Throughout the course of 5-FU-induced intestinal mucositis, signs of morbidity may be absent (**a**) or mice may exhibit piloerection, hunched posture, and inactivity (**b**). These signs can be present only as piloerection (being assigned a score of 1) or piloerection and hunched posture (score 2) or piloerection, hunched posture, and inactivity (score 3), as shown in Tables 1 and 2

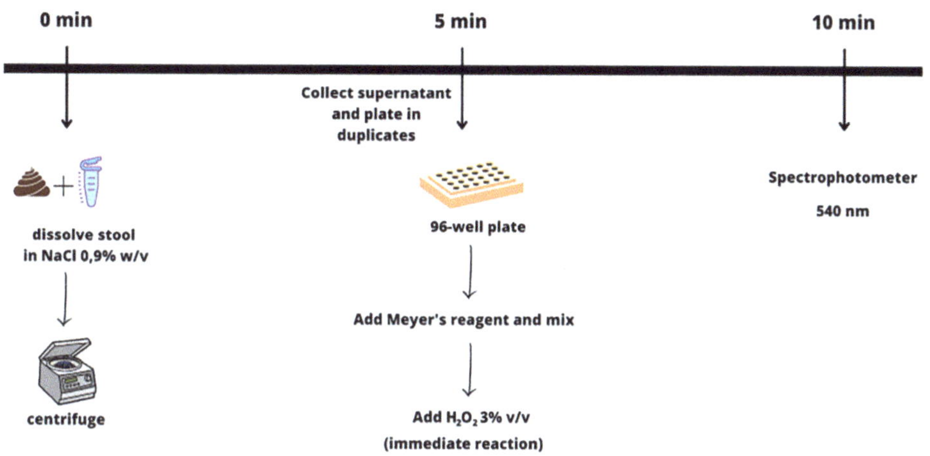

Fig. 3 Experimental protocol for fecal occult blood analysis in mice submitted to intestinal mucositis, using Meyer's reagent. Feces from mice submitted to the intestinal mucositis induction protocol by 5-FU (75 mg/kg/day; 100 μL i.p.; 3 days) or PBS (100 μL i.p.; 3 days) were collected at the times indicated in Fig. 1, to determine the stool score and, subsequently, the clinical score. Then, feces were diluted in 0.9% w/v NaCl and centrifuged (1000×*g*; 5 min). 50 μL of the supernatant was pipetted, in duplicate, into a 96-well plate and 5 μL of Meyer's reagent was added and mixed to the sample. Then, 5 μL 3% v/v H_2O_2 was added. The reaction is immediate, so the absorbance reading must be performed within 5 min, at 540 nm

1. Weigh mice using the same weight scale used in Subheading 3.1.2 (Fig. 1) (*see* **Note 3**).
2. Determine the percent of initial weight using the records obtained in Subheading 3.1.2.

3. Place the mouse in the cage wire lid, keeping the animal immobile for a few moments by softly holding the animal's tail.
4. Note if there is any alteration indicative of piloerection (ruffled fur) and take notes (Fig. 2).
5. Look at the animal's back and take note if there is any sign of hunched posture.
6. Release the animal at the cage wire lid, observe the exploratory behavior, and take notes.

Examining Fecal Consistency

1. Place the mouse in an empty cage and wait until it defecates.
2. Examine the fecal consistency of one pellet by softly smashing the fecal pellet against the cage floor using the wooden rod (*see* **Note 4**).
3. With the help of a forceps, place mouse feces in the previously weighted sterile microtubes.
4. Place the animal back in the original cage with the rest of its group.

Detecting Occult Blood in Feces Using Meyer's Reagent

The experimental protocol for the determination of occult blood in feces using Meyer's reagent is shown in Fig. 3. The steps are described below:

1. Determine pellet feces weight (*see* **Note 5**).
2. Suspend the fecal pellet in NaCl 0.9% w/v at the proportion of 1 mg of feces per 10 μL of saline solution and macerate feces using sterile pipette tips.
3. Centrifuge the feces suspension at 1000×*g* for 5 min at room temperature.
4. Collect the supernatant and plate 50 μL in duplicates in a flat-bottomed 96-well microplate.
5. Add 5 μL of Meyer's reagent and mix.
6. Add 5 μL of H_2O_2 (3% v/v) and mix.
7. Note the reaction (up to 5 min), characterized by the appearance of a reddish color.
8. Read absorbance at 540 nm on a spectrophotometer within 5 min (*see* **Note 6**).

Figure 4 shows the results of the analysis of body weight and clinical score recorded during the course of a representative 5-FU-induced intestinal mucositis experiment.

3.2 Determination of Intestinal Permeability Using Evans' Blue Dye

Intestinal permeability of mice after treatment with 5-FU might be analyzed using the Evans' blue assay, following the protocol illustrated in Fig. 5.

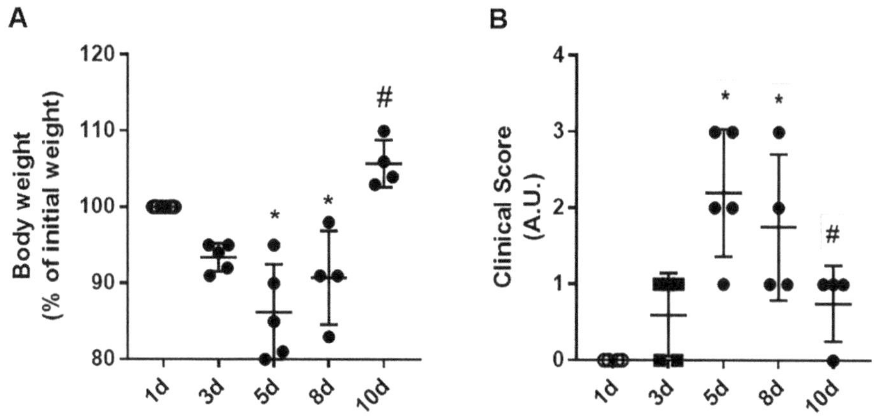

Fig. 4 Assessment of body weight and clinical score of mice subjected to intestinal mucositis. Mice treated with 5-FU (75 mg/kg/day; 100 μL i.p.; 3 days) were evaluated for percentage of initial weight (**a**) and clinical score (**b**), at the following timepoints: 1 day, 3 days, 5 days, 8 days, and 10 days after intestinal mucositis induction. The clinical score is calculated as shown in Tables 1 and 2. There is a body weight reduction and an increase in clinical score on days 5 and 8 of the protocol. At 10 days, there is an improvement in these parameters. [$n = 4$–5. $^*p < 0.05$ vs. 1d; $^{\#}p < 0.05$ vs. 5d (one-way ANOVA with Kruskal-Wallis post-test)]

3.2.1 Evans' Blue Dye Preparation and Administration

1. Dilute Evans' blue dye powder with water to a final concentration of 125 mg/mL [19].
2. Filter the solution with a syringe and a syringe filter (0.45 μm), storing in a 15 mL sterile tube protected from the light (*see* **Note** 7).
3. Administer 400 μL of Evans' blue dye to mice by oral gavage, 24 h before euthanasia (*see* **Note 8**).

3.2.2 Euthanasia and Collection of Blood and Small Intestine Samples

1. Euthanize mice according to a method approved by the animal ethics committee.
2. To collect blood, dissect the skin, and cut the muscle on the brachial plexus, exposing the vascular bed.
3. Heparinize pipette tips and tubes by filling and emptying them three times with a 1 mL heparin solution (15 IU/mL in PBS).
4. Cut the brachial plexus and transfer the collected blood to the heparinized 1.5-mL microtubes using the heparinized tips.
5. To obtain the small intestine, dissect the peritoneum from the skin, making an incision through the Alba line.
6. Remove the intestine by cutting the extremities at the pylorus of the stomach and the extremity before the cecum.
7. Collect a 2-cm-long fragment of the small intestine, open it lengthwise, and remove the luminal contents with the aid of the forceps.
8. Place the clean fragment in a previously weighed Petri dish.

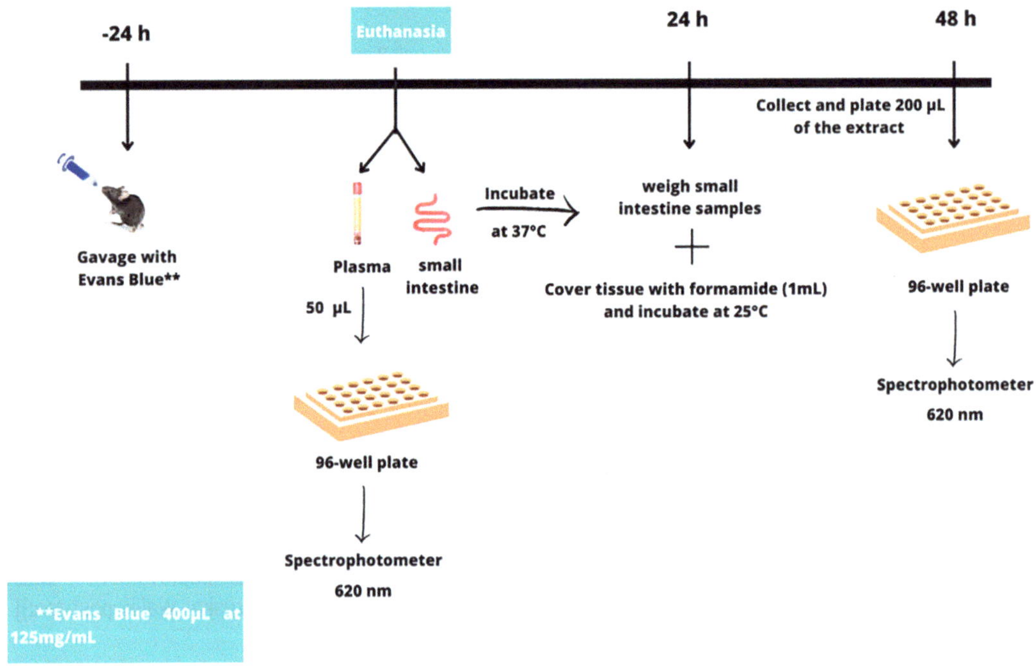

Fig. 5 Schematic protocol for Evans's blue dye assay for analyzing intestinal permeability in mice with intestinal mucositis induced by 5-FU. Mice submitted to the intestinal mucositis induction protocol (5-FU (75 mg/kg/day; 100 μL i.p.; 3 days) or PBS) received a solution of Evans' blue dye (125 mg/mL diluted in filtered water; 400 μL/animal; by oral gavage). After 24 h, plasma and fragments of the small intestine were collected to proceed with the assay. Plasma was diluted and added to a 96-well plate for determination of the dye concentration in a spectrophotometer at 620 nm. The intestine fragments were longitudinally opened and placed in previously weighed Petri dishes, and then incubated at 37 °C. After 24 h, the Petri dishes containing the intestine fragments were weighed to obtain the dehydrated tissue weight, and then 1 mL of formamide was added to extract the dye. Plates were stored protected from light at room temperature (25 °C) for 24 h, and then Evans' blue dye concentration was determined as described above for plasma

3.2.3 Obtaining Plasma and Extracting Evans' Blue Dye from the Small Intestine

1. Centrifuge blood samples at 3000×*g* for 10 min at 4 °C and transfer plasma (the clear supernatant) to another 0.5-mL microtube.
2. Incubate the open Petri dishes with tissue samples in a 37 °C incubator for 24 h.
3. Weight plates to obtain the dry weight of the tissue fragment and record them.
4. Add 1 mL of formamide to the Petri dish, covering the tissue fragment, and incubate it in the dark at room temperature for 24 h.

3.2.4 Standard Curve Preparation and Dye Quantification

1. The standard curve is prepared from the dilution of the stock solution (125 mg/mL; administered to the animals by oral gavage), to obtain the first point of the curve, corresponding to 20 μg/mL for plasma samples or 200 μg/mL for tissue

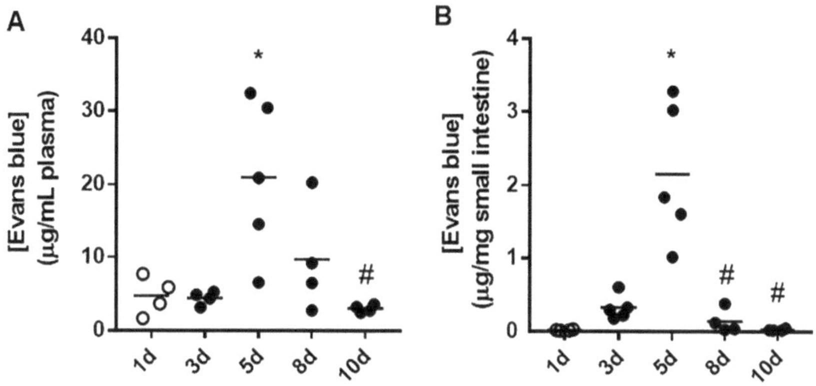

Fig. 6 Assessment of Evans' blue dye concentration in plasma and small intestine fragments. The intestinal permeability of mice after treatment with 5-FU (75 mg/kg/day; 100 µL i.p.; 3 days) or PBS was evaluated using the Evans' blue dye assay, according to the experimental protocol shown in Fig. 5. Plasma and fragments of the small intestine of these animals were collected at the following time points: 1d, 3d, 5d, 8d, or 10d. There is an increase in Evans' blue dye concentration at 5 days after mucositis induction in both analyzed sites. [$n = 4$–5. $^{*}p < 0.05$ vs. 1d; $^{\#}p < 0.05$ vs. 5d (one-way ANOVA with Tukey post-test)]

samples. A 1:2 serial dilution is then performed in order to obtain the other points of the curve. Generally, preparing 8 points of the standard curve is sufficient.

2. Add 50 µL of the standard curve and blanks, in duplicates, to a flat-bottomed 96-well microplate.
3. Dilute plasma samples (1:2) and add 50 µL, in duplicates, to the microplate.
4. For tissues, add 200 µL of Evans blue extracted by formamide, in duplicate, to the microplate (*see* **Note 9**).
5. Determine Evans' blue dye concentration by reading the absorbance at 620 nm using a plate-adapted spectrophotometer.

Figure 6 shows the concentration of Evans' blue dye in plasma and intestine from mice of a representative experiment using the experimental protocol described in Fig. 1.

3.3 Infection of 5-FU-Treated Mice with *C. albicans*

3.3.1 *C. albicans* Inoculum Preparation

1. Add 100 mL of autoclaved 2% w/v Sabouraud Dextrose Broth in a sterile Erlenmeyer flask (250 mL) (*see* **Note 10**).
2. Add 100 µL of the stocked *C. albicans* BSC1 strain into the same Erlenmeyer.
3. Keep the Erlenmeyer in the shaking incubator (37 °C at 140–160 rpm) for approximately 18–24 h.
4. After this period, pour the liquid from the Erlenmeyer into a centrifuge tube (50 mL).
5. Centrifuge at 3000×g for 10 min at 4 °C.
6. Discard the supernatant.

7. Resuspend the pellet retained on the bottom of the tube with 10 mL of sterile PBS.
8. Centrifuge again at 3000×*g* for 10 min at 4 °C.
9. Discard again the supernatant.
10. Resuspend the pellet in 5 mL of sterile PBS.
11. Adjust the inoculum by estimating the number of cells in a cell counting chamber (Neubauer) and diluting it with sterile drinking water to reach 1×10^7 cells/mL (*see* **Note 11**).

3.3.2 Exposure of Animals to C. albicans *Inoculum*

The exposure of mice to the inoculum should be done for 5 days with daily renewal of the inoculum at days 0–4 of the experimental protocol. All material and reagents used for this exposure must be sterile and should be handled in an environment with sterility control. Mice should be exposed to the inoculum 1 day before starting the protocol with 5-FU (Day 0), as depicted in Fig. 7.

1. Transfer the equivalent to 40×10^7 cells to a 50 mL tube.
2. Dilute the volume of the tube to reach 40 mL of final volume using sterile drinking water (the same offered for the control group) (*see* **Note 12**).
3. Cap the 50 mL tube with the stainless steel caps for drinking bottles and place the tube in the corresponding cage.

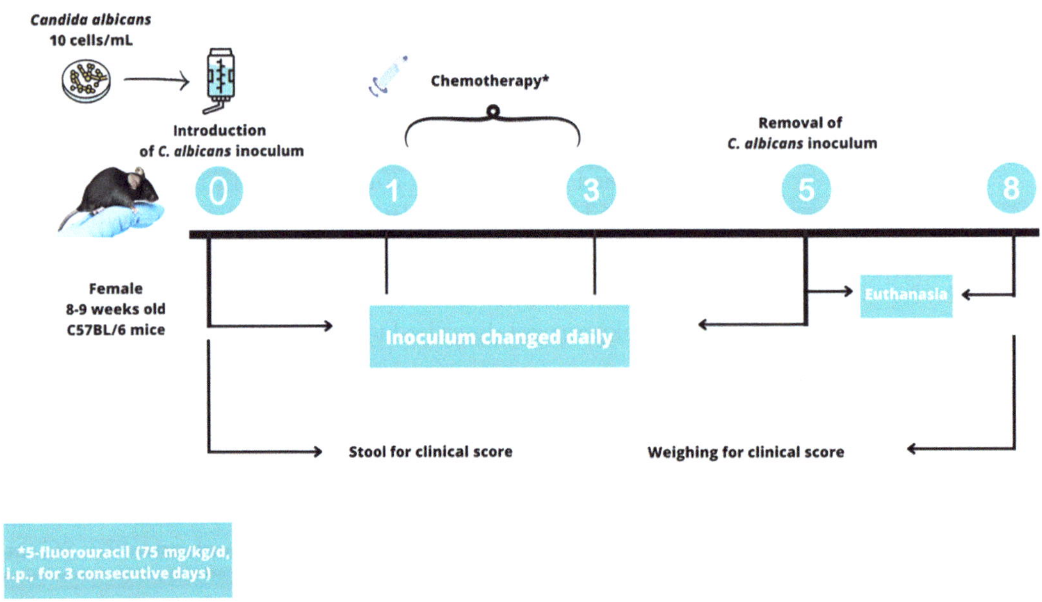

Fig. 7 Schematic protocol for exposure of mice to *C. albicans* in drinking water during the process of mucositis induction. The inoculum of *C. albicans* is added to the drinking water for 5 days, but the inoculum must be changed daily. Treatment with chemotherapy (with 5-FU (75 mg/kg/day) or PBS (both 100 μL i.p.)) is done for 3 days. In this experimental protocol, there were two different groups of mice, one submitted to euthanasia on day 5 and another one on day 8

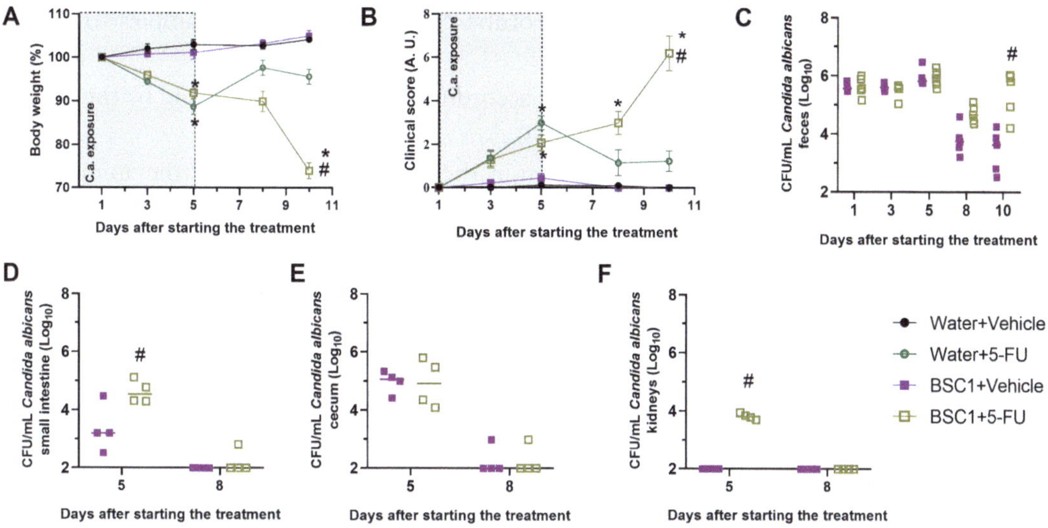

Fig. 8 *C. albicans* exposure leads to more severe disease and fungal translocation to the kidneys in mice treated with 5-FU. Mice treated with 5-FU (75 mg/kg/day; 100 μL i.p.; 3 days) or PBS were evaluated for the percentage of body weight loss (**a**), clinical scores (**b**) and fungal loads in feces (**c**), small intestine (**d**), cecum (**e**) and kidneys (**f**) at 1, 3, 5, 8, and 10 days after induction of intestinal mucositis and exposure to *C. albicans* in the drinking water. The clinical score is calculated as shown in Tables 1 and 2. There is a reduction in body weight and an increase in clinical score on day 10 in *C. albicans*-exposed, 5-FU-treated mice. These animals also presented greater fungal loads in stool, small intestine, and kidney, but not in cecum. [$n = 4$–10. *$p < 0.05$ vs. water+vehicle control; #$p < 0.05$ C. albicans+vehicle control (one-way ANOVA with Tukey post-test)]

4. Renew the inoculum daily from days 0–4 using the same procedures.
5. At day 1 of the protocol, start chemotherapy following the steps described in Subheading 3.1.
6. Record body weight changes and clinical scores as described in Subheading 3.1 (Fig. 8).

3.3.3 Euthanasia and Fungal Burden Evaluation (Stool, Small Intestine, Cecum, and Kidneys)

In this evaluation, animals should be euthanized following ethical protocols. All tissues should be collected in a sterile environment and with sterile materials. All collected material must be stored on ice until the dilution and plating procedures.

1. In a sterile environment, place 1.5 mL microtubes in racks (keep closed until use).
2. Record the weight of each microtube before adding the sample.
3. Place the mouse in an empty sterile cage and wait until it defecates.
4. Store 1 to 3 stool pellets in the previously weighted sterile 1.5 mL microtube.

5. Weigh the microtube containing the stool samples to obtain stool weight.
6. Euthanize mice according to a method approved by the animal ethics committee.
7. Wet fur and skin in the abdominal region of the animal with 70% v/v ethanol with the help of a cotton stick.
8. Proceed with laparotomy with the aid of scissors and surgical forceps, to expose the internal organs.
9. Identify the beginning of the intestine in order to guide the removal of the other organs.
10. Remove a portion of the small intestine and cecum and place them in an empty sterile Petri dish.
11. Remove all the luminal contents of the intestine and cecum and store each tissue in a previously weighed sterile microtube in ice.
12. Reach the kidneys behind the intestines and remove the entire organ, storing it in a previously weighed sterile microtube in ice.
13. Suspend the stool sample with the adequate volume to obtain a 10^{-1} w/v suspension in 0.9% w/v NaCl solution (1 mg of feces per 10 μL of diluent).
14. With the help of a micropipette tip, macerate the stool until obtaining an even homogenate.
15. Vortex the homogenate for 15 s.
16. Transfer 20 μL of the 10^{-1} suspension to 180 μL of 0.9% w/v solution to obtain the 10^{-2} homogenate.
17. Vortex the new homogenate for 15 s.
18. Repeat steps 16 and 17 twice to obtain the 10^{-3} and 10^{-4} homogenates from the 10^{-2} and the 10^{-3} dilutions, respectively.
19. After obtaining the tissue sample weight, grind the tissue sample using a pestle and mortar.
20. Recover the ground tissue by adding the adequate volume of 0.9% w/v NaCl solution to the mortar (10 μL of solution to 1 mg of tissue) in order to obtain the 10^{-1} dilution. Transfer the homogenate to a new sterile microtube.
21. Dilute the 10^{-1} tissue suspension according to steps 15 to 18 to obtain 10^{-2}, 10^{-3}, and 10^{-4} dilutions (*see* **Note 13**).
22. Using a marker pen, mark four parts of similar area in the Petri dish, one for each dilution (10^{-1} to 10^{-4}).
23. Vortex each dilution and, with pipettes and sterile tips, transfer three 10 μL replicates of the dilution to the respective marked

area in the agar plate, keeping an even distance between each drop.

24. Wait for the drops to dry and incubate plates in the incubator at 37 °C for 24 h.
25. For the kidneys, transfer 100 μL of the dilution to the Petri dish and spread it through the entire media surface area with the help of a sterile Drigalski handle.
26. Incubate plates in the incubator at 37 °C for 24 h.
27. Count the colony-forming units in the adequate dilution (from 3 up to 30 colonies per drop or from 30 to 300 colonies per plate inoculated with kidney extracts).
28. Adjust the counts to the corresponding dilution to obtain the total amount of CFUs per mL of sample (*see* Fig. 8) (*see* **Note 14**).

4 Notes

1. Sterilizing cages, shaving beds, and water is strongly recommended in order to reduce variability between experiments.
2. Different chemotherapy regimens will lead to different disease outcomes. The regimen presented here results in a mild mucositis, peaking on the 5th day after initiation of chemotherapy, despite moderate clinical scores and 50% lethality rates. Reducing the chemotherapy dosage will produce smaller death rates, but even milder disease. On the other hand, increasing the 5-FU dosage will produce a more severe phenotype, leading to 100% death before the 10th day of the experimental protocol.
3. Body weight assessment and feces collection throughout the experimental protocol should be carried out at the same time of the day, as body weight and fecal outputs change throughout the day and could bias analysis.
4. Stool consistency is classified as normal, pasty, or liquid. The "normal" parameter includes dry or soft stools, not considering the differences between these alterations. Stools characterized as "pasty" have an intermediate consistency between normal soft stools and liquid stools, without a defined shape like stools of normal consistency, but still not completely liquid.
5. Before collecting mice's feces, microtubes should be weighed. The feces weight is determined by subtracting the weight of microtubes containing feces from the weight of empty microtubes.

6. The appearance of a reddish color is considered as a positive test for occult blood in the stool. In the absence of reddish coloration, the result is considered negative. The detected absorbances are also taken in account to determine if the test is positive or not. Usually, PBS-treated mice present absorbance values below 0.250. Therefore, if the absorbance is above 0.300 absorbance units, the fecal occult blood result is considered positive, assuming a value equal to 1 in the clinical score.
7. Evans' blue dye solution can be stored in the refrigerator (4 °C) for up to 15 days after preparation.
8. Other protocols present on the literature assess vascular permeability by injecting smaller amounts of Evans' blue dye into the bloodstream and quantifying the extracted dye after perfusion of mice tissues to remove excess blood.
9. Evans blue concentration in the intestine of animals is normalized by tissue weight, which is determined before adding formamide.
10. For this step, *C. albicans* growth must be conducted with fresh cells, and it should be used in 24 h maximum. After this period, the yeast tends to change its morphology into hyphae and pseudohyphae, which can make the blastoconidia counting difficult for inoculum adjustment.
11. In the end of the process for fungi growth here described, you should obtain 10^9 cells/mL, thus, in the 11th step, the fungi suspension should be diluted into the suggested media (sterile drinking water) to reach the inoculum of 1×10^7 cells/mL.
12. Once the inoculum should be renewed daily, the volume of 40 mL should be sufficient for 24 h in a cage with 5 mice. It is recommended to use the 50 mL conical tube adapted as a drinker because of the constant renewal of the inoculum and the ease of obtaining and preparing the inoculum of *C. albicans* at 1×10^7 cells/mL in this smaller volume.
13. Organ homogenates should be obtained in a sterile environment. Using the same pestle and mortar for every mouse in the same experimental group might be possible if these instruments are thoroughly washed with sterile PBS between grounding each sample.
14. The average between the 3 micro drops of each sample should be made, respecting the variation of the count limit, between 3 and 30 colonies in each microdrop. The mean should be multiplied by the inverse value of the dilution, and then this result should be multiplied by 100 to obtain the number of CFU/mL. For the plates inoculated with kidney extracts, plates containing between 30 and 300 colonies are considered adequate. Just multiply the number of colonies by 10 to obtain

the number of CFU/mL. Values below the ones outlined above are considered below the detection limit, and the number of CFU is considered undetectable. Values above the ones outlined are considered above the detection limit and the number of CFU is considered uncountable.

Funding

The authors would like to thank Coordenação de Aperfeiçoamento de Pessoal de Nível Superior (CAPES) (process 88882.348380/2010-1), Conselho Nacional de Desenvolvimento Científico e Tecnologico (CNPq) (grant 421 465425/2014 and 304490/2014-8), Fundação de Amparo à Pesquisa do Estado de Minas Gerais (FAPEMIG) (grant APQ-00897-22) and Instituto Nacional de Ciência e Tecnologia (INCT) em dengue e interação microrganismo hospedeiro for the financial support.

References

1. Villa A, Sonis ST (2015) Mucositis: pathobiology and management. Curr Opin Oncol 27(3): 201–203
2. Ribeiro RA, Wanderley CW, Wong DV, Mota JM, Leite CA, Souza MH, Cunha FQ, Lima-Júnior RC (2016) Irinotecan- and 5-fluorouracil-induced intestinal mucositis: insights into pathogenesis and therapeutic perspectives. Cancer Chemother Pharmacol 78(5):881–893
3. Sonis ST (2009) Mucositis: the impact, biology and therapeutic opportunities of oral mucositis. Oral Oncol 45(12):1015–1020
4. Yan XX, Li HL, Zhang YT, Wu SY, Lu HL, Yu XL, Meng FG, Sun JH, Gong LK (2020) A new recombinant MS-superoxide dismutase alleviates 5-fluorouracil-induced intestinal mucositis in mice. Acta Pharmacol Sin 41(3): 348–357
5. Segers C, Mysara M, Claesen J, Baatout S, Leys N, Lebeer S, Verslegers M, Mastroleo F (2021) Intestinal mucositis precedes dysbiosis in a mouse model for pelvic irradiation. ISME Commun 1(1):1
6. Guimarães GM, Costa K, da Silva Santana Moura C, Moreira SED, Marchiori JM, de Menezes Santos ACP, Batista RRA, Queiroz-Junior CM, Raposo JDA, Braga FC, Caliari MV, Nunes AC, Fagundes CT, Neumann E (2024) Influence of tryptophan metabolism on the protective effect of Weissella paramesenteroides WpK4 in a murine model of chemotherapy-induced intestinal Mucositis. Probiotics Antimicrob Proteins. https://doi.org/10.1007/s12602-024-10413-1
7. Fijlstra M, Ferdous M, Koning AM, Rings EH, Harmsen HJ, Tissing WJ (2015) Substantial decreases in the number and diversity of microbiota during chemotherapy-induced gastrointestinal mucositis in a rat model. Support Care Cancer 23(6):1513–1522
8. Pedroso S, Vieira A, Bastos R, Oliveira J, Cartelle C, Arantes R, Soares P, Generoso S, Cardoso V, Teixeira M, Nicoli J, Martins F (2015) Evaluation of mucositis induced by irinotecan after microbial colonization in germ-free mice. Microbiology 161(10):1950–1960
9. Vanlancker E, Vanhoecke B, Stringer A, Van de Wiele T (2017) 5-fluorouracil and irinotecan (SN-38) have limited impact on colon microbial functionality and composition in vitro. PeerJ 5:e4017
10. Wardilla HR, Tissing WJE, Kissow H (2019) Animal models of mucositis: critical tools for advancing pathobiological understanding and identifying therapeutic targets. Curr Opin Support Palliat Care 13(2):2
11. Anbazhagan AN, Priyamvada S, Alrefai WA, Prasad PD (2018) Pathophysiology of IBD-associated diarrhea. Tissue Barriers 6(2): e1463897
12. Gürtler RE, Oneto ML, Cecere MC, Castañera MB, Canale DM (2001) A simple method to identify triatomine (Hemiptera: Reduviidae) feces in sensing devices used in vector

surveillance programs. J Med Entomol 38(2): 249–251
13. Radu M, Chernoff J (2013) An in vivo assay to test blood vessel permeability. J Vis Exp 73: e50062
14. Bertolini M, Sefik E, Abboud G, Merad M (2019) Candida albicans induces mucosal bacterial dysbiosis that promotes invasive infection. PLoS Pathog 15(4):e1007717
15. Teoh F, Pavelka N (2016) How chemotherapy increases the risk of systemic candidiasis in cancer patients: current paradigm and future directions. Pathogens 5(1):6
16. Kam AP, Xu J (2002) Diversity of commensal yeasts within and among healthy hosts. Diagn Microbiol Infect Dis 43(1):19–28
17. Xu H, Dongari-Bagtzoglou A (2015) Shaping the oral mycobiota: interactions of opportunistic fungi with oral bacteria and the host. Curr Opin Microbiol 26:65–70
18. Bertolini M, Sefik E, Abboud G, Merad M (2021) Mucosal bacteria modulate Candida albicans virulence in oropharyngeal candidiasis. MBio 12(4):e01220–e01221
19. Li J, Lv J-l, Cao X-y, Zhang H-p (2022) Gut microbiota dysbiosis as an inflammaging condition that regulates obesity-related retinopathy and nephropathy. Front Microbiol 13:855

Chapter 3

Direct Inoculation of Pathogenic Yeasts into Mouse Lung by Trans-tracheal Instillation

Patrícia Campi Santos

Abstract

Pathogenic dimorphic fungi are endemic in certain regions and can cause from subclinical infections to systemic mycoses. These pathogens are associated with high mortality and morbidity rates and represent emerging infectious threats to human populations worldwide. Because most fungal pathogens cause lung infection or use the lung as a route to disseminate to other organs, mammalian pulmonary infection models are crucial tools in the advancement of medical mycology. Many different types of animal models of fungal infection have been developed, with murine models being hailed as the gold standard for studies of pathogenesis, to compare virulence between isolates, preclinical evaluation of vaccines and therapies, and host antifungal immune responses. The ability to control numerous variables in performing the model allows for mimicking human disease states and quantitatively monitoring the course of the disease. To model lung inflammation and injury, fungal infectious propagules can be inoculated via intranasal delivery, intratracheal, or aerosolization approaches. The protocol in this chapter details a method for intratracheal delivery of a fungal suspension, where fungal cells are administered directly into the lungs to initiate infection. The aim is to provide a comprehensive description of techniques required to perform mouse pulmonary infection, such that reproducible results are attained, allowing for the use of this in vivo approach for high-quality studies.

Key words Fungal infection, Pulmonary disease, Mouse model, Intratracheal instillation, Immune response, Pathogenesis

1 Introduction

Fungal diseases have been on the rise worldwide since the mid-twentieth century, posing an emerging infectious threat to human populations. While fungi have intrinsic characteristics that contribute to disease emergence [1, 2] recent scenarios have mainly been driven by human modern activities, including medication (e.g., corticosteroids), the progressive devastation of tropical forests and urbanization, domestication, population growth, tourism, migration, and climate change. The increase in temperature or other changes in environmental niches may lead to the selection

Lucas Secchim Ribeiro (ed.), *Models in Inflammation and Immunopharmacology*, Methods in Molecular Biology, vol. 2993,
https://doi.org/10.1007/978-1-0716-5019-6_3,
© The Author(s), under exclusive license to Springer Science+Business Media, LLC, part of Springer Nature 2026

of thermotolerant strains with enhanced virulence characteristics and directly influence the ability of fungi to damage human hosts and infection prevalence [3–7]. The mortality rate associated with systemic mycosis remains unacceptably high due to the complexity of treatment, including the toxicity and limited efficacy of antifungal drugs, the long periods, and the high costs, particularly in developing countries. Furthermore, no vaccines have been approved for human use to prevent or treat fungal infections [8–10].

The majority of fungi associated with life-threatening infections are inhaled as infectious propagules found in the environment, including dimorphic fungi (i.e., members of the genus *Histoplasma*, *Paracoccidioides*, *Coccidioides*, *Blastomyces*, *Talaromyces*, and *Emergomyces*) and non-*Aspergillus* molds (i.e., *Penicillium marnefeii*, Mucorales order; agents of mucormycosis). The dimorphic fungi are primary pathogens generally found in the soil, plant remains, and/or animal excreta, growing at 22–25 °C as mycelia that generate conidia, which are released into the air by wind and soil disruptions along with hyphal fragments. Transition into the yeast phase occurs at 37 °C upon inhalation by the host and invasion of pulmonary tissues, which can lead to infection. The symptoms of infection can be mild and undetected or develop into more serious conditions such as pneumonia, acute respiratory distress syndrome, and disseminated disease [11].

Mice and other laboratory animals proved to be useful in mimicking clinical manifestations associated with superficial and life-threatening invasive mycoses. Fungal infections acquired through the inhalation of infectious propagules can be reproduced by intranasal or intratracheal injection of a liquid fungal suspension or inhalation of dry fungal cells for the development of mouse models in clinical mycology. The use of genetically defined inbred mouse strains, humanized mice, and gene knockout mice has allowed studies to analyze the evolution and progression of fungal diseases (pathogenesis), aspects of innate and acquired immunity, and explore therapeutics and diagnostics approaches that can improve patient care and outcome [12, 13]. Indeed, different indicators related to the infection can be assessed by mouse models in respiratory preclinical research, including mortality, microbial loads in tissues, detection of secreted cytokines and other mediators in bronchoalveolar lavage fluids, as well as the number and composition of immune cells infiltrating the lung and histological score. Among the challenges associated with the development of respiratory models is ensuring that substances are administered to the lungs in the most accurate and reproducible manner, so that a particular pathology may be induced or a novel therapeutic approach and compound can be evaluated for efficacy and safety during the preclinical phase of development.

In most models developed to date, pulmonary infection with viruses, bacteria, and other pathogens is induced either through intranasal administration (IN), oropharyngeal aspiration (OP), or by direct instillation into the trachea (IT) of a microbial suspension. In particular, IN inoculation, which consists of dropping the microbial suspension into the nostrils of anesthetized mice, is considered a rapid and easy technique suitable for repeated administrations. Intranasal instillation may be the ideal alternative for experiments designed for studying infection by airborne pathogens or inhaled drugs because it mimics the pathway made by the administered volume from the nasal passages to the lungs [14, 15]. However, given the high viscosity of the nasal mucosa, and depending on the size of the microbial particles, a part of the inoculum remains in the upper respiratory tract even after the instillation of larger volumes, preventing the reproducibility of the technique and leading to microbial spread to extrapulmonary tissues, causing off-target complications [16–18].

Instead, the intratracheal (IT) instillation has become widely used as a result of its accuracy and low risks [19]. Many techniques for this approach include tracheal intubation through the mouth cavity or surgical tracheal exposure to directly access the lung. In order to carry out the oropharyngeal aspiration (OP) intubation technique, the opening of the trachea is visualized using a laryngoscope and a catheter or needle is inserted directly into the tracheal lumen of anesthetized animals to deliver the stimuli. Although rapid, this procedure requires high technical expertise and specific instrumentation, which is often difficult to obtain or costly [20–22]. As an alternative delivery method, transtracheal instillation is performed by surgically exposing the trachea on the ventral side in the neck and inserting a needle into the lumen just below the larynx [23–25].

This chapter will describe a detailed protocol for performing mouse intratracheal instillation (IT) of pulmonary fungal infection, which is commonly used to investigate host immune responses that occur during infections caused by pathogenic fungi. Although this guideline concentrates on methods and techniques required for modeling invasive *P. brasiliensis* infection as a prototype of dimorphic fungi, the principles are the same for pulmonary infection with other airborne pathogenic fungi.

2 Materials

2.1 Fungal Inoculum Preparation

1. Frozen stock of virulent yeast cell from *P. brasiliensis* strain (Pb18).
2. Yeast Extract–Peptone–Dextrose (YPD) broth medium: 1% (w/v) yeast extract, 2% (w/v) peptone, and 2% (w/v) glucose.

3. Brain heart infusion (BHI) agar plates supplemented with 4% fetal bovine serum, 40 mg/L gentamycin, and 5% conditioned medium as growth factor (made from 7-day cultures of *P. brasiliensis* grown in YPD broth).
4. Drigalski spatulas, sterilized.
5. Parafilm.
6. BOD (Biochemical Oxygen Demand) incubator.
7. Disposable plastic inoculation loops.
8. 250-mL Erlenmeyer flasks.
9. Orbital shaking incubator.
10. 15-mL and 50-mL sterile conical centrifuge tubes.
11. Serological pipette.
12. Centrifuge.
13. Sterile Phosphate-Buffered Saline (PBS; pH 7.2).
14. 5-mL disposable syringes.
15. Hypodermic Needles (22-Gauge, 30-Gauge).
16. 2-mL microcentrifuge tubes.
17. Neubauer hemocytometer.
18. Janus green dye: prepare working solution of 0.05% w/v Janus green B in distilled water and sterilize by filtration through a 0.22 μm nitrocellulose membrane. Store at room temperature.
19. Light microscope with 10x and 20x objective lenses.

2.2 Infection of Experimental Animals

1. Male C57BL/6 or BALB/c mice, 6–8 weeks old (specific pathogen-free).
2. Ketamine hydrochloride (80 mg/kg) and xylazine (15 mg/kg) diluted in sterile PBS.
3. 25 G needles.
4. 1-mL tuberculin syringes.
5. Surgery Board.
6. Surgical tape strips.
7. Hamilton Gas-tight syringe (Hamilton, Bonaduz, Switzerland) 50 μL, Model 705 LT.
8. Silk suture 5–0.
9. Sterilized surgical instruments: dissection scissors, straight forceps, curved forceps, and tweezers.

3 Methods

3.1 Fungal Strain and Growth Conditions

All steps should be carried out under a Biosafety Level 2 (BSL2) cabinet and ensure to autoclave and discard contaminated waste according to institute guidelines.

1. Yeast cells of a highly virulent *P. brasiliensis* strain (Pb18) are stored in YPD broth medium (1% w/v yeast extract, 2% w/v peptone, 2% w/v glucose) with 25% glycerol at −80 °C. To recover the cells, remove a cryovial containing the fungal cells from freezer storage and immediately place it into ice (*see* **Note 1**).
2. After thawing, decontaminate the outside of the cryovial with 70% ethanol and bring the vial to the laminar airflow cabinet. Transfer 100 μL of the suspension to BHI agar supplemented with 4% fetal bovine serum (FBS) and gently distribute inoculum over the surface using a cell spreader until the liquid is mostly absorbed. The plates are sealed with a small strip of Parafilm and incubated at 37 °C for 7 days (*see* **Note 2**).
3. After fungal growth, collect 3 isolated fungal colonies on BHI agar using a sterile inoculating loop and transfer into 100 mL YPD broth in a 250 mL Erlenmeyer flask. In order to promote log phase growth, incubate at 37 °C in an orbital shaker with 100 rpm agitation rate for at least 5 days and no more than 7 days. Monitor the broth for turbidity, as that is an indication for the growth of the fungus (*see* **Note 3**).

3.2 Fungal Cells Harvesting and Counting Procedure

1. On the day of infection, transfer 30 mL of log-phase yeast culture to sterile 50-mL conical centrifuge tube using a serological pipette and spin at 2655 × *g* for 10 min at room temperature. A clear supernatant and a white pellet containing the yeast cells should be visible (*see* **Note 4**).
2. Carefully remove the tube from the centrifuge and place into a rack. Discard the supernatant from 50-mL conical tubes by gently and slowly tilting the tube and pouring the liquid into a 250 mL glass vial (leave 2–3 mL at the bottom to ensure the pellet is not disturbed).
3. Resuspend cells in 30 mL of sterile PBS using a serological pipette and homogenize the pellet thoroughly by pipetting up and down to disperse the yeast cells. Centrifuge again at 2655 × *g* for 10 min at room temperature.
4. Perform the wash step described in the previous item at least twice, bringing up the final, washed pellet in 10 mL of sterile PBS. Then, homogenize the fungal cell pellet thoroughly by pipetting up and down several times. If there is more than one

tube, pool the yeast cells from multiple tubes, adding sterile PBS to the pellets and resuspend them in 10 mL of PBS.

5. It is important to eliminate cell-clumping as it may interfere with the cell count and lead to clogging of the needle during the animal inoculation process. Remove the cell clumps by allowing the highly concentrated suspension to spontaneously floc and settle for 2 min, then transfer 3 mL of the upper suspension using a 1 mL pipette into a sterile 15 mL tube (*see* **Note 5**).
6. A reliable method for ensuring a proper single-cell suspension is to pass the suspension through different gauge needles multiple times, which eliminates clumps. First, load a sterile 5 mL syringe attached to a 22-gauge hypodermic needle with the highly concentrated suspension. Push the suspension through the needle back and forth into the same tube at least three times (*see* **Note 6**).
7. Then, collect the suspension again using the same 22-gauge hypodermic needle and switch to a sterile 30-gauge hypodermic needle. Dispense the cell suspension into the same tube using the thinner needle and repeat this procedure two more times.
8. Prepare 1 mL of a 1:50 dilution of the yeast cell suspension with sterile PBS in a microcentrifuge tube. The viability and cell counting are determined by the modified Janus green staining method. By this method, viable cells remain bright (unstained) and dead cells are stained dark-blue. In brief, 20 μL of cell suspension is incubated with 0.05% Janus green dye (1:1, v/v) for 10 min at room temperature. Using a pipette, take 10 μL of dye-treated cell suspension and apply to a hemocytometer chamber. Under the microscope, count the total number of cells in the four corner quadrants and calculate the concentration of the original cell suspension by the formula: (total cell number/4) $\times$ (dilution factor) $\times$ 10^4 (chamber dilution constant). The proportion of viable cells is assessed by using the formula: cell viability (%) = viable cells/total cells $\times$ 100. Only fungal preparations with viability greater than 80% are used for infection.
9. Dilute the original yeast suspension with sterile PBS to obtain the desired concentration, 4.0×10^7 yeasts/mL for infection with 1×10^6 yeasts/30 μL. This is the fungal suspension that will be used to infect the mice (*see* **Note** 7).

3.3 Mice

Due to the fact that this process involves using mouse models, it is essential to secure the necessary authorization for animal testing from the institutional ethical committee before beginning any research.

1. Isogenic C57BL/6 or BALB/c mice (male, 6–8 weeks old, 20–22 g) are maintained in the animal facility under pathogen-free conditions and constant temperature (22 ± 2 °C) and humidity (about 50%) on a period of 12 h light/dark cycle, and have free access to water and standard pellet food (*see* **Note 8**).

3.4 Anesthesia

1. Anesthetize mice with ketamine/xylazine mixture (80 and 15 mg per kg of body weight, respectively) in sterile phosphate buffered saline (PBS) administered by intraperitoneal injection, using 25 G needles on 1 mL tuberculin syringes. The mouse should remain for 10–15 min in an empty cage while the anesthesia takes effect (*see* **Note 9**).

3.5 Transtracheal Instillation of Fungal Suspension

Maintain aseptic conditions during all the procedure. Use sterile surgical instruments and materials, wear gloves and avoid contact with any non-sterile surface.

1. Transfer the animal to a laminar flow cabinet and place the sedated mouse in dorsal recumbency (crucifix position) on a plastic platform and hold it in place by delicately fixing its legs with surgical tape strips. The board inclination should be adjusted to an angle of 45–60° to allow clear visualization of the incision region and facilitate tracheal instillation (*see* **Note 10**).
2. Slightly extend the mouse neck while ensuring the maxillary front incisors are hooked and stretched by a 5–0 nylon silk or rubber band (80 mm × 1.5 mm) that is stably fixed to the board with adhesive tape.
3. Remove hair from the ventral neck with a razor blade and aseptically prepare the surgical area with gauze pad and 70% ethanol solution.
4. Hold the skin with dissecting forceps near the anterior portion of the neck (throat region) and perform a vertical incision through the skin using a curved scissor. The incision of the skin is about 0.5 cm in length. The skin fragment which is removed is so small that it will produce no pressure in the mouse neck.
5. To visualize and expose the trachea, the subcutaneous tissue (superficial cervical fascia) is carefully removed, and the two lobes of the salivary glands are softly pulled apart with fine tissue forceps. The trachea lies just below the salivary glands and is easily identified as a white, translucent tube (*see* **Note 11**).
6. Mix the suspension containing the fungal inoculum by vigorous vortexing to ensure fungal cells are evenly dispersed in solution.

7. Aspirate 30 μL of fungal suspension into a 50 μL gas-tight Hamilton syringe, avoiding the formation of bubbles (*see* **Note 12**).
8. Hold the syringe with the 30-gauge, 5/16-inch needle positioned down and parallel to the trachea. Insert the needle through the tracheal wall, not too deep between the tracheal rings and slowly inject the full volume of the inoculum into the trachea (*see* **Note 13**).
9. Wait a few seconds until the entire volume to flow through the needle, and then remove syringe from the trachea (*see* **Note 14**).

3.6 Sutures

1. The surgical skin incision should be closed using appropriate interrupted 5–0 sterile absorbable silk suture.
2. Grasp the thickest part of the needle where the suture material is attached with a serrated dissecting forceps.
3. Penetrate the skin with the tip of the needle at a 90° angle approximately 3 mm away from the edge of the surgical incision. The needle of a 5–0 silk suture is passed through the skin on each side of the incision and the ends tied.
4. Excess suture is cut at the knot and additional simple stitches, individually knotted, placed along the entire length of the incision will efficiently close the skin. Once complete, the incision edges should be well approximated and without any large gaps to prevent surgical-site contamination during healing (*see* **Note 15**).

3.7 Post-procedure Monitoring and Mouse Recovery

1. After the surgery is completed, remove the mouse from the inoculation board and place it on a pre-warmed, temperature-controlled heating pad (37 °C) in the prone position (warm surface in contact with the ventral surface of the animal) until fully recovered from anesthesia (*see* **Note 16**).
2. During recovery from anesthesia, sterile 0.9% NaCl eye drops should be applied to both eyes in order to maintain corneal hydration and avoid severe and permanent corneal opacity. Cataracts in mice are clearly visible as a white film in their eyes. Make sure to keep adding eye drops periodically during the recovery period.
3. Anesthetized animals must be observed continuously throughout the duration of the anesthetic procedure until they have fully recovered. Monitor the mouse breathing, responsiveness to external stimuli, and skin or mucous membrane color as seen by observing the ears, tail, and oral mucosa or foot pads.
4. Observe the mouse until it starts moving and regains full consciousness. Once it is confirmed that the mouse is in good

condition, place it back to a clean cage. Do not return it to the company of other animals in its routine housing until it has fully recovered to prevent possible injury.

3.8 Mice Evaluation

1. Examine the surgical incision to ensure that it is intact and that the area is clean and dry.
2. Monitor the mice daily for clinical signs including respiratory distress (sneezing, dyspnea), behavior abnormalities (not eating or drinking, hunched posture, lethargy or persistent recumbency), and for any sign of morbidity (alopecia, ruffled fur, lacrimation, eye/nasal discharge).
3. Weigh the mice every day during the infection period, starting on the day prior to infection procedure. Mice that show signs of suffering (determined by 20% body weight loss, slow movements, alteration of general appearance, and changes in posture and behavior) must be euthanized by overdose of ketamine (80 mg/kg) and xylazine (15 mg/kg) followed by cervical dislocation.

3.9 Expected Outcomes and Clinical Progression in Infected Mice

In experimental models of fungal infection, including those involving trans-tracheal inoculation of pathogenic yeasts, it is essential to understand the expected clinical course and measurable outcomes. In this model, body weight loss is a key parameter used to monitor disease progression and overall health status. Mice typically begin to exhibit detectable changes within the first few days post-infection, with progressive weight loss being one of the earliest and most consistent indicators.

Throughout the 30-day observation period, infected animals are expected to demonstrate a significant reduction in body weight compared to non-infected controls, especially between days 3 and 15 (Fig. 1a), depending on the virulence of the fungal strain and the host response. In parallel, clinical signs such as reduced mobility, ruffled fur, decreased grooming, hunched posture, and labored breathing may become increasingly evident as the infection advances. These signs usually correlate with increasing fungal burden (Fig. 1b) and pulmonary impairment.

Importantly, pulmonary impairment is directly associated with the host's inflammatory response to the infection. The recruitment and accumulation of immune cells—particularly neutrophils and macrophages—into the lung tissue form a dense inflammatory infiltrate (Fig. 1c–f). While this response is essential for fungal clearance, it also contributes to tissue damage (Fig. 1e) and impaired gas exchange, exacerbating respiratory distress. The inflammatory infiltrate not only limits pathogen spread but also plays a central role in the pathophysiology of lung dysfunction observed in infected animals. In advanced stages of infection, if the inflammatory response fails to resolve the infection or becomes

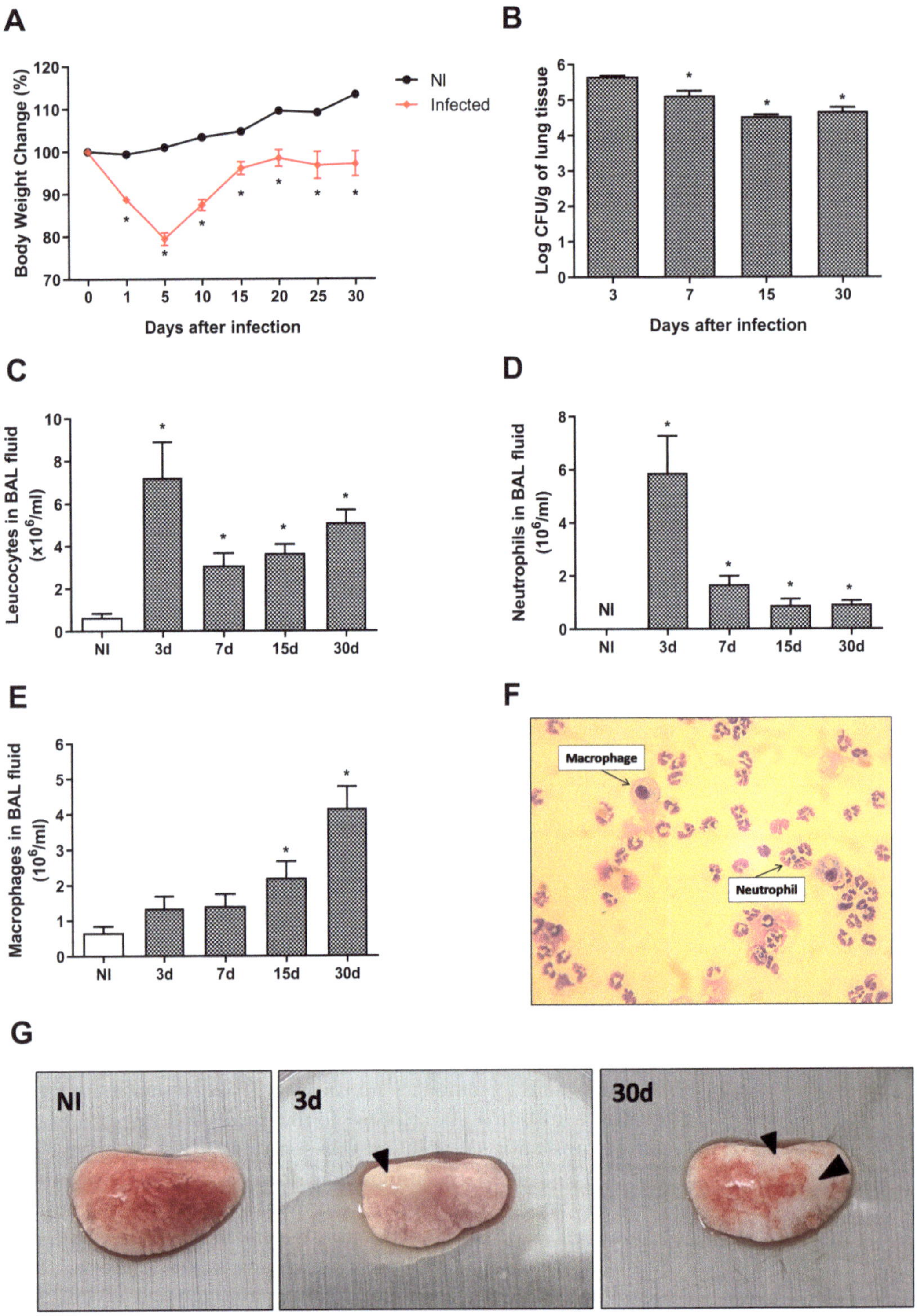

Fig. 1 Pulmonary infection of C57BL/6 mice with pathogenic yeasts induces sustained weight loss, fungal persistence, and dynamic leukocyte recruitment in mice. (**a**) Body weight change over 30 days post-infection in non-infected (NI) and infected mice. Infected animals showed a significant reduction in body weight beginning at day 3, with partial recovery but persistent impairment through day 30. (**b**) Quantification of

excessive, animals may progress to a moribund state, characterized by severe weight loss, hypothermia, and ultimately death.

Monitoring both physiological (e.g., weight loss) and clinical (e.g., breathing difficulty, behavior changes) parameters, in conjunction with histological analysis of lung inflammation, provides a comprehensive framework for assessing disease severity and host-pathogen interactions in murine models of pulmonary fungal infection.

In addition to clinical and histopathological evaluation, the characterization of the host inflammatory response can be enhanced by quantifying specific proinflammatory mediators in lung tissue or bronchoalveolar lavage fluid (BALF). Commonly assessed cytokines include tumor necrosis factor-alpha (TNF-α), interleukin-6 (IL-6), and interleukin-1 beta (IL-1β), which are key indicators of acute inflammation. The presence of neutrophilic infiltration can also be quantified through histological scoring or by measuring myeloperoxidase (MPO) activity, providing further insight into the intensity and role of the immune response during pulmonary fungal infection.

4 Notes

1. Do not let glycerol stocks thaw and refreeze. This leads to a decrease in cell viability, expression of virulence factors and their ability to infect animals.
2. Plates must be sealed with plastic wrap (Parafilm) to reduce water loss and prevent contamination due to long incubation period.
3. An orbital shaker incubation is ideal for creating a homogenous environment for microorganisms' growth and improving aeration and cellular contact with nutrients. Mixing helps the cells grow uniformly, as a short throw may reduce the yeast cells from clumping at the bottom of the flask.

Fig. 1 (continued) fungal burden in lung tissue at days 3, 7, 15, and 30 post-infection, expressed as log CFU/g of tissue. High fungal loads were observed early and persisted throughout the study period. (**c–e**) Total leukocyte counts (**c**), neutrophils (**d**), and macrophages (**e**) in bronchoalveolar lavage fluid (BALF) at different time points post-infection. A significant influx of total leukocytes and neutrophils was observed at day 3, followed by a reduction over time. Macrophage numbers increased progressively, peaking at day 30. (**f**) Representative cytospin image of BALF at 3 days post-infection showing abundant neutrophils and the presence of macrophages (Panoptical fast staining kit, 200× magnification). (**g**) Morphology of left lungs from NI and infected mice at 3 and 30 days post-infection. Infected lungs show visible inflammatory foci (black arrowheads) and consolidation areas, more prominent at day 30. Data are expressed as mean + SEM, $n = 6$ per group. * ($p < 0.05$ vs. NI)

4. Never overfill centrifuge tubes as this can lead to tube breaking or cracking and leakage of the microbial suspension during centrifugation.
5. Make sure the tip only reaches the region close to the surface of the liquid and collect the suspension slowly. Be careful not to place the tip on the bottom of the tube so that the flocculated part containing the cell clumps is not disturbed.
6. Prolonged vortexing or sonication of the cell suspension is not recommended to disrupt clumps as it may promote cell surface damage and loss of viability.
7. Prepare 30 μL of yeast suspension for two extra animals ("*n* + 2") mice to be infected to account for pipetting error. If a different inoculum is desired, adjust the cell dilution appropriately.
8. Susceptibility to infection with the pathogenic fungus *P. brasiliensis* is sex-dependent. Sexual hormones had a profound effect on the biology of immune cells, and estradiol favors protective responses to *P. brasiliensis* infection [26].
9. Anesthesia is considered adequate when the animal stays still quietly and has constant heart and respiratory rates. The toe-pinch reflex is the most effective way to check if an appropriate depth of anesthesia has been reached prior to incision. The operator's fingernails should be firmly pressed against the toe of the hindlimb. If the mouse tries to remove its limb, it is not properly sedated, and a top-up dosage (10–20% of the initial dose) should be administered. When there is no response to the toe pinch, medium-deep anesthesia has been achieved. This dosage will induce 45–60 min of anesthesia, which is enough for the infection procedure.
10. Rotate the surgical platform to orient the mouse with its head facing the operator, so that the neck is closest to you and at eye level. This position allows the operator, during the injection, to angle the syringe so that it follows the natural path of the trachea straight to the lungs.
11. The tracheal rings can be more easily seen when you gently press down with forceps on either side of the trachea. This is important for establishing where to place the injection.
12. The volume of fungal suspension described in this protocol was standardized in previous experiments. Typically, intratracheal injection volumes are 1–3 mL/kg of body weight (approximately 30–75 μL final volume). The administration of larger volumes should generally be avoided to limit any potential choking or coughing and subsequent loss of instilled material and even death of the animals by drowning. A separate vehicle control, which involves needle insertion and instillation of

sterile PBS is generally performed in order to take into consideration any host response to the administration process itself.

13. The needle bevel should face the operator and advance until the bevel is just past the trachea wall with the needle as parallel with the trachea as possible to ensure that the lumen is reached.
14. A few seconds of apnea followed by gasping after inoculum injection can be observed, which indicates that the needle is properly inserted into the trachea and the mouse immediately inhales the entire amount of liquid into the lungs. If the mouse stops breathing, withdraw the needle right away and wait until it starts breathing normally again before re-inserting it.
15. When non-absorbable sutures are used, it is recommended to remove stitches 7–10 days post-surgery. If animals will be euthanized within 14 days following surgery, removal of sutures prior to euthanasia is not necessary.
16. Small rodents, such as mice, are vulnerable to hypothermia, which is a common side effect in laboratory animals under prolonged anesthesia. Therefore, the heat keeps mice warm as they recover to avoid a drop in body temperature due to the anesthetic.

Disclosure

The author has no conflicts of interest to declare that are relevant to the content of this chapter.

References

1. Siscar-Lewin S, Hube B, Brunke S (2022) Emergence and evolution of virulence in human pathogenic fungi. Trends Microbiol 30(7):693–704. https://doi.org/10.1016/j.tim.2021.12.013
2. Carpouron JE, de Hoog S, Gentekaki E, Hyde KD (2022) Emerging animal-associated fungal diseases. J Fungi (Basel) 8(6):611. https://doi.org/10.3390/jof8060611
3. Morens DM, Fauci AS (2013) Emerging infectious diseases: threats to human health and global stability. PLoS Pathog 9:e1003467. https://doi.org/10.1371/journal.ppat.1003467
4. Rodrigues ML, Nosanchuk JD (2020) Fungal diseases as neglected pathogens: a wake-up call to public health officials. PLoS Negl Trop Dis 14:e0007964. https://doi.org/10.1371/journal.pntd.0007964
5. Fisher MC, Gurr SJ, Cuomo CA et al (2020) Threats posed by the fungal kingdom to humans, wildlife, and agriculture. Philos Trans R Soc B Biol Sci 371:20160332. https://doi.org/10.1098/rstb.2016.0332
6. Van Rhijn N, Bromley M (2021) The consequences of our changing environment on life threatening and debilitating fungal diseases in humans. J Fungi 7:367. https://doi.org/10.3390/jof7050367
7. Nnadi NE, Carter DA (2021) Climate change and the emergence of fungal pathogens. PLoS Pathog 17:e1009503. https://doi.org/10.1371/journal.ppat.1009503
8. Armstrong-James D, Brown GD, Netea MG et al (2017) Immunotherapeutic approaches to treatment of fungal diseases. Lancet Infect Dis 17:e393–e402. https://doi.org/10.1016/S1473-3099(17)30442-5
9. Kischkel B, Rossi SA, Santos SR et al (2020) Therapies and vaccines based on nanoparticles for the treatment of systemic fungal infections.

Front Cell Infect Microbiol 10:463. https://doi.org/10.3389/fcimb.2020.00463

10. Zhang Z, Bills GF, An Z (2023) Advances in the treatment of invasive fungal disease. PLoS Pathog 19(5):e1011322. https://doi.org/10.1371/journal.ppat.1011322
11. Gonzalez A, Taborda CP (2021) Editorial: pathogenesis of dimorphic fungal infections. Front Cell Infect Microbiol 11:793245. https://doi.org/10.3389/fcimb.2021.793245
12. Capilla J, Clemons KV, Stevens DA (2007) Animal models: an important tool in mycology. Med Mycol 45(8):657–684. https://doi.org/10.1080/13693780701644140
13. Jensen HE (2022) Animal models of invasive mycoses. APMIS 130(7):427–435. https://doi.org/10.1111/apm.13110
14. Southam DS, Dolovich M, O'Byrne PM, Inman MD (2002) Distribution of intranasal instillations in mice: effects of volume, time, body position, and anesthesia. Am J Physiol Lung Cell Mol Physiol 282(4):L833–L839. https://doi.org/10.1152/ajplung.00173.2001
15. Coelho C, Camacho E, Salas A, Alanio A, Casadevall A (2019) Intranasal inoculation of *Cryptococcus neoformans* in mice produces nasal infection with rapid brain dissemination. mSphere 4(4):e00483-19. https://doi.org/10.1128/mSphere.00483-19
16. Kimura K, Miyazaki S, Tateda K, Matsumoto T, Tsujimoto S, Yamaguchi K (1999) Factors affecting the course and severity of transnasally induced *Staphylococcus aureus* pneumonia in mice. J Med Microbiol 48:1005–1010. https://doi.org/10.1099/00222615-48-11-1005
17. Mizgerd JP, Skerrett SJ (2008) Animal models of human pneumonia. Am J Physiol Lung Cell Mol Physiol 294(3):L387–L398. https://doi.org/10.1152/ajplung.00330.2007
18. Medina E (2010) Murine model of pneumococcal pneumonia. Methods Mol Biol 602:405. https://doi.org/10.1007/978-1-60761-058-8_22
19. Driscoll KE, Costa DL, Hatch G et al (2000) Intratracheal instillation as an exposure technique for the evaluation of respiratory tract toxicity: uses and limitations. Toxicol Sci 55(1):24–35. https://doi.org/10.1093/toxsci/55.1.24
20. Ikeda A, Matsushita S, Sakakibara Y (2009) A simple and reliable method of endotracheal intubation in mice: advantages of exposing the trachea. Scand J Lab Anim Sci 36(4):363–368. https://doi.org/10.23675/sjlas.v36i4.198
21. Spoelstra EN, Ince C, Koeman A et al (2007) A novel and simple method for endotracheal intubation of mice. Lab Anim 41(1):128–135. https://doi.org/10.1258/002367707779399400
22. Vergari A, Gunnella B, Rodolà F et al (2004) A new method of orotracheal intubation in mice. Eur Rev Med Pharmacol Sci 8(3):103–106
23. Liu L, Gao Z, Xia C et al (2012) Comparative study of trans-oral and trans-tracheal intratracheal instillations in a murine model of acute lung injury. Anat Rec 295(9):1513–1519. https://doi.org/10.1002/ar.22531
24. Dos Santos AR, Fraga-Silva TF, de Fátima Almeida-Donanzam D et al (2022) IFN-γ mediated signaling improves fungal clearance in experimental pulmonary mucormycosis. Mycopathologia 187(1):15–30. https://doi.org/10.1007/s11046-021-00598-2
25. Samarasinghe AE, Hoselton SA, Schuh JM (2011) A comparison between intratracheal and inhalation delivery of *Aspergillus fumigatus* conidia in the development of fungal allergic asthma in C57BL/6 mice. Fungal Biol 115(1):21–29. https://doi.org/10.1016/j.funbio.2010.09.013
26. Pinzan CF, Ruas LP, Casabona-Fortunato AS, Carvalho FC, Roque-Barreira MC (2010) Immunological basis for the gender differences in murine Paracoccidioides brasiliensis infection. PLoS One 5(5):e10757. https://doi.org/10.1371/journal.pone.0010757

Chapter 4

Murine Models of Experimental Periodontal Disease

Anna Clara Paiva Menezes dos Santos, Thales Augusto Anestino, Luiza de Almeida Queiroz Ferreira, Mariana Rates Gonzaga Santos, Lívia Maria Alves Ferreira, Pedro Augusto de Oliveira, Celso Martins Queiroz-Junior, and Mila Fernandes Moreira Madeira

Abstract

Periodontal disease (PD) is a chronic inflammatory process of infectious etiology that affects the periodontal tissues. PD is caused by the subgingival biofilm, which, in dysbiosis, leads to an uncontrolled response of the immunological system in the periodontal tissues. To further understand the mechanisms involved in PD and how it is linked to other diseases, several animal models have been developed. These models allow researchers to study the different aspects of PD in a controlled setting, such as its pathogenesis and treatment options. Oral inoculation of periodontal bacteria, such as *Aggregatibacter actinomycetemcomitans* or *Porphyromonas gingivalis*, is one of the most commonly used models for studying PD. In these methods, the bacteria are inoculated directly into the oral cavity, allowing for rapid colonization and development of the disease. Another widely used mouse model for PD involves the application of a silk ligature around the second molar, the ligature triggers oral micro-organisms accumulation inducing an inflammatory response in the surrounding tissues, leading to gingival inflammation and pocket formation. The application of mouse models of PD has several advantages, such as relatively low cost, fast results, and the possibility of performing more accurate studies. In this chapter, we will describe bacteria- and ligature-induced periodontal disease models in detailed steps.

Key words Periodontal disease, Alveolar bone, Bone loss, Dysbiosis, Inflammation, Mice

1 Introduction

The periodontal health of an individual is dependent on the homeostasis between biological and inflammatory markers. The absence of an acute inflammatory process in the gingiva and periodontium is a sign of health, and is characterized by a physiological bone level of 1.0–3.0 mM to the apical margin until the

Authors Celso Martins Queiroz-Junior and Mila Fernandes Moreira Madeira have equally contributed to this chapter.

Lucas Secchim Ribeiro (ed.), *Models in Inflammation and Immunopharmacology*, Methods in Molecular Biology, vol. 2993,
https://doi.org/10.1007/978-1-0716-5019-6_4,
© The Author(s), under exclusive license to Springer Science+Business Media, LLC, part of Springer Nature 2026

cementoenamel junction [1–3]. The initial phase of periodontal disease (PD) is a reversible inflammatory process that occurs with an interplay of the accumulation of a dental biofilm and the host's immune-inflammatory response, and it is limited to the gingiva [2, 4]. When untreated, gingivitis becomes a chronic and multifactorial process leading to the destruction of the tooth-supporting tissues. The loss of the connective tissue and the alveolar bone defines periodontitis. PD results from a chronic inflammatory process that initiates with dysbiosis of the symbiotic oral microbial community. The involved microorganisms are able to subvert the host's immune response to their own benefit, leading to a persistent infection that can lead to severe damage of the periodontal structures [5]. Moreover, individual risk factors such as genotype, smoking, systemic diseases, and diet can influence disease progression [2].

Periodontal health is a complex issue, influenced by a wide range of factors, especially the microorganisms of the supra and subgingival biofilm. In the PD scenario, bacteria such as *Porphyromonas gingivalis*, *Treponema denticola*, *Tannerella forsythia*, *Fusobacterium* spp., *Prevotella* spp., and *Campylobacter* spp. can act as potential pathogens. *P. gingivalis* is a keystone pathogen and its virulence is associated with the production of enzymes such as gingipains, lipid A phosphatases, and other factors associated with pain and alveolar bone loss [5–8]. On the other hand, *Aggregatibacter actinomycetemcomitans* is a Gram-negative bacteria that has been implicated in aggressive periodontal tissue destruction. It is believed to mediate this destruction through apoptosis of the host's cells induced by the release of toxins, such as leukotoxin and cytolethal-distening toxins and also the release of enzymes that degrade collagen and other extracellular matrix components, as well as the modulation of cytokines, prostanoids, and host inflammatory mediators [8–10]. The microorganisms involved in PD can disrupt the host's immune response to damage the periodontal tissues and create a hyperinflammatory stage. Moreover, the products of the tissue destruction are used as nutrients by the microorganisms, which maintain oral dysbiosis [5, 11].

Although it is widely accepted that PD is progressive and cumulative, it is difficult to precisely know the timeline of the disease progression and isolate some variables that are associated with the patients, such as smoking, diet, stress, and chronic inflammation [12]. Thus, studying human PD is challenging. In this way, experimental PD models can provide valuable knowledge of the biological mechanisms of each phase of the disease and allow the testing of novel therapeutic approaches [12–14]. Nevertheless, as any experimental model, experimental mouse models of PD do not reflect the whole clinical setting of a polymicrobial human infection, but tend to be closer to the human condition than in vitro experiments [12].

Experimental PD has been studied mainly using murine models, which allow researchers to mimic the acute and chronic phases of the disease. Common bacteria used in these models are *A. actinomycetemcomitans* and *P. gingivalis* [15–17]. The *A. actinomycetemcomitans*-induced PD model is one of the widely used models to induce experimental PD in different animals, including mice, rats, and non-human primates [18–24]. *A. actinomycetemcomitans* is a periodontopathogen with great destructive capacity and can be easily cultivated in laboratory microaerophilic conditions [25]. This model results in an inflammatory response, with consequent alveolar bone loss, by the upregulation of genes related to bone resorption and activation of osteoclasts [17, 21–24]. *P. gingivalis* is also used experimentally. These bacteria can colonize the periodontal tissues, inducing dysbiosis of the oral microbiota, resulting in alveolar bone loss [15]. Different strains of *P. gingivalis* or mice background can be used for this purpose [16]. Another way in which PD can be studied is by the ligature model [25–27]. This model involves the placement of a sterile silk ligature around the second upper left molar tooth of a mouse or rat. The ligature is left in place for up to 14 days and then removed [28]. The ligature model mimics the acute phase of PD, leading to aggressive alveolar bone loss and it is currently used to study the pathogenesis of PD and the regeneration of the supporting tissues [29]. More recently, models associating a bacterial inoculum with the ligature model have also been developed to allow the colonization of human microorganisms and the host's immune response in vivo [29, 30].

In this chapter, we will describe the experimental mouse model of PD induced by *P. gingivalis* ATCC 33277, an obligated anaerobic pathogen, which is inoculated in mice. Also, we will detail the *A. actinomycetemcomitans* (Aa)-induced PD model, in which the strain ATCC 43718 is inoculated in the oral cavity of mice to promote colonization, facilitate the development of a dysbiotic oral microbiota, and damage periodontal tissues [31]. Finally, the ligature model of PD will be unraveled and the applications of all these models will be discussed.

2 Materials

2.1 Mouse Model of Periodontal Disease Induced by *Aggregatibacter actinomycetemcomitans*

2.1.1 *A. actinomycetemcomitans* Culture

1. Vial of *A. actinomycetemcomitans* strain ATCC 43718.
2. Solid culture media: to perform an experiment with 20 mice, 16 plates with 4% (w/v) Tryptic Soy Agar (TSA) growth medium supplemented with 0.5% (w/v) yeast extract should be prepared. The medium should be suspended in distilled or deionized water, heated until boiling and dissolved, then sterilized in an autoclave at 121 °C for 15 min. The mixture should be cooled between 45 and 50 °C and distributed in 15–20 mL

sterile Petri dishes (90 × 15 mm), dried in the laminar flow hood, and stored at 4 °C until use. Additionally, a bacteria-free control plate should be used at each experimental period to rule out any possible contamination.

3. Liquid culture media: to perform an experiment with 20 mice, 3% (w/v) Trypticase soy broth (TSB) growth medium supplemented with 0.5% (w/v) yeast extract should be suspended in distilled or deionized water, heated to boiling, and mixed until completely dissolved. 5 mL of the solution should be dispensed into sterile test tubes/culture tubes (16 × 100 mm) and sterilized in an autoclave at 121 °C for 15 min. After allowing the tubes to cool, they should be stored at 4 °C until use; 3 tubes will be needed for each bacterial spike, and a bacteria-free control tube should be used at each experimental period to rule out any possible contamination.
4. Laminar flow hood.
5. Jar for microaerobic environment.
6. Pipette (100–1000 μL and 20–200 μL), and sterile tips.
7. Bacteriological loop.
8. Petri dishes (90 × 15 mm).
9. Test tubes/culture tubes (16 × 100 mm).
10. Biochemical oxygen demand (BOD) incubators for 37 °C under conditions of 5–10% CO_2.

2.1.2 Inoculum Preparation

1. Tubes containing TSB with grown *A. actinomycetemcomitans*.
2. Gram staining battery.
3. Sterile 50 mL tubes.
4. Centrifuge.
5. Phosphate Buffered Saline (1× PBS): 8 g/L NaCl, 0.2 g/L KCl, 1.44 g/L Na_2HPO_4, 0.24 g/L KH2PO4 in ddH_2O. Adjust the pH to 7.4, and then add H_2O to complete the solution up to 1 L. Dispense the solution into sterile 50 mL tubes and sterilize them by autoclaving at 121 °C for 20 min. Finally, store PBS at room temperature.
6. Flat-bottomed 96-well plate.
7. Sterile tips and properly calibrated pipettes.
8. Laminar flow hood.
9. Inoculum plus 1.5% (w/v) carboxymethylcellulose (CMC): to prepare the inoculum (100 μL per mouse) for a total of 10 mice, add 15 mg of CMC to 1 mL of the bacterial suspension containing 1×10^9 CFU/mL.

10. PBS-1.5% (w/v) CMC solution: to prepare the PBS-CMC solution (100 μL per mouse) for a total of 10 mice, add 15 mg of CMC to 1 mL of 1× PBS.

2.1.3 Periodontal Induction

1. C57BL6/J male mice from 6 to 8 weeks old.
2. Bioterism facility, duly sanitized micro isolators, and ventilated rack with duly inspected filters.
3. Duly sanitized bottles for filtered water and good quality murine commercial chow.
4. Sterile pipette and tips.
5. Prepared bacterial inoculum plus 1.5% CMC.
6. Prepared PBS-CMC solution.

2.2 Mouse Model of Periodontal Disease Induced by Porphyromonas gingivalis

2.2.1 Porphyromonas gingivalis Culture

1. Vial of *P. gingivalis* strain ATCC 33277.
2. Solid culture media: To perform an experiment with 20 mice, 16 plates of 5.6% (w/v) Brain Heart Infusion Agar (BHI agar) supplemented with 5% (v/v) sheep blood, 1 mg/mL of hemin, and 5 mg/mL of menadione must be prepared. The medium should be suspended in distilled or deionized water, heated until boiling and then sterilized in an autoclave at 121 °C for 15 min. After cooling to 45–50 °C, the mixture should be mixed well and the sheep blood added to the 5.6% w/v BHI agar, followed by the addition of hemine and menadione. Then, 15–20 mL of the mixture should be distributed into sterile Petri dishes (90 × 15 mm) and left to dry in the laminar flow hood. To prevent contamination, a bacteria-free control plate should be used at each experimental period, and the plates should be stored at 4 °C until use.
3. Liquid culture media: to perform an experiment with 20 mice, prepare 5 tubes of 3.7% (w/v) Brain Heart Infusion broth (BHI broth) growth medium suspended in distilled or deionized water, supplemented with 1 mg/mL of hemin, and 5 mg/mL of menadione. Heat to boiling and mix until completely dissolved, then dispense 5 mL into sterile test tubes/culture tubes (16 × 100 mm).
4. Laminar flow hood.
5. Chamber or 2.5 L jar for anaerobic environment.
6. Pipettes (100–1000 μL and 20–200 μL), and sterile tips.
7. Bacteriological loop.
8. Petri dishes (90 × 15 mm).
9. Test tubes/culture tubes (16 × 100 mm).
10. Anaerobic generator.
11. Distilled or deionized water.

12. Carbon dioxide indicator tube.
13. Anaerobiosis indicator strip.
14. Biochemical oxygen demand (BOD) incubators for 37 °C under conditions of 5–10% CO_2.

2.2.2 Inoculum Preparation

1. Sterile tubes contain BHI broth and grown bacteria.
2. Gram staining battery.
3. Sterile 50-mL tubes.
4. Centrifuge.
5. 1× PBS: *see* Subheading 2.1.2, **item 5**.
6. Flat-bottomed 96-well plate.
7. Inoculum plus 1.5% CMC: *see* Subheading 2.1.2, **item 10**.
8. PBS-CMC solution: *see* Subheading 2.1.2, **item 9**.

2.2.3 Periodontal Induction

1. C57BL6/J male mice from 6 to 8 weeks old.
2. Animal facility, duly sanitized micro isolators, and ventilated rack with duly inspected filters.
3. Duly sanitized bottles for filtered water and good-quality murine commercial chow.
4. Sterile pipette and tips.
5. Inoculum plus CMC; *see* Subheading 2.1.2, **item 10.**
6. PBS-CMC solution: *see* Subheading 2.1.2, **item 9.**

2.3 Mouse Model of Ligature-Induced Periodontal Disease

1. C57BL6/J male mice from 6 to 8 weeks old.
2. Anesthetic solution: add 1.8 mL of 10% (w/v) ketamine hydrochloride and 1.5 mL of 2% (w/v) xylazine to 6.7 mL of sterile 0.9% saline solution, so that, when injecting 100 μL of the solution intraperitoneally, a mouse weighing 20 g receives 90 mg/kg ketamine hydrochloride and 15 mg/kg xylazine.
3. Operating table that allows mouse mouth opening for visualization of the upper molars (made by hand or 3D printed—Fig. 1a, c).
4. Sterile 1 mL syringe and needle (0.3 × 13 mm).
5. Sterile Silk thread (6-0).
6. Two ophthalmic needle holder tweezers (Fig. 1b).
7. Ophthalmic scissors.
8. Lighting.

2.4 Quantification of Alveolar Bone Loss

1. Container to identify and accommodate the sectioned maxillae. (Optional: Reusable 24-well tissue culture plates, each well containing a properly identified hemimaxilla).

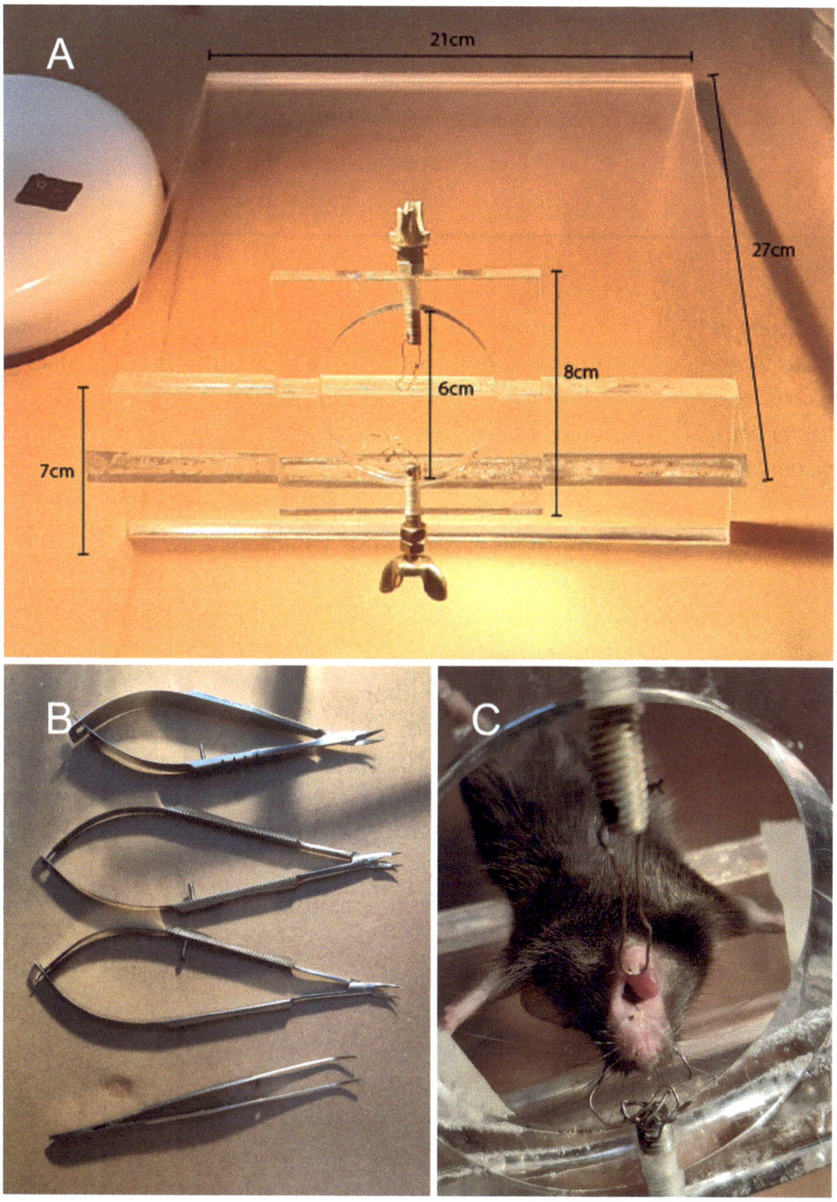

Fig. 1 Surgical field. (**a**) The surgical table was used for positioning the mouse during the procedure. (**b**) Tweezers and scissors are essential tools for the surgical procedure. (**c**) Demonstration of the animal positioning for the procedure. The mouse's incisors will be secured in the metal arches of the surgical table, keeping the mouth wide open for the passage of the silk thread

2. 500 μL of 30% hydrogen peroxide for each hemimaxilla, a total of 10 mL for 20 hemimaxilla.
3. Scissors and tweezers.
4. 200 μL of 0.3% methylene blue for each hemimaxilla, a total of 4 mL.

5. Dental pink utility wax and millimeter ruler.
6. Stereomicroscope and a digital camera.
7. A computer with ImageJ/Fiji, freeware program available at https://imagej.net/Fiji and a software for creating graphs and tables (e.g., GraphPad Prism).

3 Methods

3.1 Mouse Model of Periodontal Disease Induced by *Aggregatibacter actinomycetemcomitans*

3.1.1 A. actinomycetemcomitans *Culture*

1. To ensure the success of every experiment, it is essential to properly prepare the TSA and TSB culture media for bacteria culture at least 24 h in advance.
2. Let a − 80 °C frozen vial of *A. actinomycetemcomitans* strain ATCC 43718 (*see* **Note 1**) at room temperature.
3. Once the bacterial vial has been completely thawed inside the laminar flow hood, use a micropipette to gently homogenize the contents and aliquot 100 μL of it onto the surface of a Petri dish containing the TSA solid medium.
4. Spread the bacteria over the entire surface of the medium with a bacteriological loop.
5. Incubate the Petri dishes in a microaerophilic atmosphere at 37 °C, under 5–10% of CO_2 using a glass candle jar (a lit candle is carefully placed inside the jar before sealing the airtight lid to create a microaerobic atmosphere) in a BOD incubator, for 2 days.
6. Two days after incubation, the bacterial growth can be observed by the formation of translucent colonies with irregular edges and rough surfaces strongly adhered to the agar (Fig. 2a).
7. Then, at this point perform:
   (a) Bacterial culture in TSB: Using a bacteriological loop, transfer as many colonies as possible from the TSA plates to the TSB and place the TSB in the glass candle jar. Label the glass jar with the strain name, date, and time. Place the jar in the BOD incubator, set the incubation conditions (microaerophile, 37 °C, 5–10% CO_2), and incubate it for 20–24 h. Once the incubation period is complete, observe the broth for bacterial growth. If the bacterial growth is visible, then you can use the bacterial culture for the first time of infection.
   (b) Bacterial culture in TSA to expand the culture: Bacterial colonies can be collected from the TSA plates using a bacteriological loop and plated on fresh Petri dishes containing TSA culture medium. The Petri dishes should then be incubated under a microaerophilic atmosphere

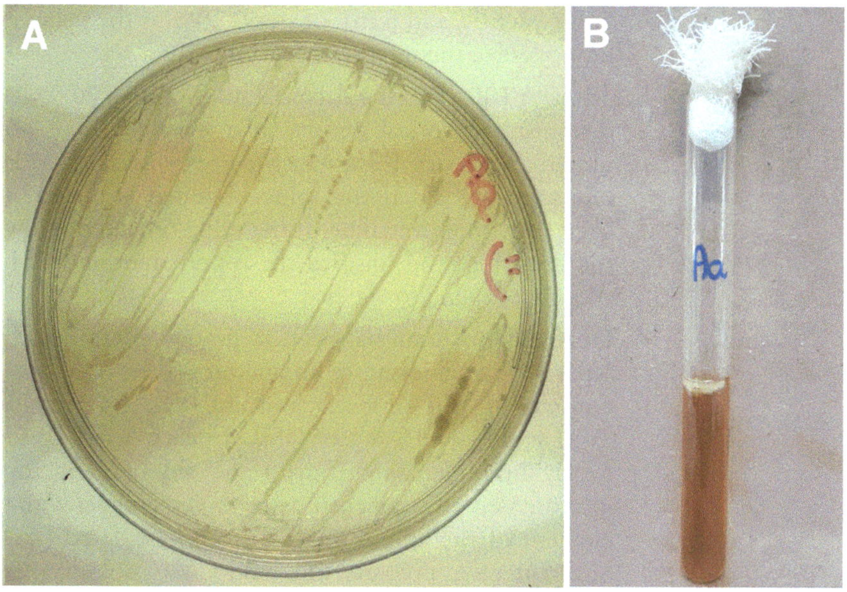

Fig. 2 Cultivation of *Aggregatibacter actinomycetemcomitans*. (**a**) Translucent colonies with irregular edges and rough surfaces strongly adhered to the agar. (**b**) TSB media with an expressive whitish, irregular, and floating aggregation of bacteria

with 5–10% of CO_2 at 37 °C in a BOD incubator for 48 hours. The newly formed colonies can be used for the next cycle of infection.

8. Four days after initial incubation (Subheading 3.1.1, **items 3** and **4**), repeat the instructions in Subheading 3.1.1, **item** 7, line "a" for the second cycle of infection and line "b" for the third cycle of infection.
9. Six days after initial incubation (Subheading 3.1.1, **items 3** and **4**), repeat the instructions in Subheading 3.1.1, **item** 7, line "a" for the third cycle of infection.

3.1.2 Inoculum Preparation

1. After 3 days of initial incubation of the bacteria (2 days in TSA and 1 day in TSB) in a properly prepared laminar flow hood, a slight turbidity of the TSB media can be observed, demonstrating an expressive whitish, irregular, and floating aggregation of bacteria (Fig. 2b). To verify the purity of the culture, a Gram stain should be performed (*see* **Note 2**).
2. Transfer the TSB with bacteria growth into a sterile 50 mL tube, and centrifuge it at 750 g for 5 min at 24 °C.
3. At the end of the centrifugation, observe the formation of a bacterial pellet at the bottom of the tube. Carefully decant the supernatant, discarding it properly, taking care not to disturb the pellet.

4. Subsequently, add 5 mL of sterile PBS to the pellet and gently homogenize the mixture with a pipette.
5. Transfer 100 μL of the bacterial solution to a 96-well plate, and measure the optical density at a wavelength of 560 nM to obtain an A_{560} of 1.20, which corresponds to 1×10^9 CFU/ml (*see* **Note 3**). If the value obtained is higher, dilute the bacteria with more saline. If the value obtained is lower, repeat the centrifugation, discard the supernatant and add a smaller volume of saline to the pellet.
6. Once the inoculum has been accurately assessed and prepared, separate the amount of inoculum to be used for all mice, considering 100 μL for each mouse (e.g., for 10 mice separate 1 mL) and add 15 mg of CMC to each 1 mL of suspension, gently homogenizing it with the aid of a micropipette to increase viscosity and ensure prolonged contact with the dental element of the animal that is to be inoculated (*see* **Note 4**). Subsequently, prepare the PBS-CMC solution by adding 15 mg of CMC to 1 mL of sterile PBS, and again homogenizing it gently with the aid of a micropipette.
7. After 2 days of first cycle of infection (as per new TSA plates and TSB broth made according to Subheading 3.1.1 **item** 7 line "b", and **item 8**, respectively), repeat the instructions from Subheading 3.1.2 **items 2–6** for a second cycle of infection.
8. After 4 days of the first cycle of infection (according to new TSB broth made as described in Subheading 3.1.1 **item 9**), repeat the instructions from Subheading 3.1.2 **items 2–6**, for the third cycle of infection.

3.1.3 Periodontal Infection

1. On the third day after the initial incubation of the bacteria (2 days in TSA and 1 day in TSB), the first cycle of infection will be carried out once the inoculum has been prepared.
2. After inoculum preparation, mice should be correctly grouped according to their identifiers.
3. Take the group that will receive the inoculum and manually restrain each mouse carefully. With the aid of a sterile pipette and tip, aliquot 100 μL of the inoculum containing the bacteria plus 1.5% CMC and then carefully dispense it into the mouse's oral cavity (Fig. 3).
4. Once the inoculum is dispensed into the oral cavity of each mouse in the group, carefully return the animal to its micro-isolator. For mice in the control group, be sure to add 100 μL of the PBS-CMC solution to their oral cavity. Preferentially, start the experiment with the control group, to avoid any possible contamination.

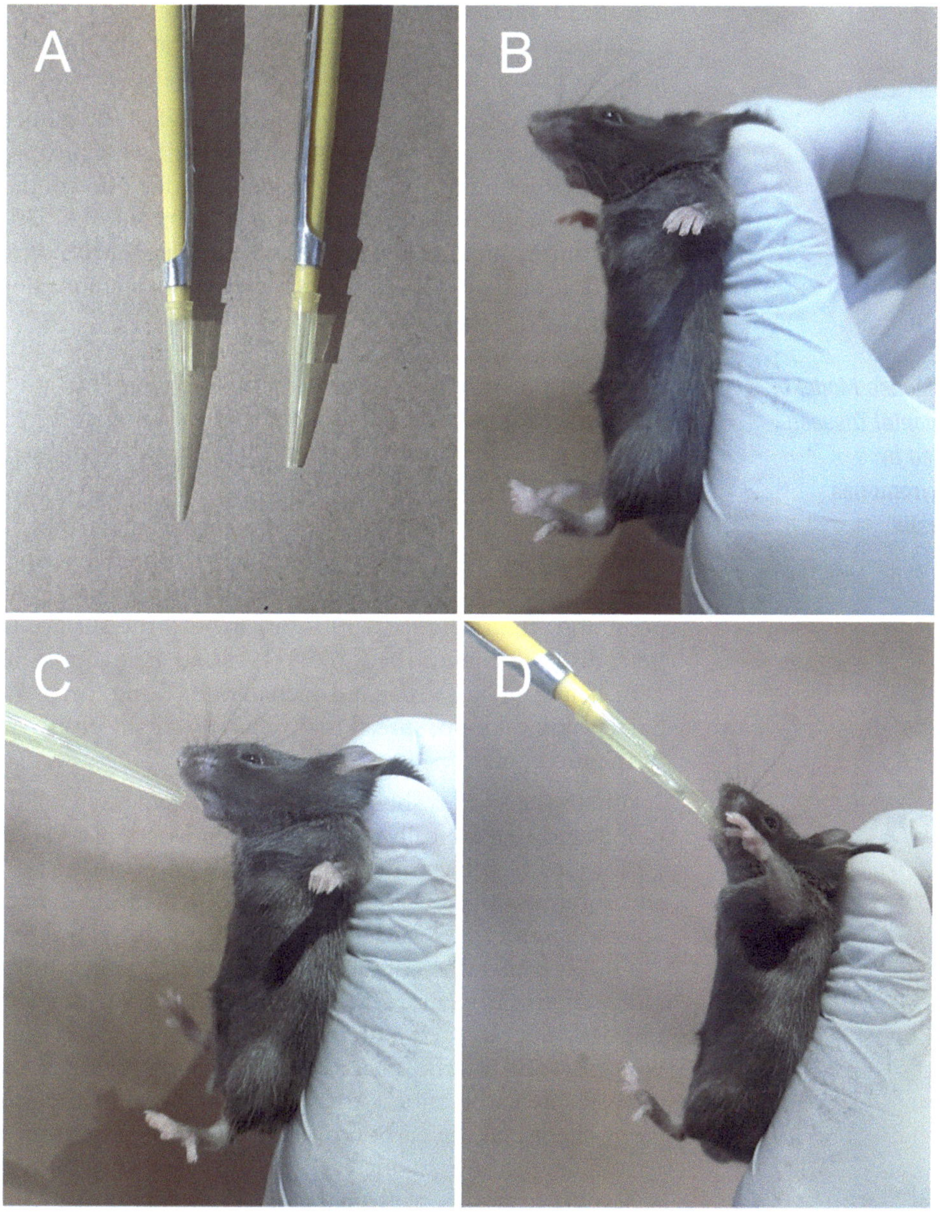

Fig. 3 Periodontal induction. (**a**) Micropipette and tips duly trimmed for inoculation. (**b**) Manual immobilization of the animal. (**c, d**) Bacterial inoculation in the oral cavity of the animal

5. Following inoculation, the experimental groups are subjected to 30 minutes of water deprivation and 4 hours of feed deprivation to prevent any mechanical interference that might impair the successful colonization of the bacteria.
6. The second cycle of infection is carried out 2 days after the first infection cycle. For this, repeat the instructions from Subheading 3.1.3 **items 2–5**.

7. The third cycle of infection is carried out 2 days after the second cycle of infection. For this, repeat the instructions from Subheading 3.1.3 **items 2–5**.
8. The primary phenotype of the mouse model of PD induced by *A. actinomycetemcomitans* is alveolar bone loss, which is evaluated by obtaining maxillary tissues. Under our conditions, significant alveolar bone resorption is detected after 30 days of the third round of infection. The euthanasia procedure and collection of the samples will be described in Subheading 3.4 (*see* **Note 5**).

3.2 Mouse Model of Periodontal Disease Induced by Porphyromonas gingivalis

3.2.1 P. gingivalis Culture

1. Before using a cryopreservation tube containing a vial of *P. gingivalis* ATCC 33277, ensure that the laminar flow hood or the anaerobic chamber is properly prepared with the necessary material (*see* **Note 6**).
2. Inside the laminar flow hood or the anaerobic chamber, after visualizing the complete thawing of the bacterial vial, gently homogenize the bacterial content with the aid of a micropipette and, then, aliquot 100 μL of it onto the surface of the Petri dish containing the 5.6% BHI blood agar supplemented with hemin (1 mg/mL) and menadione (5 mg/mL). With a bacteriological loop, spread the bacteria over the entire surface of the medium.
3. Incubate the Petri dishes containing supplemented solid BHI blood agar anaerobically at 37 °C, under 5–10% of CO_2 using a jar for an anaerobic atmosphere, in a BOD incubator or in an anaerobic chamber for 7–11 days. To generate an anaerobic atmosphere, a 2.5 L anaerobic jar is required, along with an anaerobic generator composed of diatomite, iron, citric acid, and sodium carbonate, which, when in contact with 15 mL of distilled or deionized water, release CO_2. Measure the CO_2 concentration in the anaerobic atmosphere using a carbon dioxide indicator tube. After 15 min, observe the gradual color change of the carbon dioxide indicator tube from beige to red, indicating an increase in CO_2 concentration in the anaerobic atmosphere from 4% to 10%. Additionally, use an anaerobiosis indicator strip to measure the absence of O_2 in the atmosphere of the jar, which should become white in 6–10 h of incubation.
4. After 7–11 days of incubation, bacterial growth can be observed as a mat or isolated dark-colored colonies, often with a translucid halo around them and a characteristic odor from the production of volatile sulfur compounds, such as hydrogen sulfide (*see* **Note 7**) (Fig. 4a).
5. Then, at this point perform:

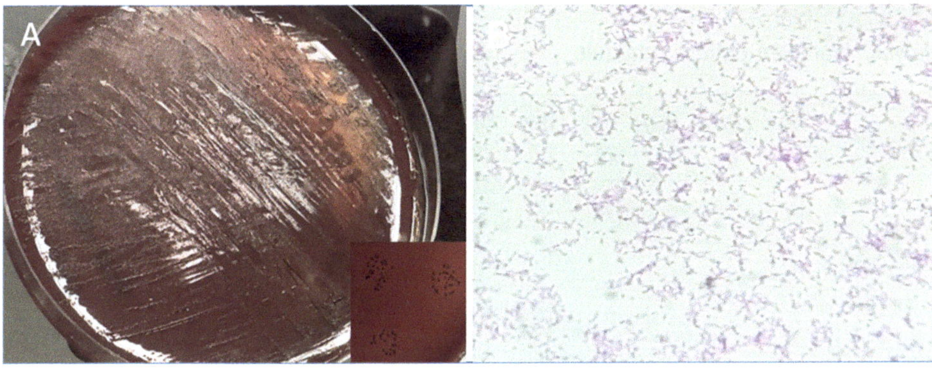

Fig. 4 Cultivation of *Porphyromonas gingivalis*. (**a**) Isolated dark-colored colonies, often with a translucid halo around them (**b**) Gram stain demonstrating morphological and staining characteristics of *P. gingivalis* evidencing the presence of Gram-negative coccobacilli

(a) Bacterial culture: using a bacteriological loop, transfer as many colonies as possible from the BHI blood agar plates to the BHI broth and place it in anaerobic conditions, at 37 °C, with 4–10% CO_2 using an anaerobic jar in a BOD incubator or anaerobic chamber. Label the anaerobic jar with the strain name, date, and time and incubate the bacterial culture for 48 h. After incubation, observe the broth for bacterial growth. If bacterial growth is observed, then use the culture for the first cycle of infection.

(b) Bacterial culture in supplemented solid BHI blood agar to expand the culture: bacterial colonies are collected with a bacteriological loop from the BHI blood agar and plated on new Petri dishes with supplemented solid BHI blood agar culture medium, incubated under anaerobic conditions, at 37 °C, 4–10% of CO_2 using an anaerobic jar in a BOD incubator or anaerobic chamber, for another 7 days (the fresh colonies will be used for the next infection cycle). Initially, it is important to produce at least 4 BHI blood agar plates with bacteria so that they can be used to maintain the culture of the bacteria for the days of infection, according to the instructions in the image (Table 1).

3.2.2 Inoculum Preparation

1. In a laminar flow hood, take the BHI broth with the bacterial growth (turbidity visualization), perform the Gram stain (*see* **Note 2**) to further check the purity of the culture (Fig. 4b). Then, transfer the contents of the medium tubes to sterile 50 mL tubes, close these tubes tightly, and centrifuge them at 750 g for 6 min at 24 °C.
2. After centrifugation, observe the formation of a bacterial pellet at the bottom of the tube. Discard the supernatant, carefully avoiding mixing the pellet with the now translucent supernatant.

Table 1
***P. gingivalis* inoculum culture schedule**

	Day 0	Day 7	Day 14	Day 21	Day 27	Day 28	Day *29*	Day 30	Day 31	Day 32	Day 33	Day 34	Day 36	Day 38
Initial culture on BHI blood agar	X													
Repeat onto a new BHI blood agar plate		X	X	X	X	X	X		X		X			
Replicate to a tube of BHI broth								X		X		X	X	
Inoculation in the oral cavity of animals										X		X	X	X

Each row of the table describes a step of the re-spikings that must be done to obtain the necessary *P. gingivalis* inoculum. The columns show the days when these steps must be performed

3. Next, add 1 mL of sterile 1× PBS to the bacterial pellet and homogenize it mechanically with a pipette.
4. Transfer 100 μL of the bacterial solution to a flat-bottomed 96-well plate and quantify it at a wavelength of 600 nm using a microplate reader. An A600 of 0.9 should be obtained, which corresponds to 1×10^9 CFU/mL. If values higher than this are obtained, dilute the inoculum by adding more 1× PBS. Conversely, if values lower than this are obtained, repeat the centrifugation, discard the supernatant, and add a smaller volume of 1× PBS to the pellet.
5. Once the inoculum has been correctly obtained, add 1.5% CMC and homogenize it using a pipette. Additionally, prepare another tube containing only sterile 1× PBS and 1.5% CMC. This CMC helps promote bacterial adhesion to the teeth of the inoculated mice, thereby allowing for a longer contact time.

3.2.3 Periodontal Induction

1. After grouping and identifying the mice, use a micropipette to pipette 100 μL of PBS-CMC solution and manually immobilize the mouse. Carefully dispense the solution into the mouse's oral cavity, and once finished, ensure that all mice are returned to their microisolator compartments.
2. Subsequently, perform the inoculation in the oral cavity of mice belonging to the infected group. With the aid of a micropipette, pipette 100 μL of the inoculum plus 1.5% CMC,

manually immobilize the mouse, and carefully dispense the solution into the mouse's oral cavity. Make sure that all mice are returned to their microisolator compartments afterward (Fig. 3).

3. Following inoculation, all experimental groups are subjected to 30 min of water deprivation and 4 h of feed deprivation to prevent any mechanical interference that might impede the successful colonization of the bacteria.
4. Repeat the inoculation process 4 times every 48 h, for 8 days, taking care to use fresh bacterial inoculum each time. Follow the instructions in Subheading 3.2.1, **items 5** and **6**. To do this, it is important to organize yourself to subculture the bacteria in BHI agar medium and BHI broth medium, following a routine of *P. gingivalis* culture and inoculum preparation (*see* **Note 8**).
5. The primary phenotype of the mouse model of PD induced by *P. gingivalis* is alveolar bone loss. Under our conditions, significant alveolar bone loss is observed 45 days after infection, when mice are euthanized for collection of the samples, as described in Subheading 3.4.

3.3 Mouse Model of Ligature-Induced Periodontal Disease

1. Prepare the surgical field and separate the materials necessary for placing the ligature (Fig. 1).
2. Mice should be anesthetized with an intraperitoneal injection of a solution of xylazine and ketamine (*see* Subsection 2.3, **item 2**). Using appropriate personal protective equipment (PPE), take the anesthetic solution with a sterile 1 mL syringe and needle. Remove any bubbles. Manually restrain the mouse with your thumb and forefinger. The mouse's tail should be between the pinky and the palm of the hand. Inject the anesthetic intraperitoneally and withdraw the needle. Return the mouse to the cage and wait for it to fall to start the procedure. To place the ligature, the mouse cannot move and must not feel pain (*see* **Note 9**).
3. Place the mouse facing up on the operating table with its mouth wide open. Attach the mouse's paws to the table with adhesive tape. Position the mouse's tongue outside the mouth, with the tweezers, to facilitate visualization of the upper molar teeth (Fig. 1c) (*see* **Note 10**).
4. Using two ophthalmic tweezers, hold the ligature thread 2 mm apart between the mouse's first and second molars, positioning the tweezers in the buccal vestibule and on the palate. A reference point is the distal cusp of the first upper molar (Fig. 5). The thread should be positioned distally to the distal cusp of the first molar. Gently make back-and-forth movements with the ligature (*see* **Note 11**) while applying some pressure toward

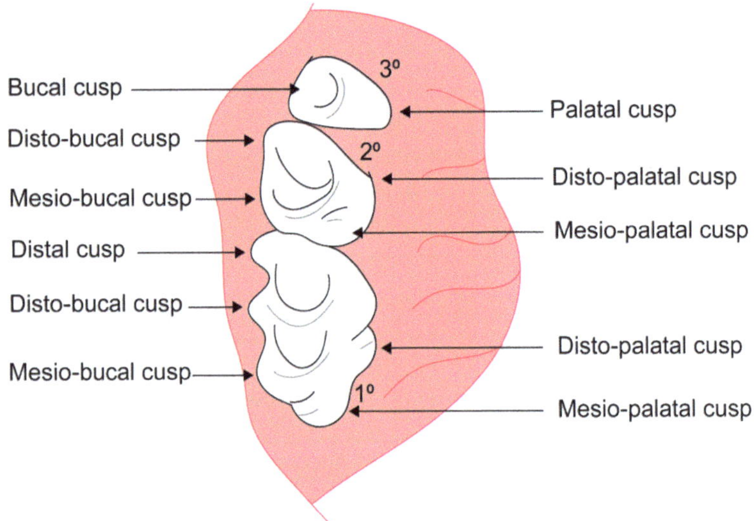

Fig. 5 Illustration of mouse molars. Drawing to help identify the cusps of mouse molars. Anatomical landmarks can help with ligature placement

the space between the teeth, being mindful not to injure the soft tissues of the cheek or palate (Fig. 6b).

5. After positioning the ligature between the first and second molars (Fig. 6b), the thread should be pulled from the buccal region. Then, the tweezers should be positioned about 2 mm of distance, in the section of the thread that was in the palatal region. The procedure described in the previous topic should be repeated to position the ligature between the second and third molars. A reference point is the disto-palatal cusp of the second upper molar. The thread should be positioned distally to the disto-palatal cusp of the second molar (Fig. 5).
6. After placing the ligature between the second and third molars (Fig. 6c), the tip of the thread should be pulled toward the buccal region, leaving the ligature next to the gingival sulcus of the second molar.
7. Finally, make a simple surgical knot joining the parts of the thread in the buccal region (Fig. 7). Hold one end of the ligature in your hand and pass the thread that will be on the hand around a tweezer twice (Fig. 7a). Then, hold the other end of the ligature with the tweezer and pull both sides of the thread (one with your hand for one side, and the other with the tweezer for the opposite side) (Fig. 7b). This will cause the loops around the forceps to slide closer to the molars, tightening the knot.
8. To finish off the knot, hold one end of the ligature in your hand and pass the thread that will be on the hand around the tweezer, just once. Hold the other end of the ligature with the tweezer and pull both sides of the thread, as done before.

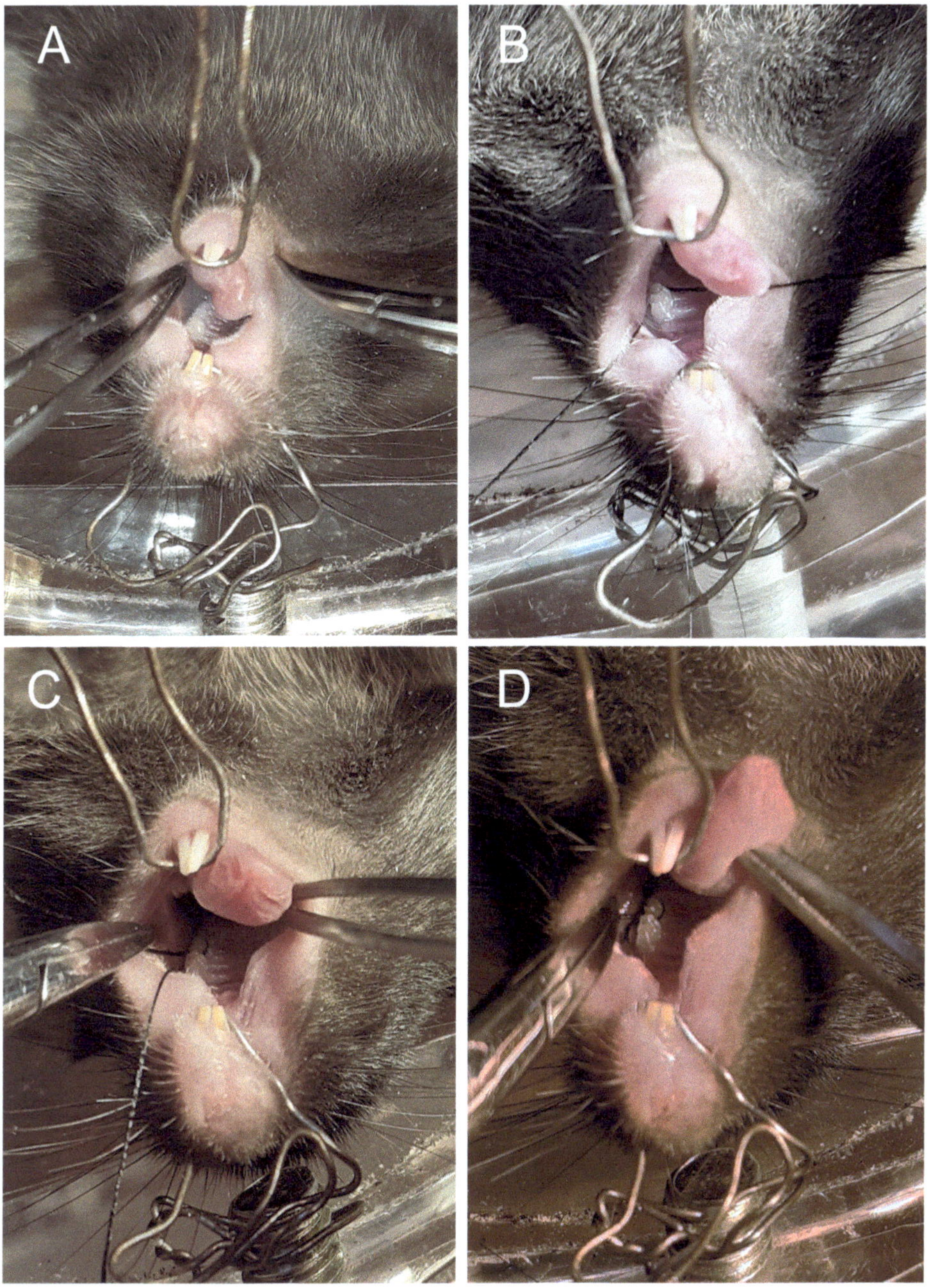

Fig. 6 Thread passing steps. (**a**) The mouse's mouth wide open to visualize the molars. (**b**) Thread placed between the first (1°) and second (2°) molars. (**c**) Thread placed between second and third (3°) molars. (**d**). How the ligature should look after the knot

9. Cut the excess thread, taking care not to undo or cut the knot, but snip off any large end of the thread that is in the mouse's mouth (Fig. 6d). It is safe to leave a tip of up to 5 mm after the ligature knot.

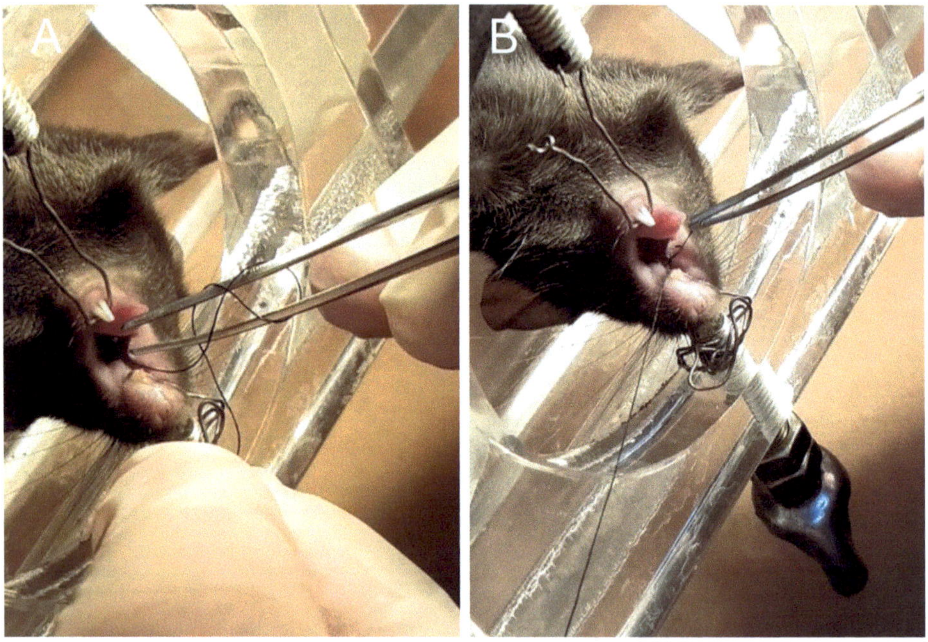

Fig. 7 Surgical knot. (**a**) How to pass the thread around the collet twice. (**b**) Finish the knot by pulling the ends of the thread, with the tweezers on one side and with your hand on the other

10. The experiment time for evaluating the PD phenotype may vary, depending on a characterization curve of parameters such as alveolar bone loss and the number of osteoclasts (*see* **Note 12**). In our research group, we standardized performing euthanasia 4 days after the ligation.

3.4 Quantification of Alveolar Bone Loss

1. After the trial period for each mouse model of PD, 300 μL of anesthetic (described in Subheading 2.3, **item 1**) must be intraperitoneally injected with a syringe and needle. For this, the animal must be immobilized by the back with the ventral surface facing upwards, and the head tilted downwards at a 45-degree angle in order to position the intestines cranially and avoid perforation. The needle should be inserted into the lower right quadrant of the abdomen at a 30–45-degree angle and the anesthetic administered in a dose of 250–300 mg/kg of ketamine and 25–30 mg/kg of xylazine, after which the animal must be euthanized.
2. After euthanasia, mouse maxillae should be carefully obtained with the aid of scissors and tweezers by first cutting the skin and soft external tissues. The entire maxilla can then be removed by either (i) cutting the bone from the nose and keeping both incisors or (ii) obtaining the hemimaxilla, breaking the front portion of both incisors and the jointed region of anterior maxilla between the two incisors and the vomer bone. The

hemisected maxilla should then be placed in a 24-well plate, each well containing a correctly identified hemimaxilla, and left to soak overnight in 500 μL of 15% v/v hydrogen peroxide each well.

3. Using tweezers, mechanically dissect the hemimaxilla to remove all the soft tissue, exposing the bone tissue and teeth. Subsequently, wash each well containing the hemimaxilla three times with distilled water and allow it to dry overnight.
4. Stain the mouse maxilla with 0.3% methylene blue for 30 min to clearly identify the cement-enamel junction (CEJ).
5. Wash off the excess dye with tap water and allow the hemimaxilla to dry overnight.
6. For each hemimaxilla sample of the infected and control group, position the buccal side of the crowns of the teeth parallel to the horizontal plane and fix the hemimaxilla on wax. Then, using a stereomicroscope and a digital camera, take photographs of the palatal surface at 20× magnification. Additionally, associate the sample with a millimeter ruler fixed inferiorly to the position of the maxilla to enable measurements.
7. Using ImageJ software, quantitative analysis is employed to measure the palatal area between the CEJ and the alveolar bone crest (ABC) in the first molar (*see* **Note 13**). Install and prepare ImageJ Software for analysis:
   (a) Download ImageJ software from the NIH website (http://rsbweb.nih.gov/ij).
   (b) Install and open the ImageJ software.
   (c) Drag the image to be analyzed (TIFF or DM3 file) directly into ImageJ. Alternatively, click File Open to open the selected image. For accuracy and reproducibility, ensure that each image contains a millimeter ruler. This is important for setting the appropriate units within the ImageJ settings.
   (d) After opening the image, select the icon Straight, Segmented or Freehand lines, or Arrows, on the toolbar, and mark the position 0 on the millimeter ruler present at the bottom edge of image and drag it to the position of 1.00 mM, used to record and keep track of measurements. Next select Analyze Set scale Know distance, fill in the Known distance field with 1.00 and the length Unit Field with millimeters. Check the Global box to standardize the scale for all images.
   (e) Then, click on Freehand Selections on the toolbar to access the Freehand tool and trace the outline of the entire area between the CEJ and the alveolar bone crest (ABC) in the first molar. Click on Analyze Measure, to take the

measurements that ImageJ will make. Two independent measurements of the area (square millimeters) between the CEJ and the ABC are taken and the average of these measurements is used for each animal.

8. Organize the measurements of each group of animals in tables. Use this data for plotting graphs.

4 Notes

1. *A. actinomycetemcomitans* is a Gram-negative bacterium that presents different serotypes based on lipopolysaccharide (LPS) antigenicity [1]. The *A. actinomycetemcomitans* serotypes differ in their ability to induce response in host immune cells; of the seven *A. actinomycetemcomitans* serotypes, the serotype b, in particular, has shown a higher immunogenicity, and it is isolated most often from patients affected with periodontitis when compared with the other serotypes [1].
2. Before the preparation of the inoculum, it is necessary to perform the Gram stain of the slide, respecting the protocol time for each dye/solution: crystal violet, iodine, decolorizer solution, and fuchsin. After staining the slide, it should be placed to dry. As *A. actinomycetemcomitans* is a Gram-negative coccobacillus and *P. gingivalis* is a Gram-negative coccobacilli, they should appear red in color when viewed under a microscope.
3. From the standardization of "optical density x colony forming units (CFU)" it is possible to quantify the number of bacteria (CFU/mL) in your inoculum. As previously verified by the microdrop culture technique, the absorbance of 1.20 at an A_{560} corresponds to 10^9 CFU/mL of bacteria, as defined by previous experiments carried out in our laboratory, through the microdrop technique.
4. Colonization is a key component of survival for microorganisms in the hazardous environment of the oral cavity, and it is well-established *that A. actinomycetemcomitans* colonization of teeth is a critical step in the infectious process [6, 7]. The use of CMC contributes to tooth adherence and biofilm persistence, which are prerequisites for survival in the oral domain.
5. In a mouse model of PD induced by *A. actinomycetemcomitans*, in our conditions, a significant alveolar bone loss is not observed before 30 days after infection [2].
6. If the microorganism will be cultured in an anaerobic jar, thawing should be carried out in the laminar flow hood. Additionally, it is important to properly prepare the anaerobic generators and indicators supplied and guided by the

manufacturer. Every process carried out in aerobiosis must be done efficiently and quickly; however, it is important to remember that *P. gingivalis* is an obligate anaerobic microorganism, which can survive in aerobic conditions for up to 1 h of exposure. Therefore, it is advisable to consider working within this time frame or even shorter intervals.

7. The number of replications performed with the periodontopathogen can significantly affect the phenotype and pathogenicity potential. To ensure the reproducibility of the experimental model, it is essential to carefully consider the exact number of 4 replications.
8. Assessing the interval of 7 days in culture conditions is crucial for planning an effective bacterial culture routine and assembling the inoculum, as colonies of *P. gingivalis* typically develop well within that timeframe.
9. It is important to be extremely cautious when administering an anesthetic to each animal; mice should be well anesthetized, but administering an anesthetic concentration higher than recommended can be fatal for the animal.
10. Visualization of molars with the naked eye can be challenging for those without training. To make the process easier, a head magnifying glass or a table adapted to a stereo microscope can be used. This will allow for a more detailed and accurate observation of the molars.
11. To place the ligature between the mouse's molar, it is necessary to make, with the tweezers, a back-and-forth motion similar to that used for flossing teeth in our daily lives, while pressing firmly.
12. The ligature model induces an increase in alveolar bone loss from day 3 to day 9 (remaining stable from day 9 to day 18) with a notable rise in the number of osteoclasts in the bone crest of the second molar. This amount drops between days 9 and 18 [29].
13. Microtomography (MicroCT) scanning of maxillae obtained from control and infected mice can be used to quantitatively assess alveolar bone loss by measuring the area (μm^2) between the cement-enamel junction (CEJ) and the alveolar bone crest (ABC) in three-dimensional images (Fiji—National Institutes of Health, USA) of the first, second, and third molars.

Acknowledgments

We would like to thank Pró-Reitoria de Pesquisa/UFMG (PRPq/UFMG), Conselho Nacional de Desenvolvimento Científico e Tecnológico—CNPq—(408433/2018-3) and Fundação de Amparo a

Pesquisa de Minas Gerais (FAPEMIG), Coordenação de Aperfeiçoamento de Pessoal de Nível Superior (CAPES) (Brazil), and the National Institute for Science and Technology in Dengue and Host-microbial interactions (APQ-03606-17).

Funding Mila Madeira has received a research grant from FAPEMIG (APQ-01790-22). Celso Queiroz-Junior has received a research grant from a FINEP-funded project (MCTI/FINEP-MS/SCTIE/DGITIS/CGITS, grant number 6205283BBB28-4F9C-AA65-808FE4450542). The authors have no conflicts of interest to declare that are relevant to the content of this chapter.

References

1. Chapple ILC, Bouchard P, Cagetti MG et al (2017) Interaction of lifestyle, behaviour or systemic diseases with dental caries and periodontal diseases: consensus report of group 2 of the joint EFP/ORCA workshop on the boundaries between caries and periodontal diseases. J Clin Periodontol 44(18):S39–S51. https://doi.org/10.1111/jcpe.12685
2. Chapple ILC, Mealey BL, Van Dyke TE et al (2018) Periodontal health and gingival diseases and conditions on an intact and a reduced periodontium: consensus report of workgroup 1 of the 2017 world workshop on the classification of periodontal and peri-implant diseases and conditions. J Periodontol 89(S1):S74–S84. https://doi.org/10.1002/jper.17-0719
3. Papapanou PN, Sanz M, Buduneli N et al (2018) Periodontitis: consensus report of workgroup 2 of the 2017 world workshop on the classification of periodontal and peri-implant diseases and conditions. J Periodontol 89(S1):S173–S182. https://doi.org/10.1002/jper.17-0721
4. Murakami S, Mealey BL, Mariotti A, Chapple ILC (2018) Dental plaque-induced gingival conditions. J Periodontol 89(S1):S17–S27. https://doi.org/10.1002/jper.17-0095
5. Hajishengallis G (2015) Periodontitis: from microbial immune subversion to systemic inflammation. Nat Rev Immunol 15(1):30–44. https://doi.org/10.1038/nri3785
6. Socransky SS, Haffajee AD, Cugini MA, Smith C, Kent-Jr RL (1998) Microbial complexes in subgingival plaque. J Clin Periodontol 25(2):134–144. https://doi.org/10.1111/j.1600-051x.1998.tb02419.x
7. Pérez-Chaparro PJ, Gonçalves C, Figueiredo LC, Faveri M, Lobão E, Tamashiro N, Duarte P, Feres M (2014) Newly identified pathogens associated with periodontitis: a systematic review. J Dent Res 93(9):846–858. https://doi.org/10.1177/0022034514542468
8. Darveau RP, Hajishengallis G, Curtis MA (2012) Porphyromonas gingivalis as a potential community activist for disease. J Dent Res 91(9):816–820. https://doi.org/10.1177/0022034512453589
9. Hiyoshi T, Domon H, Maekawa T, Nagai K, Tamura H, Takahashi N, Yonezawa D, Miroshi T, Yoshida A, Tabeta K, Terao Y (2019) *Aggregatibacter actinomycetemcomitans* induces detachment and death of human gingival epithelial cells and fibroblasts via elastase release following leukotoxin-dependent neutrophil lysis. Microbiol Immunol 64(3–4):100–110. https://doi.org/10.1111/1348-0421.12672
10. Kachlany SC, Schwartz AB, Balashova NV, Hioe CE, Tuen M, Le A, Kaur M, Mei Y, Rao J (2010) Anti-leukemia activity of a bacterial toxin with natural specificity for LFA-1 on white blood cells. Leuk Res 34(6):777–785. https://doi.org/10.1016/j.leukres.2009.08.022
11. Hajishengallis G, Chavakis T (2021) Local and systemic mechanisms linking periodontal disease and inflammatory comorbidities. Nat Rev Immunol 21(7):426–440. https://doi.org/10.1038/s41577-020-00488-6
12. Rojas C, García MP, Polanco AF, González-Osuna L, Sierra-Cristancho A, Melgar-Rodríguez S, Cafferata EA, Vernal R (2021) Humanized mouse models for the study of periodontitis: an opportunity to elucidate unresolved aspects of its immunopathogenesis and analyze new immunotherapeutic strategies. Front Immunol 12:663328. https://doi.org/10.3389/fimmu.2021.663328
13. Oz HS, Puleo DA (2011) Animal models for periodontal disease. J Biomed Biotechnol

2011:754857. https://doi.org/10.1155/2011/754857

14. Struillou X, Boutigny H, Soueidan A, Layrolle P (2010) Experimental animal models in periodontology: a review. Open Dent J 4:37–47. https://doi.org/10.2174/1874210601004010037
15. Kimura S, Nagai A, Onitsuka T, Koga T, Fujiwara T, Kaya H, Hamada S (2000) Induction of experimental periodontitis in mice with *Porphyromonas gingivalis*-adhered ligatures. J periodontal 71(7):1167–1173. https://doi.org/10.1902/jop.2000.71.7.1167
16. Hariyani N, Halimah AN, Al-Junaid M, Fadhila O, Budhy TI (2021) Mouse periodontitis models using whole *Porphyromonas gingivalis* bacteria induction. Saudi Dent J 33(8):819–825. https://doi.org/10.1016/j.sdentj.2021.08.001
17. Izawa A, Ishihara Y, Mizutani H, Kobayashi S, Goto H, Okabe E, Takeda H, Ozawa Y, Kamiya Y, Sugita Y, Kubo K, Kamei H, Kikuchi T, Mitani A, Hayashi J, Nishihara T, Maeda H, Noguchi T (2014) Inflammatory bone loss in experimental periodontitis induced by *Aggregatibacter actinomycetemcomitans* in interleukin-1 receptor antagonist knockout mice. Infect Immun 82(5):1904–1013. https://doi.org/10.1128/iai.01618-13
18. Sedgh LM, Bacino M, Kapila YL (2021) Periodontal diseases: the good, the bad, and the unknown. Front Cell Infect Microbiol 11:766944. https://doi.org/10.3389/fcimb.2021.766944
19. Highfield J (2009) Diagnosis and classification of periodontal disease. Aust Dent J 54(1):S11–S26. https://doi.org/10.1111/j.1834-7819.2009.01140.x
20. Queiroz-Junior CM, Santos ACPM, Galvão I, Souto GR, Mesquita RA, Sá MA, Ferreira AJ (2019) The angiotensin converting enzyme 2/angiotensin-(1-7)/Mas Receptor axis as a key player in alveolar bone remodeling. Bone 128:115041. https://doi.org/10.1016/j.bone.2019.115041
21. Queiroz-Junior CM, Madeira MFM, Coelho FM, de Oliveira CR, Cândido LCM, Garlet GP et al (2012) Experimental arthritis exacerbates *Aggregatibacter actinomycetemcomitans*-induced periodontitis in mice. J Clin Periodontol 39(7):608–616. https://doi.org/10.1111/j.1600-051x.2012.01886.x
22. Madeira MFM, Queiroz-Junior CM, Cisalpino D, Werneck SMC, Kikuchi H, Fujise O, Ryffel B, Silva TA, Teixeira MM, Souza DG (2013) MyD88 is essential for alveolar bone loss induced by *Aggregatibacter actinomycetemcomitans* lipopolysaccharide in mice. Mol Oral Microbiol 28(6):415–424. https://doi.org/10.1111/omi.12034
23. Madeira MFM, Queiroz-Junior CM, Montero-Melendez T, Werneck SMC, Corrêa JD, Soriani FM, Garlet GP, Souza DG, Teixeira MM, Silva TA, Perretti M (2016) Melanocortin agonism as a viable strategy to control alveolar bone loss induced by oral infection. FASEB J 30(12):4033–4041. https://doi.org/10.1096/fj.201600790r
24. Madeira MFM, Queiroz-Junior CM, Costa GM, Santos PC, Silveira EM, Garlet GP et al (2012) MIF induces osteoclast differentiation and contributes to progression of periodontal disease in mice. Microbes Infect 14(2):198–206. https://doi.org/10.1016/j.micinf.2011.09.005
25. D'Ângelo MQ, Queiroz-Junior CM, Maltos KLM, Ferreira AJ, Pacheco CMF, Soares RV (2020) The blockade of kappa opioid receptors exacerbates alveolar bone resorption in rats. Arch Oral Biol 120:104923. https://doi.org/10.1016/j.archoralbio.2020.104923
26. Queiroz-Junior CM, Maltos KLM, Pacheco DF, Silva TA, Albergaria JDS, Pacheco CMF (2013) Endogenous opioids regulate alveolar bone loss in a periodontal disease model. Life Sci 93(12–14):471–477. https://doi.org/10.1016/j.lfs.2013.08.007
27. Queiroz-Junior CM, Pacheco CMF, Maltos KLM, Caliari MV, Duarte IDG, Francischi JN (2009) Role of systemic and local administration of selective inhibitors of cyclo-oxygenase 1 and 2 in an experimental model of periodontal disease in rats. J Periodontal Res 44(2):153–160. https://doi.org/10.1111/j.1600-0765.2007.01069.x
28. Sallay K, Sanavi F, Ring I, Pham P, Behling UH, Nowotny A (1982) Alveolar bone destruction in the immunosuppressed rat. J Periodontal Res 17(3):263–274. https://doi.org/10.1111/j.1600-0765.1982.tb01153.x
29. Marchesan J, Girnary MS, Jing L, Miao MZ, Zhang S, Sun L, Morelli T, Schoenfisch MH, Inohara N, Offenbacher S, Jiao Y (2018) An experimental murine model to study periodontitis. Nat Protoc 13(10):2247–2267. https://doi.org/10.1038/s41596-018-0035-4
30. Chun J, Kim KY, Lee J-H, Choi Y (2010) The analysis of oral microbial communities of wild-type and toll-like receptor 2-deficient mice using 454 GS FLX Titanium pyrosequencer. BMC Microbiol 6(10):101. https://doi.org/10.1186/1471-2180-10-101
31. Kinane DF, Stathopoulou PG, Papapanou PN (2017) Periodontal diseases. Nat Rev Dis Primers 3:17038. https://doi.org/10.1038/nrdp.2017.38

Chapter 5

Infectious Models in Mice for Study of Gut Inflammation: Primary Cultures of Myenteric Neurons

Mayra Fernanda Ricci and Fabiana Simão Machado

Abstract

The human gut has 200–600 million neurons coordinated by an integrative system called the enteric nervous system (ENS). The ENS forms a vast network that controls all the complex activities of the gastrointestinal tract (GIT). In some intestinal alterations, this complex control is impaired by the loss or reduction of enteric neurons/glial cells due to inflammation, which affects neurotransmitter and motility activities and leads to the development of disease. Elucidating the mechanism of intestinal inflammation-induced dysfunction of GIT is one of the major gaps in inflammatory disease research. In this chapter, we present a detailed protocol that opens up the study of enteric neurons, glial cells, and smooth muscle cell activity in infectious models of murine intestinal inflammation. Primary myenteric culture mimics aspects of in vivo tissue and is an excellent platform for studying intestinal disease.

Key words Enteric nervous system, Infectious models, Myenteric plexus, Tissue culture, Gut inflammation

1 Introduction

The gastrointestinal tract (GIT) is a complex system responsible for several processes, including digestion, absorption, excretion, and protection. The GIT regulates itself from the transition of the striated skeletal muscles in the pharynx to the smooth muscles. Originating from the smooth muscle, a network organized in two layers houses 200–600 million neurons [1], sometimes referred to as "little brains" [2], consisting of a network of preganglionic parasympathetic fibers and postganglionic neurons that receive sympathetic nerve fibers and form synapses with sensory axons. This integrative system has been termed the enteric nervous system (ENS), a division independent of the autonomic nervous system (ANS). The ENS controls motility of the small and large intestine, except during defecation, in which the lumbosacral spinal cord acts

Lucas Secchim Ribeiro (ed.), *Models in Inflammation and Immunopharmacology*, Methods in Molecular Biology, vol. 2993, https://doi.org/10.1007/978-1-0716-5019-6_5,
© The Author(s), under exclusive license to Springer Science+Business Media, LLC, part of Springer Nature 2026

under central control, and is organized into specific circuits that control smooth muscle and mucosal function [3–6].

The proper function of GIT is vital; therefore, the mucosal, muscular, and nerve innervation alterations observed from the esophagus to the colon in some intestinal diseases are common and, at the same time, extremely complex. One of the most effective responses to intestinal diseases is the activation of the immune system, which has developed and established mechanisms to fight infectious and non-infectious diseases. The neuro-immune connection has important regulators of physiology in the GIT, but also across multiple organs such as the lungs, brain, and other tissues [7–10].

It is important to know that the balance between the pro- and anti-inflammatory responses of the immune system, as exacerbation of inflammation can compromise the immunomodulatory response and lead to the development of disease. The inflammatory processes in intestinal diseases such as Chagas disease cause damage to the neurons in the myenteric plexus, resulting in a decrease in the number of neurons [11, 12]. This loss and/or damage alters the neural control of muscle contractility, or control of transmural fluid movements, and can be highly morbid and even fatal [13]. In addition, other infectious diseases such as toxoplasmosis have been associated with intestinal alterations and inflammation in individuals with psychiatric disorders [14, 15], suggesting that part of the parasite's effects on behavior and cognition may be related to changes in the gastrointestinal tract. In addition, loss of ganglion cells in the colon is associated with constipation in mice models of colitis and aging [16, 17]. Also, intestinal inflammation leading to motility disorders has been linked in part to an oxidative stress mechanism. However, it is important to emphasize that it is not known how autonomic neuronal antioxidant mechanisms contribute to the susceptibility of enteric neurons to oxidative stress in inflammatory diseases [18].

The aim of this chapter is to present a method of using mice primary cultures of myenteric neurons, whose importance and relevance in the context of studying the structural and cellular aspects of gut inflammation represents a powerful approach to elucidate gaps in the mechanism of intestinal disease development. The method presented here results in a mixed culture of enteric neurons, glia, and muscle cells. The ENS culture described in this chapter yields several cell types that were prompt for investigation.

2 Materials

2.1 Mice

For this technique, mice from 6 to 8 weeks (sex and breed are indifferent) are needed to remove the full length of the large intestine. The segment is processed, as explained below, and the cells isolated from the colon muscle wall are used for many in vitro assays (electrophysiology, immunohistochemistry, immunofluorescence, single-cell PCR, and other methodologies). Appropriate national and institutional ethics should be in place before performing animal experiments.

2.2 Cell Culture

1. Media: Dulbecco's Modified Eagle Medium (DMEM). Complete media can be obtained by supplementation with 10% v/v fetal bovine serum and antibiotics, e.g., penicillin/streptomycin at 2% (w/v), prepare 0.2 g of antibiotic mixture in 10 mL and use for all the experiment.
2. DMEM without supplementation (*see* **Note 1**).
3. H/H solution (Hanks/HEPES solution): Hanks' Balanced Salt Solution (HBSS) and 1 M HEPES buffer, maintaining pH at 7–7.4.
4. Sterile flasks, petri dishes, plates for cell culture, and 9 mm glass coverslips (*see* **Note 2**).
5. Krebs solution: 126 mM NaCl, 5 mM KCl, 1.25 mM $MgSO_4$, 1.25 mM KH_2PO_4, 2.5 mM CaCl, 25 mM $NaHCO_3$, 11 mM Glucose in 200 mL ddH_2O. Refrigerate to 4 °C.
6. Enzymatic solutions: 1 mg/ml of collagenase type II-S and 0.25% of Trypsin from porcine pancreas (*see* **Note 3**).
7. Matrigel®.
8. Facility with a laminar flow hood, binocular stereoscopic loupe, centrifuge, heat water bath with shaker to 37 °C, and incubator for 37 °C, 5% CO_2 cell culture.

3 Methods

3.1 Preparation of Matrigel® Glass Coverslips in 48-Well Plates

All the procedures are performed in sterile conditions in a laminar flow hood.

1. Glass coverslips are cleaned in PBS (pH 7.4) for 15 min.
2. The glass coverslips are transferred to dishes with 70% ethanol for 15 min.
3. The glass coverslips are dried with soft paper and exposed to 15 min of UV radiation.
4. Use sterile forceps to place the glass coverslips in 12 wells of a 48-well plate (*see* **Note 4**).
5. Coat coverslips with Matrigel®; pipette 2.5 μl of matrigel (dilution 1:1 with DMEM) stock on each coverslip (*see* **Note 5**).
6. Incubate on coverslips for 1 h in laminar flow hood.
7. Perform this procedure only on the day of the experiment.

3.2 Preparation of Large Intestine Tissue

In this procedure, tissues from two mice are used for the isolation of enteric, glia, and muscle cells (*see* **Note 6**).

1. Prepare the surgical area. Clean the area with 70% ethanol and cover it with sterile paper. Prepare small surgical scissors, tweezers, syringe and needle, entomological pins, and three 50 mL glass beakers to clean the tissue.
2. Place 30 mL ice-cold Krebs solution in two 50 mL glass beakers labeled "dirty" and "clean".
3. Euthanize mouse by cervical dislocation (*see* **Note** 7).
4. Remove the large intestine and place in Krebs solution (dirty).
5. Using the syringe and needle to create a tool to clean; fill the syringe with ice-cold Krebs solution and clean the feces in the intestine.
6. The intestinal fragments are sectioned along the mesenteric border and the edges are carefully stretched open and fixed with entomological needles.
7. Use the binocular stereoscopic loupe to delaminate the mucosa and circular layer; then the longitudinal layer containing the Auerbach's plexus was exposed.
8. Cut the intestinal fragment into smaller pieces and place in 50 mL glass beakers (clean) and maintain on ice-cold Krebs solution (*see* **Note 8**).

3.3 Primary Cultures of Mouse Myenteric Neurons

1. Place the fragments in the tube (15 mL) and spin for 5 min at 200 × g in a centrifuge cooled to 4 °C. Carefully remove the supernatant with a pipette and discard the ice-cold Krebs solution.
2. Add 2 mL of Collagenase type II-S solution (1 mg/mL) under shaking (water bath at 37° C) for 15 min and manually homogenize by inversion every 5 min.
3. Centrifuge for 5 min at 200 × g (4° C). Carefully remove the supernatant with a pipette and discard the collagenase type II-S solution.
4. The fragments are washed with Hanks/HEPES buffer solution for 5 min under stirring.
5. Centrifuge for 5 min at 200 × g (4° C). Carefully remove the supernatant with a pipette and discard the Hanks/HEPES buffer solution.
6. Add 2 mL of Trypsin solution (0.25%) in water bath at 37°C for 10 min (*see* **Note 9**).
7. Stir the tube (water bath at 37 ° C) for 15 min and manually homogenize by inversion every 5 min.

2. Using 48-well plates for cell culture assays at a minimum density of 2×10^4 cells per well guarantees the homogenous distribution and interaction of the cells in culture.
3. This type of collagenase is important for the success in the enzymatic digestion.
4. The number of wells may vary slightly, this is an average obtained from each experiment.
5. Gradually defrost the Matrigel® (from the freezer to the refrigerator for 10 min), then dilute 1:1 in DMEM medium.
6. The number of animals was patronized to obtain 2×10^4 cells per well, ensuring the ideal number and density of cells for culture.
7. According to the animal model and in accordance with national and institutional ethical guidelines, euthanasia, if possible, was performed by cervical dislocation under general anesthesia.
8. The small fragments of the intestine remain in the backers with ice-cold Krebs solution until the beginning of enzymatic digestion.
9. Do not shake in these first 10 minutes of incubation.
10. Do not generate air bubbles.
11. The 30-minute incubation is very important for the adherence of the cells.

Acknowledgments

Fabiana Simão Machado has received research grant from Conselho Nacional de Desenvolvimento Científico e Tecnológico (CNPq; 307,828/2022–0, Fundação de Amparo à Pesquisa do Estado de Minas Gerais (FAPEMIG; APQ-01816-23, RED-00096-22, and RED-00067-23).

References

1. Jänig W (2006) The integrative action of the autonomic nervous system. Neurobiology of Homeostasis, Cambridge. https://doi.org/10.1017/9781108778411
2. Armour JA (2008) Potential clinical relevance of the "little brain" on the mammalian heart. Exp Physiol 93:165–176. https://doi.org/10.1113/expphysiol.2007.041178
3. de Groat WC, Nadelhaft I, Milne RJ, Booth AM, Morgan C, Thor K (1981) Organization of the sacral parasympathetic reflex pathways to the urinary bladder and large intestine. J Auton Nerv Syst 3:135–160. https://doi.org/10.1016/0165-1838(81)90059-X
4. Hasler W (2003) Physiology of gastric motility and gastric emptying. Lippincott, Philadelphia. https://doi.org/10.1111/j.1532-950x.1977.tb01640.x
5. Furness J (2006) The enteric nervous system. Blackwell, Oxford. https://doi.org/10.1002/9780470988756
6. Schubert ML, Peura DA (2008) Control of gastric acid secretion in health and disease. Gastroenterology 134:1842–1860. https://doi.org/10.1053/j.gastro.2008.05.021

7. Huh JR, Veiga-Fernandes H (2020) Neuroimmune circuits in inter-organ communication. Nat Rev Immunol 20:217–228. https://doi.org/10.1038/s41577-019-0247-z
8. Delves PJ, Roitt IM (2000) The immune system. N Engl J Med 343:37–49. https://doi.org/10.1056/NEJM200007063430107
9. Parkin J, Cohen B (2001) An overview of the immune system. Lancet 357:1777–1789. https://doi.org/10.1016/S0140-6736(00)04904-7
10. Maloy KJ, Powrie F (2011) Intestinal homeostasis and its breakdown in inflammatory bowel disease. Nature 474:298–306. https://doi.org/10.1038/nature10208
11. Koberle F (1968) Chagas' disease and Chagas' syndromes: the pathology of American Trypanosomiasis. Adv Parasitol 6:63–116. https://doi.org/10.1016/s0065-308x(08)60472-8
12. De Oliveira RB, Troncon LEA, Dantas RO, Meneghelli UG (1998) Gastrointestinal manifestations of Chagas' disease. Am J Gastroenterol 93:884–889. https://doi.org/10.1111/j.1572-0241.1998.270_r.x
13. Furness J (2012) The enteric nervous system and neurogastroenterology. Nat Rev Gastroenterol Hepatol 9:286–294. https://doi.org/10.1038/nrgastro.2012.32
14. Severance EG, Alaedini A, Yang S et al (2012) Gastrointestinal inflammation and associated immune activation in schizophrenia. Schizophr Res 138:48–53. https://doi.org/10.1016/j.schres.2012.02.025
15. Severance EG, Gressitt KL, Yang S, Stallings CR, Origoni AE, Vaughan C, Khushalani S, Alaedini A, Dickerson FB, Yolken RH (2014) Seroreactive marker for inflammatory bowel disease and associations with antibodies to dietary proteins in bipolar disorder. Bipolar Disord 16:230–240. https://doi.org/10.1111/bdi.12159
16. Takahashi T, Qoubaitary A, Owyang C, Wiley JW (2000) Decreased expression of nitric oxide synthase in the colonic myenteric plexus of aged rats. Brain Res 883:15–21. https://doi.org/10.1016/S0006-8993(00)02867-5
17. Pelletier A-M, Venkataramana S, Miller KG, Bennett BM, Nair DG, Lourenssen S, Blennerhassett MG (2010) Neuronal nitric oxide inhibits intestinal smooth muscle growth. Am J Physiol Gastrointest Liver Physiol 298:G896–G907. https://doi.org/10.1152/ajpgi.00259.2009
18. Bubenheimer RK, Brown IAM, Fried DE, McClain JL, Gulbransen BD (2016) Sirtuin-3 is expressed by enteric neurons but it does not play a major role in their regulation of oxidative stress. Front Cell Neurosci 10:73. https://doi.org/10.3389/fncel.2016.00073

Chapter 6

Intravital Microscopy of Mesenteric and Brain Vessels: A Valuable Tool for Leukocyte Recruitment Evaluation

William Antonio Gonçalves, Lara Penna, Isadora Oliveira Gondim, Vinicius Leal Pitcella, Sabrina Berger da Silva, Luan Lopes Menezes, Vanessa Pinho, Mauro Martins Teixeira, and Barbara Maximino Rezende

Abstract

Graft-versus-host disease (GVHD) is the most serious complication limiting the clinical utility of allogeneic hematopoietic stem cell transplantation (HSCT), in which lymphocytes of donors (graft) are activated in response to the host antigen. This disease is related to increased inflammatory response through releasing inflammatory mediators and recruiting defense cells, such as leukocytes. Intravital microscopy is a technique that allows the observation of leukocyte interactions in vivo. It can be performed using conventional light microscopy in thin and transparent tissues or epifluorescence microscopy in solid and non-transparent tissues. Here, we describe the procedure to execute this method at the mesenteric and brain vessels of mice subjected to GVHD.

Key words Intravital microscopy; graft-versus-host-disease, Inflammation, Mesenteric vessels, Brain vessels, Leukocyte recruitment, Fluorescent labeling

1 Introduction

Hematopoietic stem-cell transplantation (HSCT) is a procedure required for the treatment of different forms of cancer and hematological diseases [1]. The unhealthy bone marrow cells from the patient are destroyed using chemotherapy and/or radiotherapy. This process of destroying cells is named *conditioning regimen.* After this, the hematopoietic stem cell from the donor replaces the host one [2]. Graft-versus-host disease (GVHD) is a systemic

Supplementary Information The online version contains supplementary material available at https://doi.org/10.1007/978-1-0716-5019-6_6.

Lucas Secchim Ribeiro (ed.), *Models in Inflammation and Immunopharmacology*, Methods in Molecular Biology, vol. 2993,
https://doi.org/10.1007/978-1-0716-5019-6_6,
© The Author(s), under exclusive license to Springer Science+Business Media, LLC, part of Springer Nature 2026

inflammatory syndrome derived from allogeneic HSCT and affects 30–60% of patients who receive these cells [3, 4].

The development of acute GVHD can be divided into 3 phases. The first one is a consequence of the conditioning regimen since it causes damage to tissues of organs that have extensive cell proliferation, including the intestine, skin, liver, and lungs. These injuries mediate the release of damage-associated molecular patterns (DAMPs) and proinflammatory cytokines that activate the antigen-presenting cells (APCs) [3, 5]. The activation of those cells leads to the second phase, characterized by donor-T-cell activation, differentiation, and migration. APCs-activated cells interact with donor T cells that respond to the host human leukocyte antigens (HLA). The difference between HLA from donor and receptor is responsible for activating the donor T cells. After this, these cells proliferate into different types of T cells, such as T-helper 1 (Th1, Th2, Th9, Th17, Th22), and cytotoxic T cells. In the effector phase, the third one, these T cells migrate and damage the main GVHD target organs: the liver, skin, and intestine [6]. It is responsible for the common clinical manifestations related to acute GVHD, including skin rash; decreased appetite, nausea, diarrhea, weight loss, mucosal damage, and jaundice [7]. Furthermore, the involvement of the central nervous system in graft-versus-host disease has been recently explored [8–10]. Some studies have shown that patients subjected to allogeneic transplantation could develop neurological complications, suggesting that the alloreactivity observed in GVHD may be responsible for these complications [8–10].

The diagnosis of acute GVHD is made by analyzing the clinical manifestations that usually appear in the first 100 days after the transplant and their severity. Histological confirmation using skin or gastrointestinal tract biopsy can also be performed [11]. Current GVHD prophylaxis is based on glucocorticoids and immunosuppressive strategies. However, they are partially effective, showing high risks of side effects such as infections, loss of beneficial graft-versus-leukemia effect, and other long-term outcomes [12–14].

The study of microcirculation in the organism through intravital microscopy (IVM) is a refined research tool to investigate cellular events in physiological and pathological processes, including the leukocyte rolling and adhesion in vessels of inflamed tissues of different pathologies, such as GVHD [15, 16].

Some IVM models can be performed using conventional light microscopy in thin and transparent tissues such as the cremaster muscle, liver edge, and mesentery. In this case, it is possible to observe the leukocyte rolling, adhesion, and migration through the blood vessel. On the other hand, epifluorescence microscopy is necessary for solid and non-transparent tissues such as the brain and intestinal wall. In that case, fluorescent labeling is necessary to study the leukocyte traffic and vascular alterations. The leukocytes can be labeled with fluorophores such as rhodamine 6G or fluorochrome-conjugated antibodies through an intravenous injection [17].

One of the challenges of IVM is to excite the fluorescence in the tissue of interest. For better visualization and image capture, it is important to perform minimal damage to the tissue and the fluorophore, with constant attention to spatial and temporal resolution, so that it is sufficient to observe the behavior of individual cells [17]. In the multiphoton microscope, in which intravital visualization is performed, support for multiphoton excitation lasers is crucial with attention to the wavelengths that are going to be used and the fluorophore. Excitations are usually obtained using Ti:Sapphire tunable lasers, emitting NIR light (wavelengths between 680 and 1080 nm) in very short pulses (in the order of 80–140 fs) and at high repetition rates (80–100 MHz) [18].

The most common way to accomplish the IVM is to anesthetize small animals and perform a short surgery to gently expose the tissue of interest to a light source or laser and gather the images using a microscope. A great variety of animals can be used for this technique; however, mice are the most commonly used due to the possible genetic modifications [17, 18].

In this chapter, we describe the procedure to perform the in vivo intravital microscopy technique at the mesenteric and brain vessels of mice subjected to GVHD. It is interesting to execute this method in the model of GVHD to evaluate leukocyte recruitment and adhesion because the intestine is one of the most affected organs in this disease model. The analysis of the cell trafficking in the mesentery thus contributes to analyzing, for example, the efficiency of new therapeutic strategies in reducing the inflammation related to GVHD.

While leukocyte entry into tissues is governed by chemokines and adhesion molecules, the molecular mechanisms for entry into different regions of the CNS, including the spinal cord, are not fully understood [19–21]. In this regard, intravital microscopy has emerged as a valuable tool for studying this crucial process in the inflammatory response.

In summary, this chapter will be useful to applying this technique in other experimental models involving leukocyte recruitment to the intestine and central nervous system.

2 Materials

2.1 Surgery Equipment (See Note 1)

1. Petri plate.
2. Surgical scissors.
3. Centrifuge.
4. Tweezers.

5. Nylon strainer of 70 um.
6. Falcon tube.
7. Neubauer chamber.
8. Scalpel blade.
9. Cotton swab.
10. Equine Straight Short Dental Pen.
11. Plastic stage.
12. 20× objective and 10× eyepiece.
13. Heated plate at 37 °C.

2.2 Solutions

1. Anesthesia: 200 mg/kg of ketamine (2-(2-chlorophenil)-2-(methylamine)cyclo-hexanone) and 10 mg/kg of xylazine (N-(2,6-Dimethylphenyl)-5,6-dihydro-4H-1,3-thiazin-2-amine)].
2. 1 mg/mL of Rhodamine 6G chloride.
3. Sterile 0.9% saline (sodium chloride (NaCl) solution [0.9 g per 100 ml of solution].
4. 70 mg/L of Ciprofloxacin.
5. Incomplete RPMI medium.
6. Trypan Blue dye.

2.3 Intravital Microscopy (Fig. 1)

1. Nikon SMZ800 Stereo Microscope (**Note 2**).
2. Nikon High Intensify illuminator N.I 150.
3. Nikon Thermo Plate.

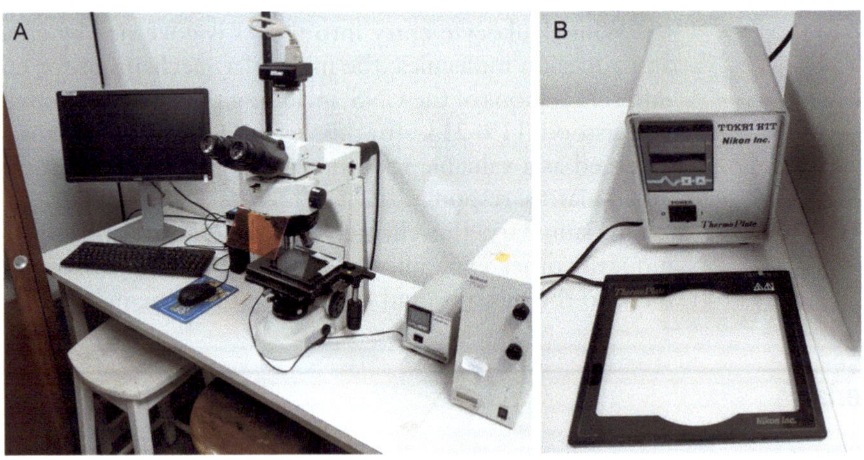

Fig. 1 Equipment for intravital microscopy technique. (**a**) Software NIS-Elements was installed in a computer for image capture, video recording, and leukocyte recruitment analysis. Nikon SMZ800 Stereo Microscope and Nikon High Intensify illuminator N.I 150 was used to visualize the mesenteric vessels. (**b**) Nikon Thermo Plate was used to maintain the body temperature of the mice at 37 °C

4. Software NIS-Elements D 3.0.
5. Mercury HBO of 103 W/2 lamp.

2.4 Mice

8–12-week-old male C57BL/6 J and BALB/c mice are used. All mice are housed under standard conditions in a temperature-controlled room (23 ± 1 °C) on an automatic 12 h/12 h light/dark cycle. The mice must have free access to commercial rodent food and water.

3 Methods

3.1 GVHD Induction

1. C57BL/6 J recipient mice will be irradiated with 9 Gy total body radiation (source: Cobalt-60) in two doses at 2 h intervals to minimize gastrointestinal toxicity.
2. To perform the transplantation, a cell suspension will be prepared from a pool of spleen cells added to a pool of femoral and tibial bone marrow cells.
3. The spleen from the donor mice will be removed and gently pelleted in a Petri plate with the use of a Nylon strainer of 70 um in 5 mL of ice-cold incomplete RPMI medium.
4. Subsequently, the cells will be placed in a Falcon tube and decanted for 3 min to remove the lumps. The cell preparation will be centrifuged at 350 g for 5 min.
5. After centrifugation, the supernatant will be discarded and the pellet resuspended in 10 mL of incomplete RPMI medium for counting the number of cells in a Neubauer chamber, using Trypan dye at a dilution of 1:200. This counting will provide the viability and number of cells that will be injected into the recipient mice.
6. To isolate the cells from the femoral and tibial bone marrow, both femurs and tibias of mice will be removed and their extremities cut with a scalpel blade, allowing the insertion of a 30G needle to wash the cells with 3 ml of cold incomplete RPMI medium.
7. Subsequently, the liquid obtained from the lavage containing the cells will be placed in a Falcon tube and centrifuged at 350 g for 5 min.
8. After centrifugation, the supernatant will be discarded and the pellet resuspended in 10 mL of incomplete RPMI medium for counting the number of cells in a Neubauer chamber, as mentioned above.
9. A total of 3×10^7 splenocytes + 1×10^7 bone marrow cells from BALB/c donors will be injected intravenously (i.v.) into C57BL/6 J recipient mice. C57BL/6 J mice that receive

splenocytes from C57BL/6 J mice do not develop any disease and are considered the control group.

10. Due to the toxicity of the high level of body irradiation, recipient mice receive an oral suspension of ciprofloxacin (70 mg/l) in their drinking water from 1 day before to 15 days after transplantation.
11. The following clinical parameters are evaluated every 2 days after induction: body weight change, piloerection, activity, trunk flexion posture, and occurrence of diarrhea. Each clinical parameter receives a score from 0 to 2 depending on presence (2)/absence (0) or intensity: mild (0.5)/moderate (1)/intense (2). The sum of all parameters can reach a maximum of 12 points. Mice are also monitored daily for mortality. Mice are analyzed by intravital microscopy 10 days after transplantation, when they have a clinical score between 6 and 8 points. Animals with clinical score higher than 8 points usually do not survive the surgical procedure.

3.2 Intravital Microscopy of Mesenteric Vessels (Fig. 2)

1. To maintain stable hemodynamic parameters and good observation of the leukocyte recruitment process, the animals are anesthetized with an intraperitoneal injection of 100 μl of a mixture of xylazine hydrochloride (10 mg/kg) and ketamine (200 mg/kg). After a midline incision using surgical scissors, the intestinal loop of interest is gently exposed with cotton swabs and mounted on a plastic stage. It is important to keep the vital signals of the animal stable to have a good visualization of leukocyte rolling. To ensure this, the use of the thermal plate prevents the body temperature to drop and harm the animal. It is not necessary to fixate the mice (**Notes 3–6**).
2. To measure the leukocyte-endothelial cell interactions, the fluorescent marker rhodamine 6G [1 mg/mL] injected

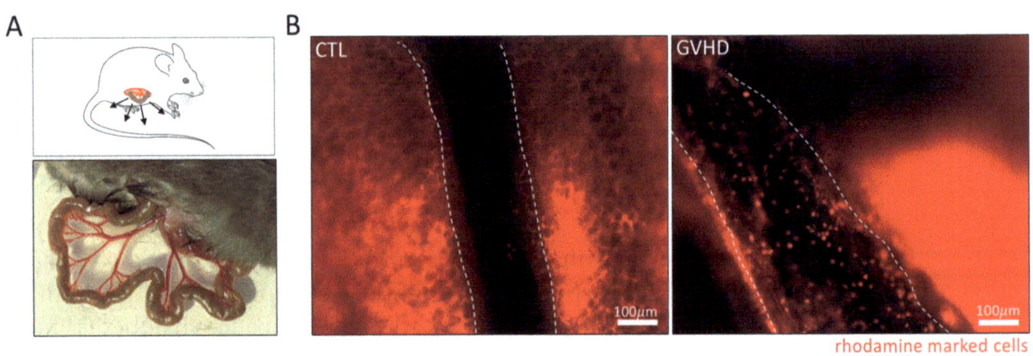

Fig. 2 Illustration of surgical incision and visualization of rhodamine-marked cells. (**a**) Representative images of the surgical incision in the peritoneum (top) and exposition of intestinal loops (bottom), and (**b**) the leukocyte recruitment in mesenteric vessels of the control mice (CTL) and GVHD mice. The intravital microscopy images were artificially colored to represent the rhodamine filter color seen in the microscope objective

intravenously will be used via the caudal vein as a single bolus of 100 ul immediately before measurements. Rhodamine epi-illumination is achieved with a variable mercury HBO of 103 W/2 lamp together with a set of filters supplied by the Nikon High Intensify illuminator N.I 150 (**Note** 7).

3. 2 to 4 regions of interest will be selected, looking for unbranched venules with 70 to 120 μm in diameter in each mouse, using a 20× objective and a 10× eyepiece. The number of rolling and adherent leukocytes will be determined offline using NIS ELEMENT imaging software for data analysis. Rolling leukocytes are defined as those cells that move at a velocity less than that of the erythrocytes within a given vessel. The flux of rolling cells is measured as the number of rolling cells that pass by a given point in the venule per minute (Video 1). A leukocyte is considered to be adherent if it remains stationary for ≥30 s, and total leukocyte adhesion is quantified as the number of adherent cells in the intravascular space within a 100 μm length of venule in 5 min. It is expected that GVHD mice show increased number of rolling and adherent leukocytes compared to the control group (**Note 4**).

3.3 Intravital Microscopy of Brain Vessels (Fig. 3)

1. To analyze cerebral microcirculation, the animals are anesthetized with an intraperitoneal injection of 100 μl of a mixture of xylazine hydrochloride (10 mg/kg) and ketamine (200 mg/kg).
2. After the onset of anesthesia, the mouse will receive an intravenous injection of Rhodamine 6G [1 mg/mL] (**Note** 7) and then placed in a prone position on a heated plate at 37 °C (**Note 6**).
3. The skin over the cranial region will be removed using surgical scissors and tweezers, exposing the superior aspect of the skull. Then, a window of approximately 1 cm in length will be opened in the parietal bone, exposing the brain tissue using a dental pen with a grinding disc (**Note 8**) (Fig. 3).

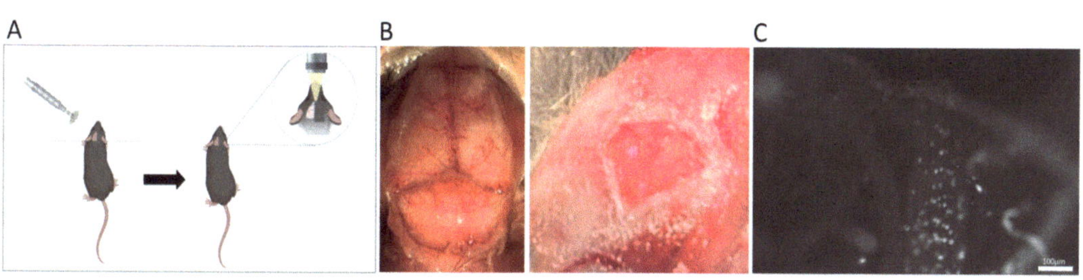

Fig. 3 Illustration of intravital microscopy in the mouse brain. (**a**) Representative image of the surgical procedure in the calvaria. (**b**) The visualization of venules and cellular recruitment analysis were realized after careful unilateral remotion of parietal bone fragment. (**c**) Leukocyte recruitment in a brain vessel of GVHD mice

4. Once the brain tissue is exposed, the microvasculature will be observed, and cellular recruitment will be analyzed (**Note 9**) (Video 2). Typically, only venules with a thickness of approximately 60 to 80 μm are evaluated.
5. The counting of rolling and adherent cells and image acquisition occur in the same manner as previously described for mesenteric microvasculature analysis (See the Subheading 3.2, **step 3**).

4 Notes

1. Usually, after intravital microscopy, the animals are euthanized, so there is no need to use sterile material. If the researcher would like to keep the animals alive after the procedure, sterile surgical material is required.
2. Other more complex and higher cost microscopes can be used, for example, the Nikon Eclipse Ti Inverted microscope. Here, we mention a relatively cheaper and more accessible model of microscope. Each microscope model may require adjustments in surgical technique and positioning of the mice for visualization of the tissue of interest.
3. In the process of making the cut, a lot of care and attention is needed to maintain the mice's homeostasis conditions, without excessive blood loss. The ideal is to make a small incision to avoid the exposure of other organs and harmful bleeding.
4. Regarding the recruitment of leukocytes, it will occur immediately after the mesenteric cells are activated by the surgical procedure and consequently, more leukocytes are recruited and may interfere with the results, making it necessary to improve the operative techniques to avoid the trauma greater than necessary. In order to keep the integrity of the mesentery, a way to perform the exposure of the organ is by taking it using tweezers covered with cotton.
5. During the experimental protocol, especially when the peritoneal cavity is exposed, changes such as dehydration, loss of volume, and hypothermia may occur. Therefore, it is crucial to promote the maintenance of the mice's physiological conditions. Practices such as covering the exposed organs that will not be observed with the gauze soaked with Sterile 0.9% saline solution can reduce the loss of water and temperature, and even slow down the movement of organs.
6. Room temperature is usually lower than that of mice (25–30 °C), so there is a recommendation to use warmer lamps, such as infrared, to avoid hypothermia, especially when

the peritoneum is open and the organs are exposed. The use of the Thermo Plate also helps to avoid the drop of the body temperature.

7. The evaluation of the stability of the fluorophore must be carried out according to the experimental protocol. Light microscopy illuminates a translucent tissue and makes it possible to observe the migration of defense cells. In the case of solid organs and visualization of cells in extracellular spaces, there is a greater difficulty, with reduced visualization of cells outside the microcirculation and alterations in the leakage process of rhodamine from the microcirculation to the extravascular space.
8. The surgery for removing a portion of the cranial vault should be conducted in a manner that ensures the disk does not come into direct contact with the brain tissue. Therefore, after thinning the targeted bone, it should be delicately lifted using forceps.
9. It is important to note that any bleeding occurring in the surgical area can potentially impact the results and should be minimized.

Funding

The authors have no conflicts of interest to declare that are relevant to the content of this chapter.

References

1. Barriga F, Ramírez P, Wietstruck A, Rojas N (2012) Hematopoietic stem cell transplantation: clinical use and perspectives. Biol Res 45:307–316. https://doi.org/10.4067/S0716-97602012000300012
2. Bazinet A, Popradi G (2019) A general practitioner's guide to hematopoietic stem-cell transplantation. Curr Oncol 26:187–191. https://doi.org/10.3747/CO.26.5033
3. Bryant A, Mallick R, Huebsch L, Allan D, Atkins H, Anstee G, Sabloff M, Scrivens N, Maze D, Bredeson C, Kekre N (2017) Low-dose antithymocyte globulin for graft-versus-host-disease prophylaxis in matched unrelated allogeneic hematopoietic stem cell transplantation. Biol Blood Marrow Transplant 23:2096–2101. https://doi.org/10.1016/j.bbmt.2017.08.007
4. Holtan SG, Yu J, Choe HK, Paranagama D, Tang J, Naim A, Galvin J, Joachim Deeg H (2022) Disease progression, treatments, hospitalization, and clinical outcomes in acute GVHD: a multicenter chart review. Bone Marrow Transplant 57:1581–1585. https://doi.org/10.1038/S41409-022-01764-W
5. Toubai T, Magenau J (2020) Immunopathology and biology-based treatment of steroid-refractory graft-versus-host disease. Blood 136:429–440. https://doi.org/10.1182/BLOOD.2019000953
6. Ghimire S, Weber D, Mavin E, Wang XN, Dickinson AM, Holler E (2017) Pathophysiology of GvHD and other HSCT-related major complications. Front Immunol 8. https://doi.org/10.3389/FIMMU.2017.00079
7. Zeiser R, Blazar BR (2017) Acute graft-versus-host disease – biologic process, prevention, and therapy. N Engl J Med 377:2167–2179. https://doi.org/10.1056/NEJMRA1609337
8. Ruggiu M, Cuccuini W, Mokhtari K, Meignin V, Peffault De Latour R, Robin M et al (2017) Case report: central nervous system involvement of human graft versus host disease: report of 7 cases and a review of

literature. Medicine 96. https://doi.org/10.1097/MD.0000000000008303

9. Chen BJ (2022) Chronic GVHD of the CNS. Blood 139:1271–1272. https://doi.org/10.1182/BLOOD.2021014079
10. Vinnakota JM, Zeiser R (2021) Acute graft-versus-host disease, infections, vascular events and drug toxicities affecting the central nervous system. Front Immunol 12. https://doi.org/10.3389/FIMMU.2021.748019
11. Ramachandran V, Kolli SS, Strowd LC (2019) Review of graft-versus-host disease. Dermatol Clin 37:569–582. https://doi.org/10.1016/J.DET.2019.05.014
12. Choi SW, Reddy P (2014) Current and emerging strategies for the prevention of graft-versus-host disease. Nat Rev Clin Oncol 11:536–547. https://doi.org/10.1038/NRCLINONC.2014.102
13. Zeiser R, von Bubnoff N, Butler J, Mohty M, Niederwieser D, Or R, Szer J, Wagner EM, Zuckerman T, Mahuzier B, Xu J, Wilke C, Gandhi KK, Socié G (2020) Ruxolitinib for glucocorticoid-refractory acute graft-versus-host disease. N Engl J Med 382:1800–1810. https://doi.org/10.1056/NEJMOA1917635
14. Rezende BM, Bernardes PTT, Gonçalves WA, De Resende CB, Athayde RM, Ávila TV, Martins DG, Castor MGM, Teixeira MM, Pinho V (2019) Treatment with Apocynin limits the development of acute graft-versus-host disease in mice. J Immunol Res 2019. https://doi.org/10.1155/2019/9015292
15. Vaghela R, Arkudas A, Horch RE, Hessenauer M (2021) Actually seeing what is going on – intravital microscopy in tissue engineering. Front Bioeng Biotechnol 9. https://doi.org/10.3389/FBIOE.2021.627462
16. Jorch SK, Deppermann C (2021) Intravital imaging allows organ-specific insights into immune functions. Front Cell Dev Biol 9. https://doi.org/10.3389/FCELL.2021.623906
17. Pinho V, Coelho FM, Menezes GB, Cara DC (2011) Intravital microscopy to study leukocyte recruitment in vivo. Methods Mol Biol 689:81–90. https://doi.org/10.1007/978-1-60761-950-5_6
18. Masedunskas A, Milberg O, Porat-Shliom N, Sramkova M, Wigand T, Amornphimoltham P, Weigert R (2012) Intravital microscopy: a practical guide on imaging intracellular structures in live animals. BioArchitecture 2:143–157. https://doi.org/10.4161/BIOA.21758
19. Campbell JJ, Hedrick J, Zlotnik A, Siani MA, Thompson DA, Butcher EC (1998) Chemokines and the arrest of lymphocytes rolling under flow conditions. Science 279:381–384. https://doi.org/10.1126/SCIENCE.279.5349.381
20. Alon R, Grabovsky V, Feigelson S (2003) Chemokine induction of integrin adhesiveness on rolling and arrested leukocytes local signaling events or global stepwise activation? Microcirculation 10:297–311. https://doi.org/10.1038/SJ.MN.7800195
21. Schenkel AR, Mamdouh Z, Muller WA (2004) Locomotion of monocytes on endothelium is a critical step during extravasation. Nat Immunol 5:393–400. https://doi.org/10.1038/NI1051

Chapter 7

Classical Histological Procedures with Murine Tissues

Anna Clara Paiva Menezes dos Santos, Antônio Felipe Silva Carvalho, and Celso Martins Queiroz-Junior

Abstract

Histology is a classic and widely used technique that allows the visualization of tissues under a microscope. It is a relatively cheap methodology that generates important visual results in the scientific field, such as the characterization of the parenchymal and stromal components of a biological sample and the morphological tissue damage induced by an insult (i.e., infection, surgery, mechanical stress). Despite these advantages, the classical histological procedures require a full-step guide in order to obtain standardized samples. Reaching high-quality slides deserves care from the sample collection to the confection of the slides. This chapter describes the detailed steps for obtaining histological slides from murine samples, including collection, fixation, processing, inclusion, and microtomy.

Key words Histology, Fixation, Processing, Embedding, Microtomy

1 Introduction

Created in the early seventeenth century, the microscope enabled a new understanding of cell biology and medical science. Antony van Leeuwenhoek perfected the optical equipment of his time and used it to observe structures invisible to the naked eye. The primitive microscope invented by the researcher in 1674, equipped with only a glass lens, allowed to increase visual perception by up to 300× and, with it, it was possible to observe bacteria with a size of 1 to 2 μm [1]. A few years later, Leeuwenhoek's primitive microscope was improved by Robert Hooke, with an extra lens and the ability to magnify the image even further. Hooke's first observations and the studies by Antony van Leeuwenhoek led to the discovery of cells [2]. In the middle of the nineteenth century, Rudolf Virchow stated that all diseases involve alterations in regular

Authors Anna Clara Paiva Menezes dos Santos and Antônio Felipe Silva Carvalho have equally contributed to this chapter.

Lucas Secchim Ribeiro (ed.), *Models in Inflammation and Immunopharmacology*, Methods in Molecular Biology, vol. 2993, https://doi.org/10.1007/978-1-0716-5019-6_7,
© The Author(s), under exclusive license to Springer Science+Business Media, LLC, part of Springer Nature 2026

cells, that is, all disease, ultimately, is a cellular disease. Diseases could be characterized not only by a group of clinical symptoms, but by typical morphological changes that could be observed through tissue analysis. Thus, understanding the histology of a healthy tissue sample is essential for histopathological interpretation [3, 4].

Histology is the study of the microscopic structures of biological cells and tissues, using light, fluorescence, or electron microscopes. It allows the examination of a representative sample that has been previously prepared using histological techniques [4, 5]. Histopathology can be used for diagnosis in different tissues because it is a fast, cheap, detailed, and reproducible technique [6]. Despite the tradition of the technique and the wide use, some limitations can be mentioned, including possible subjective interpretation errors [7], the increasing need to look at tissues from a volumetric perspective, in three dimensions (3D) [8], and the invasive analysis given that biopsies must be obtained from a tissue to be analyzed [9]. Nevertheless, the widespread use of histological methods in scientific research until today shows how this technique is still the gold standard for diagnosing tissue morphological alterations.

For the analysis of morphological microstructures of tissues by optical microscopy, it is necessary to make histological slides. The process of obtaining slides for histological analysis involves five main steps, which include organ collection, fixation, processing, embedding, and cutting [10]. All these steps will be detailed in the current chapter.

2 Materials

2.1 Organ Collection

1. Anesthetic solution for mouse euthanasia: add 4.8 mL of 10% (w/v) ketamine hydrochloride to 4.5 mL of 2% (w/v) xylazine, so that, when injecting 100 μL of the solution (intraperitoneally), a mouse weighing 20 g receives 240 mg/kg ketamine hydrochloride and 45 mg/kg xylazine, which is a dose able to induce rapid loss of consciousness and mouse euthanasia.
2. Surgical material: flat tip tweezers and sharp-edged scissors to perform the removal of mouse organs.
3. Concentrated Phosphate-buffered saline (10× PBS): 1.37 M sodium chloride (NaCl), 0.027 M potassium chloride (KCl), 0.1 M dibasic sodium phosphate (Na_2HPO_4), 0.018 M monobasic potassium phosphate (KH_2PO_4), pH 7.2–7.4.
4. Phosphate-buffered saline (1× PBS): For the final volume of 100 mL, add 90 mL of distilled water to a beaker and then add 10 mL of 10× PBS, pH 7.2–7.4.

5. Syringes (1 mL and 10 mL) and needles (30G, 0.3 × 13mm and 22G, 0.7 × 25mm).
6. Petri dish.

2.2 Fixation

1. 10% (v/v) formalin in 1× PBS: to prepare 1 L of 10% (v/v) buffered formalin, add 100 mL of 37% (w/v) formalin to 900 mL of 1× PBS.
2. 70% ethanol.
3. 1× PBS.
4. Histology tissue cassettes and a glass pot.
5. *Optional*: 8 mL glass bottles with rubber stopper or plastic conical tubes (15 or 50 mL) for storing the samples in the fixative solution.
6. Scalpel and scalpel blade.
7. Petri dish.
8. Pencil or printed paper sample identifications.
9. Tap water.

2.3 Processing

1. Glass pot to place the samples: Processing can be done in a single pot if the samples are separated into cassettes, or in small individual pots.
2. Battery for dehydration: prepare ethanol solutions diluted in distilled water in the following concentrations 70%, 80%, 90%, and three different 100% ethanol (100% ethanol I, 100% II, and 100% III).
3. For hard tissues (bone and teeth samples) more baths are performed in the dehydration phase, so it will be necessary to prepare ethanol 70% I, 70% II, 80% I, 80% II, 90%, 95%, 100% I, 100% II, and 100% III.
4. 14% EDTA: 14 g ethylenediaminetetraacetic acid in 100 mL distilled water. This is only used for hard tissues to decalcify the sample. The pH of the solution should be adjusted to 7.2–7.4 (*see* **Note 1**).
5. Xylene: it is used in clearing, with specific glass bottles for xylene baths, in an exhaust hood.
6. Paraffin: it is used for tissue impregnation that needs to be performed with liquid paraffin. Therefore, an inclusion oven at 60 °C containing pots with paraffin is required for carrying out paraffin baths.
7. Tweezers.
8. *Optional*: Automated tissue processing systems can be used, instead of the classical manual baths. Some examples are HistoCore PELORIS 3, HistoCore PEGASUS Plus from Leica Biosystems and other systems.

2.4 Embedding

1. Paraffin.
2. Hot plate: keep at 60 °C so that the paraffin remains liquid.
3. Beaker with liquid paraffin, or an electric paraffin bucket.
4. Metal molds for embedding.
5. Plate with ice.
6. Tweezers: used for positioning the samples in the metal molds. Positioning of the sample is easier if the tweezers are warm.

2.5 Microtomy

1. Microtome and a microtome razor.
2. Histological slides, pencils, ice, paper, and a brush to remove the cut.
3. *Optional:* Container with 30% ethanol in water.
4. Water bath at 40 °C.
5. Histology oven at 60 °C.

3 Methods

The methodology will be described based on soft tissues. Hard tissues, such as bone and teeth, have some methodological differences that will be depicted in the notes.

3.1 Organ Collection

1. For a tissue sample of a mouse to be analyzed histologically, it must be taken from a dead organism. Thus, in experimental conditions with rodent models, the mice must be euthanized using ethically acceptable methods of euthanasia, such as anesthetic overdose. To that end, inject the mice intraperitoneally (i.p.) with 100 μL of the anesthetic solution for mouse euthanasia (*see* **Note 2**).
2. *Optional*: In some samples, such as lung samples, removing the blood from blood vessels may improve the visualization of parenchymal components of the tissue. To that end, tissue perfusion should be performed at this step (*see* **Note 3**).
3. With the aid of scissors and tweezers, the region where the organ of interest is located should be exposed.
4. After locating the organ of interest for histological analysis, its handling must be performed delicately in order to avoid tissue laceration or any damage that can disrupt the normal morphology of the organ.
5. Preferentially, manipulate neighboring organs/structures to obtain the target organ (*see* **Note 4**). Use flat tip tweezers to pick organs next to the target organ in order to avoid direct contact with target tissue.

6. There is a specific collection technique for each type of tissue (*see* **Note 5**).
7. As soon as the sample has been removed from the mouse body, it should be gently cleaned with 1× PBS to remove superficial blood clots by carefully placing it in a Petri dish containing 1× PBS.

3.2 Fixation

1. After organ collection, certify that the obtained fragment presents thickness of up to 6 mm. This is the recommended maximum thickness that allows adequate penetration of the fixative solution into the sample (*see* **Note 6**). If an organ was collected with more than 6 mm thickness, segment the sample with the aid of a scalpel blade on a Petri dish. Gently manipulate and segment the target organ in the adequate anatomical plane to allow the better visualization of the target tissue (*see* Subheading 3.4).
2. As soon as the sample has been collected and sectioned, place it into pencil-identified histological cassettes, taking care not to cramp the sample with the cassette (*see* **Note** 7).
3. Place all identified cassettes into a glass bottle with a volume of 10% buffered formalin that is 10–20 times that of the samples, in order to allow sufficient fixation.
4. Let the samples to fixate at room temperature for 48 h.
5. Then, remove the 10% buffered formalin from the bottle with the samples and let all cassettes washing with tap water for 2 h to remove the fixative residues (*see* **Note 8**). At this step, it is possible to perform a new cleavage in the samples with a scalpel blade in order to finely adjust the plane of inclusion.
6. For hard tissue samples (bones, teeth), the decalcification step is subsequent to the washing step (*see* **Note 1**).
7. After washing, remove the water and add 70% ethanol to the samples (*see* **Note 9**).
8. After 24 h, the processing of samples for embedding in paraffin blocks can start.

3.3 Processing

1. Dehydration: In this step, the samples are immersed in ethanol with increasing concentrations, to remove water from the tissue. Samples must be soaked in new 70% ethanol for 30 min.
2. The ethanol must be removed and replaced by 80% ethanol.
3. The alcohol exchange process must be repeated until the samples undergo a 30-min bath also in ethanol 90%, 100% I, 100% II, and 100% III (*see* **Note 10**).
4. Dehydration is different for hard tissues (*see* **Note 11**).

5. Clearing: Here, the ethanol present in the tissue is replaced by a reagent that has a greater affinity for paraffin and promotes clarification of the samples (xylene, *see* **Note 12**). For this, dry off the excess of ethanol.
6. Inside an exhaust hood, the xylene must be placed in the flask with the samples until it covers them (*see* **Note 13**). Let the solution be in place for 10–20 min, when the xylene should be removed and replaced with a cleaner one.
7. This process must be repeated two more times, to complete 3 xylene baths (*see* **Note 14**).
8. The clearing step is performed in the same way for soft and hard tissues.
9. Paraffin impregnation: After the last clearing xylene bath, the cassettes with samples must be placed, with the help of tweezers, inside a pot with melted paraffin, in the histological oven (*see* **Note 15**).
10. After 20–30 min, the samples must be transferred from this pot to paraffin II, with the help of tweezers.
11. The procedure must be repeated, transferring the samples to paraffin III. In total, the samples must pass through at least 3 baths of 20–30 min each, in 3 different pots of paraffin, the last one being the cleanest one (*see* **Note 16**).
12. This procedure is different for hard tissues (*see* **Note 17**).

3.4 Embedding

1. Embedding is the step where the external part of the tissue will be wrapped in paraffin, forming a block that will later be cut to obtain the slide.
2. A metal mold must be filled with liquid paraffin and kept on the hot plate.
3. Remove a cassette with the samples from paraffin III and place it on the hot plate.
4. Open the cassette and position the sample properly in the bottom of the metal form, using tweezers.
5. Check the position of the samples with attention to which part of the sample is facing the bottom of the mold. Each sample has an inclusion plane, based on what you want to observe in each tissue. We will detail the inclusion planes of some mouse organs in the next topic (see Subheading 3.5 Inclusion planes).
6. When the sample is properly positioned in the metal mold, it must be removed from the hot plate, and the paraffin will start to solidify. Place the mold on a cold plate to accelerate the solidification.
7. Place the cassette over the mold and wait for the paraffin to solidify.

8. When the paraffin is almost solid, place more paraffin on top of the cassette to fill it and "bond" the cassette to the sample.
9. Wait for the paraffin to cool completely. When that happens, you can remove the block from the mold, slightly pushing the cassette in the opposite direction to the direction of the mold. If the paraffin is completely solid, it will come off easily.
10. The blocks must be stored for the next step in histology, which is the microtomy.

3.5 Inclusion Planes

1. In this topic, we will detail the position in which some tissues can be embedded. Each subtopic will have a description of a suggested inclusion plane of a tissue and an image to illustrate. We will address the embedding of liver, lung, brain, heart, muscles, femur, knee, maxilla, and intestine samples.
2. Liver: choose one of the liver lobes to study. Usually the median and the left lateral lobes provide wide area for evaluation, in view of their sizes. Only one lobe is needed for routine evaluation, but it is advisable to standardize which one will be evaluated. The chosen lobe should be positioned with its ventral surface, which presents a convex aspect, in contact with the bottom of the inclusion metal mold (*see* Fig. 1).
3. Lung: choose one lung lobe to perform histological evaluation. Usually, the unique left lung lobe provides a wide area for evaluation. If you deserve to essentially observe the respiratory alveoli and associated respiratory bronchioles, position the sample with its ventral surface, which presents a convex aspect, facing the bottom of the mold (*see* Fig. 2). If you deserve to observe the bronchi and larger bronchioles and blood vessels, include the sample with its dorsal surface facing the bottom of the mold. The lung is an airy tissue. Therefore, it is normal for the organ to float after placing the paraffin in the mold. It is important to reposition the lung at the bottom of the mold, gently pressing it with hot tweezers while the paraffin solidifies.

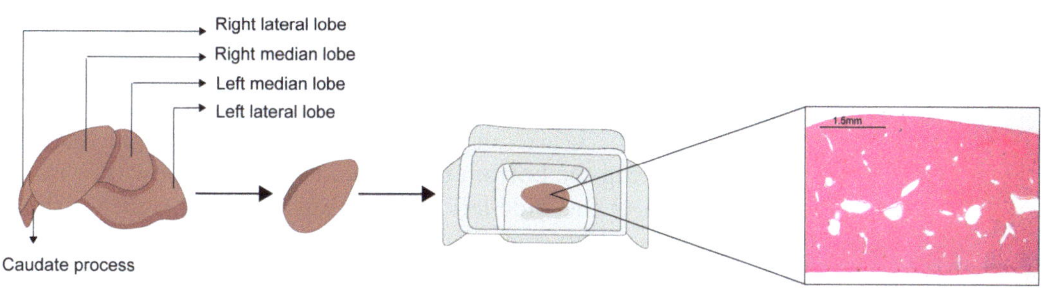

Fig. 1 Drawing representing the liver and a lobe that can be removed at the time of sample collection. To the right of the image, we see the illustration of a mold of inclusion, the sample with its convex surface facing the bottom of the mold and the resulting H&E-stained slide

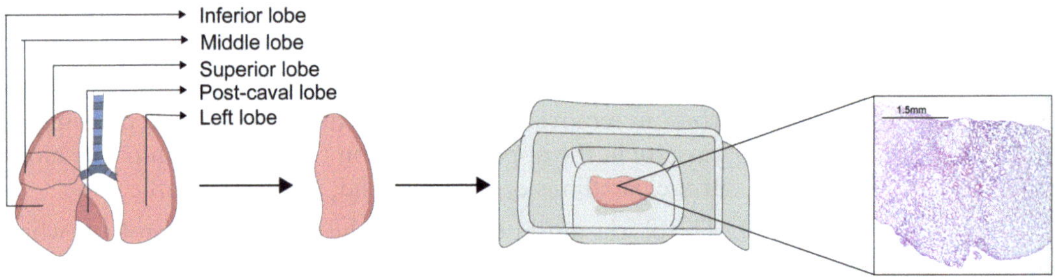

Fig. 2 Representative image of the lung and the left lobe that can be removed at the time of sample collection. To the right of the image, the illustration shows the positioning of the ventral convex surface of the sample on the bottom of the mold, which privileges the visualization of respiratory alveoli and bronchioles in the H&E-stained slide

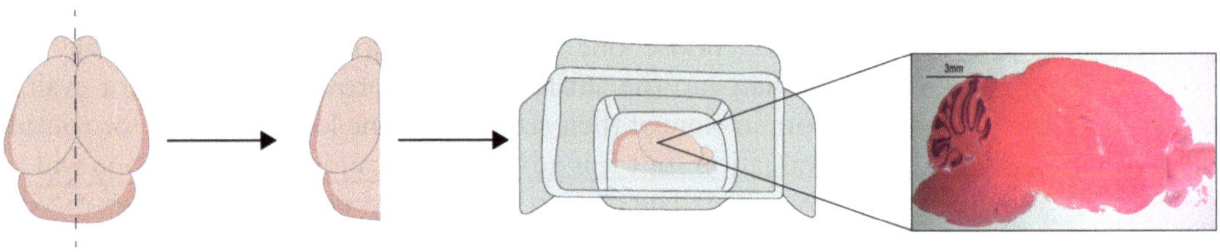

Fig. 3 Drawing representing the positioning of the brain for inclusion. The dotted line represents the region where the tissue can be cut to allow the visualization of the structures in the sagittal plane. To the right of the image, the illustration shows the brain positioning in the mold and the H&E-stained slide shows evidence of the cerebellum, cortex, and hippocampus

4. Brain: brain samples can be split in half, following the sagittal plane with the aid of a scalpel blade (*see* Fig. 3). After division, each side of the sample can be positioned in the mold with the cut region facing the bottom of the mold. This will allow the visualization of different regions of the brain cortex, hippocampus, cerebellum, besides the meninge. Other planes, such as a section in the coronal plane, can also be performed, and will generate different visualizations of the same sample.
5. Heart: To embed heart samples, it is important to perform a section in the transverse plane of the organ during the collection of the sample, to remove the veins and atriums' walls (*see* Fig. 4). This will expose the ventricles' walls and allow the fixative and solutions to penetrate properly in the sample. Then, the heart can be positioned with the region that was cut facing the bottom of the mold. This will allow the visualization of the wide left ventricle wall aside the right ventricle, and the interventricular septum.
6. Muscles: Muscle samples can be included horizontally, with their most convex surface facing the bottom of the mold, to allow the visualization of longitudinal muscle fibers. For the cross-sectional view of the muscle fibers, it is necessary to cut

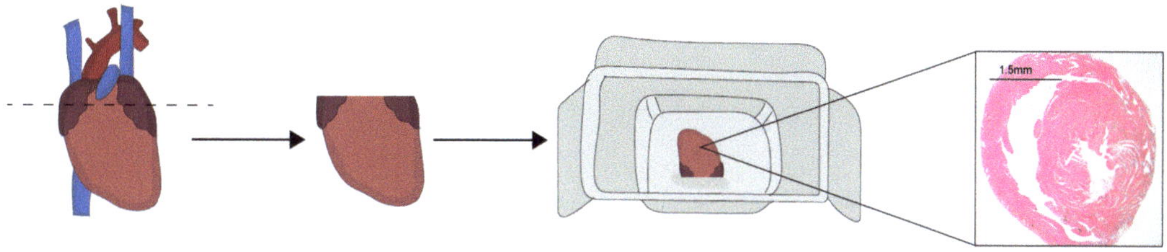

Fig. 4 Drawing representing the positioning of the heart for inclusion. The dotted line represents the region where the tissue should be cut. To the right of the image, we see the illustration of a mold of inclusion and the indicated position of the organ, showing the ventricle walls in the H&E slide

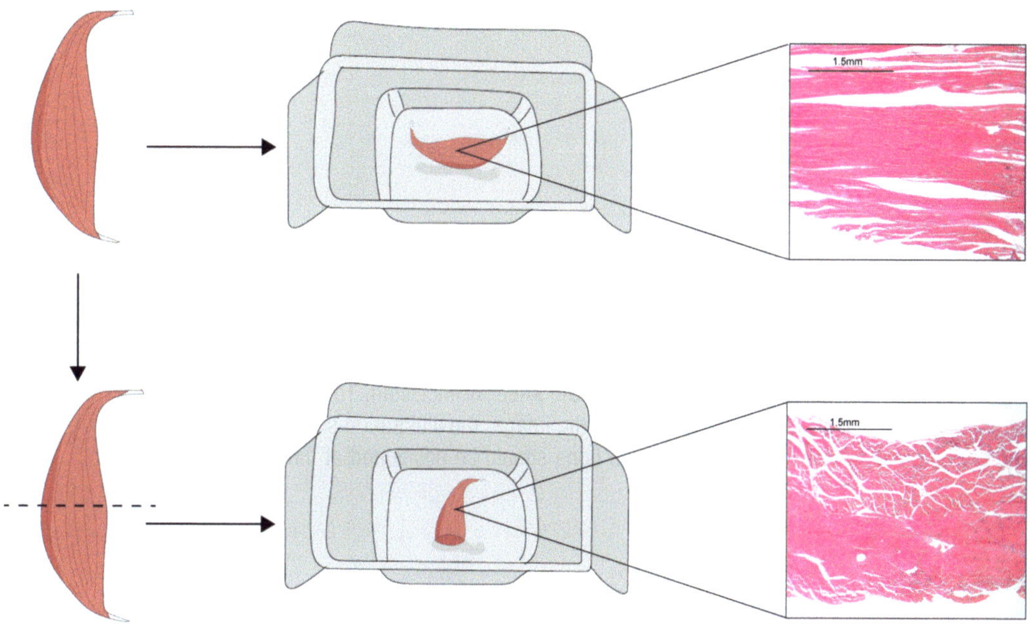

Fig. 5 Representation of muscle sample inclusion planes. This tissue can be included in two ways for different analyses. The longitudinal view of the muscle fibers is shown in the upper panels. The dotted line represents the sample cutting plane for transversal visualization of the muscle fibers

the muscle transversely (*see* Fig. 5) and include it with the cut region facing the bottom of the mold.

7. Femur: femur samples can be positioned with their patellar face facing the bottom of the mold (*see* Fig. 6). This will privilege the visualization of the distal epiphysis and associated diaphysis.
8. Knee: Knee samples can be split in half, following the sagittal plane with the aid of a scalpel blade (*see* Fig. 7). After splitting, each side of the sample can be positioned with the cut region facing the bottom of the mold. The bones should face the bottom of the mold, while the muscle should face upwards. This will allow visualization of the epiphyses of the femur and

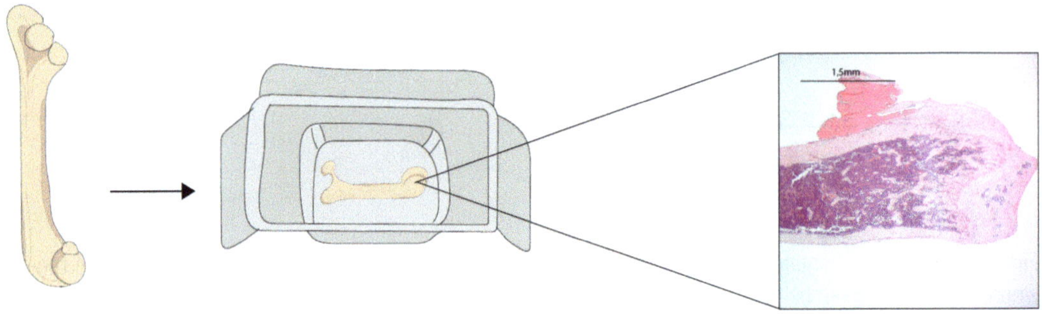

Fig. 6 Drawing representing the femur. To the right of the image, we see the illustration of a mold of inclusion and the position of the organ

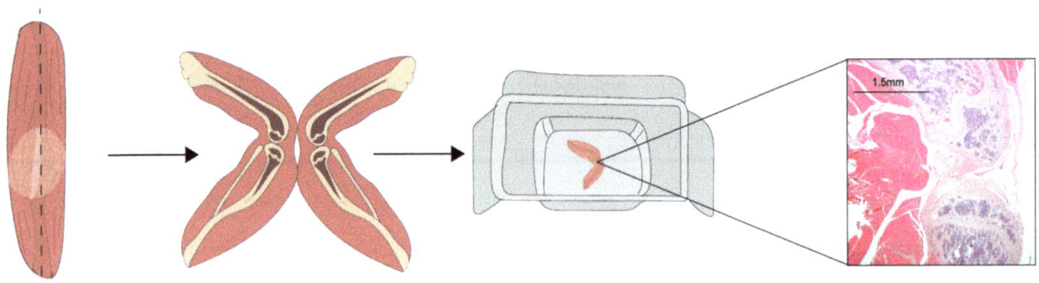

Fig. 7 Drawing representing the positioning of the knee for inclusion. The dotted line represents the region where the tissue should be cut. To the right of the image, the illustration shows the aspect of the cut sample and the embedding plane. The face with the bone should be positioned at the bottom of the mold

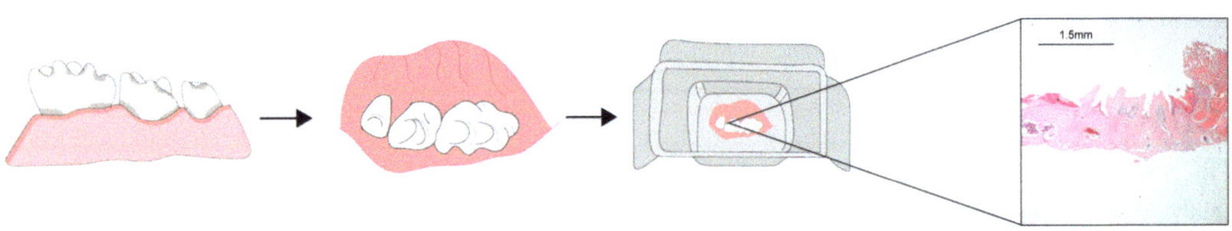

Fig. 8 Drawing representing the plane of inclusion of the maxilla. The buccal region should be facing the mold as shown in the image on the left. The H&E slide shows a region of the sample that evidences the second molar tooth

tibia involved in the knee joint, the meniscus located between the two epiphyses, the joint capsule, the synovial tissue, patellar tendon, and the skeletal striated muscle surrounding the bone.

9. Maxilla: Maxilla samples should be included with the buccal region facing down (*see* Fig. 8). In this way, it will be possible to visualize the 3 upper molars, including their dental pulp, the gingiva, the trabecular and compact bone. The embedding of this tissue is difficult because the contact region of the sample with the mold is small and it tends to lie down, changing the

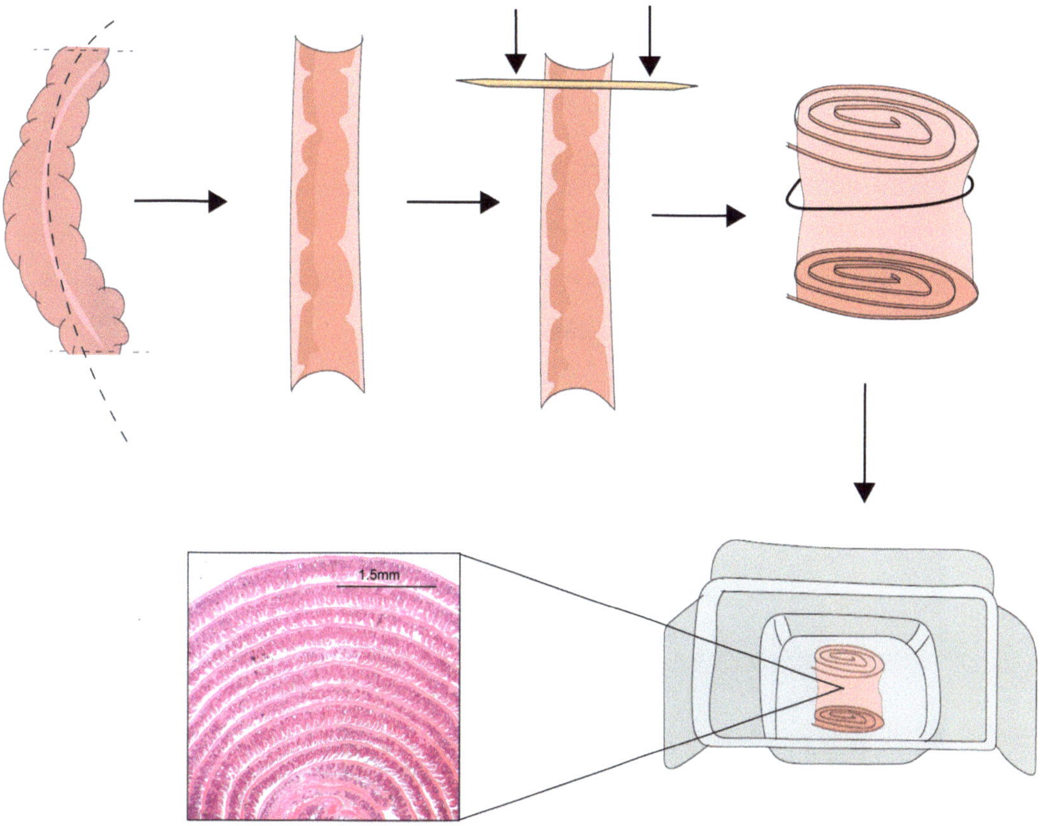

Fig. 9 Representative plane of inclusion of the intestine. The first line of the image shows how the roll should be made, after extracting the tissue. Upon inclusion, the intestine roll must be positioned with its base facing the mold. This will allow a wide area of the sample for evaluating morphological alterations of intestinal villi and inflammatory mucosal damage

angle of inclusion. Use hot tweezers to hold the sample in position until the paraffin starts to solidify.

10. Intestine: With the aid of a scalpel blade, the intestine can be sectioned into four parts (duodenum, proximal jejunum, distal jejunum, and ileum) and individually rolled into a "swiss roll" (*see* **Note 5**). The inclusion of samples should be carried out by placing the base facing downwards (*see* Fig. 9). This will provide a wide area for evaluation.

3.6 Microtomy

1. To observe the tissue under the microscope, it is necessary to cut it into 5 μm (or less) slices for light to pass through the tissue (*see* **Note 18**). The process will be described based on the levers of the microtome identified in Fig. 10.
2. With lock 1 closed, push lever 2 (*see* Fig. 10a) and position the cassette in the proper space (indicated in the image A by the number 3) (*see* Fig. 11a).

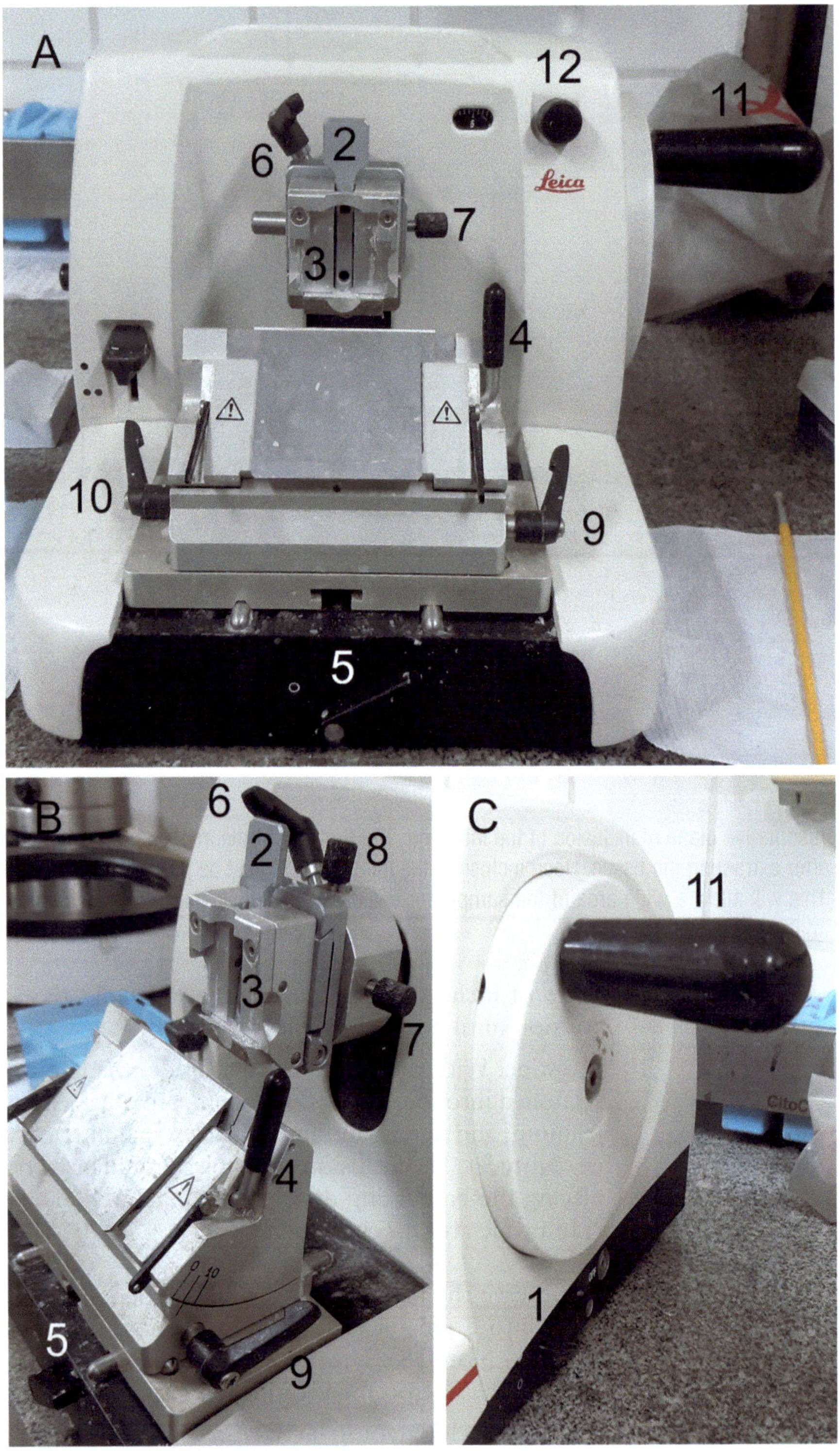

Fig. 10 Photos of a rotary microtome. The numbers indicate the locks and levers that will be described in the text. 1. Lever for activating the forward crank; 2. Block holder security lock; 3. Block holder; 4. Knife locking lever; 5. Base lever; 6. Block holder activation lever; 7. Block holder horizontal movement button; 8. Block holder vertical movement button; 9. Knife angle locking lever; 10. Base lateral movement lever; 11. Forward crank; 12. Thickness button

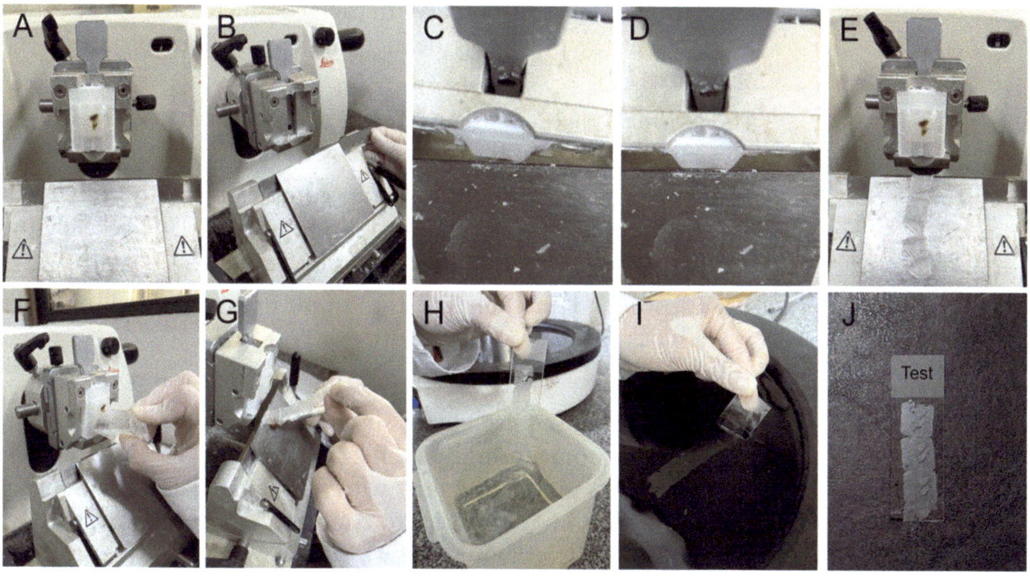

Fig. 11 Pictures taken to demonstrate the steps of microtomy. (**a**). Paraffin block with lung samples, in the correct position; (**b**) Razor being placed; (**c**) Demonstration of the crooked block, before alignment; (**d**) Block aligned with the razor, as it should be before making the cuts; (**e**) Thicker cuts (10 μm thick), to demonstrate the block roughing phase; (**f**) How to humidify the block; (**g**) How to pick up the cuts that will be used, to put them in water with ethanol; (**h**) How to take the slices from the water; (**i**) Placing the slices to open in hot water; (**j**) Ready slide after the end of this process

3. Unlock lever 4 (*see* Fig. 10b) to position the knife blade (*see* Fig. 11b). Then lock it again.
4. Lock 5 prevents movement of the microtome base. Open lock 5 to bring the knife closer to the tissue block.
5. Check that the block is straight, parallel to the knife. If not, release the lock indicated by number 6. This lock prevents the movement generated by buttons 7 and 8. Button 7 allows a horizontal movement of the platform with the block. Button 8, on the other hand, allows a vertical movement. Release lock 6 and use buttons 7 and 8 to align the block with the knife. When the block is straight, close the latch 6 (*see* Figs. 10c and 11c-d).
6. Before starting to cut. Check lock number 9. It allows changing the angle of the knife in relation to the paraffin block. Normally, we keep the indicator in 5 degrees, as shown in Fig. 10b.
7. Lock number 10 prevents the base of the microtome from moving laterally. It is a less-used lever. Keep it closed when making cuts.
8. Check all locks after positioning and aligning the block with the specimen. Open only lock 1 to allow movement of the forward crank (indicated by number 11).

9. Turn the forward crank. With each complete turn of the crank, a cut in the previously defined thickness will be produced (*see* **Note 19**).
10. Normally, we cut with a thickness of 10 μm to remove the paraffin that is in front of the tissue (*see* Fig. 11e).
11. When the sample is coming out completely, use button 12 to set the thickness needed for cutting on the blade (usually the pattern is to cut samples with 5 μm thickness).
12. Close lock 1 and humidify the sample in the paraffin block with a block of ice or a damp cloth (*see* Fig. 11f).
13. Turn the forward crank to take the cuts that will be used.
14. After obtaining the cut, it should be carefully removed from the microtome, using a brush (*see* Fig. 11g). The cut should be placed in a plastic container with 30% ethanol at room temperature. This makes it easier to collect the sections with the slide and later open them in hot water.
15. Then, the glass slide should be immersed in container and positioned close to the cut and perpendicular to it. When removing the slide from the solution, the cut will adhere to the slide and come out of the solution as well (*see* Fig. 11h).
16. Subsequently, the cut sample should be placed in a water bath at 40 °C to open the samples. Simply insert and remove the slide from the bath perpendicular to the water (*see* Fig. 11i). Hot water will cause the cut to open, undoing possible paraffin folds. Identify the slide.
17. Before carrying out staining, it is necessary to fix the sections on the slide. Let the slides for 30 minutes in the oven at 60 °C before proceeding with staining.

4 Notes

1. In tissues with large depositions of calcium, such as bones and teeth, tissue rigidity does not allow the microtomy to happen in the same way as in soft tissues. Therefore, it is necessary that the samples pass through a decalcification step after washing the fixative and before starting the processing. Slow decalcification of samples yields better results than rapid decalcification with strong acids, with regard to the morphological quality of the obtained samples. Usually, slow decalcification is obtained using a calcium chelator, such as Ethylenediaminetetraacetic Acid Disodium Salt (EDTA). To that end, it is necessary to prepare a 10–14% EDTA solution in distilled water, with a pH of 7.4. Adjusting the pH is fundamental to EDTA dissolution. After washing the fixative, the samples must be placed in a

container and covered by the EDTA solution. It is necessary to change the EDTA of the samples every 48 h. The decalcification time varies between 15 and 30 days, according to the tissue. At the end of this period, it is necessary to check whether the sample is ready for processing. Try to cut a part of the tissue that will not be used. It must have a smooth aspect during the cutting process. Another way to ensure decalcification is by trying to pierce the tissue with a small needle. If the needle passes through the tissue easily, it can proceed to process. It is necessary to carry out the calcium chelator washout before processing the tissue. To do this, wash samples under running water for 30 min.

2. The recommended lethal dose for mice is in the range 240–300 mg/kg of ketamine hydrochloride and 30–45 mg/kg of xylazine [11].
3. Before removing the target tissues for analyses, you can perform tissue perfusion. The blood present in the mouse's circulatory system may be removed with the aid of a solution with a neutral pH, such as 1× PBS or 0.9% saline solution. For this, with scissors and tweezers, the skin in the chest area must be removed. Then, with the tip of a pair of scissors, pierce the muscle in the region below the sternum and make a longitudinal cut dividing the rib cage to expose the heart. Locate and hold the right ventricle with tweezers and inject 10 mL of 1× PBS with a syringe, simulating the functioning of the circulatory system. In this process, it is possible to visualize the lungs inflating and the liver changing color from wine to light brown.
4. Tips for organ manipulation: For lung collection, remove the organ by picking the heart; for liver collection, cut the ligaments associated to the diaphragm to deliver the organ; for intestine collection, pick the stomach; for brain collection, perforate the bregma region with the tip of a sharp-edged scissor and then open the instrument, so that this movement opens the skullcap and exposes the brain. With the aid of tweezers, remove the occipital region of the skullcap and insert the tip of a spatula below the cerebellum to deliver the organ. The use of inappropriate instruments, such as toothed tweezers, for example, can damage the tissue and generate artifacts that can alter the morphological structure of the tissue.
5. The intestine, for example, can be sectioned into four parts: duodenum, proximal jejunum, distal jejunum, and ileum. Soon after collection, the intestine fragments must be rolled individually into a "swiss roll". To do this, place the extracted portion of the intestine on brown paper soaked in formalin. With a syringe, clean the lumen of the intestine with 1x PBS. Then, make a longitudinal cut, opening the intestinal tube. The organ

should look like a leaf. Be very careful not to press and damage the intestinal villi. With a small wooden rod, roll up the sample with the villi facing inwards. Tie it with a suture thread and straighten the rod from the center of the roll (*see* Fig. 9). Another example is the liver, a heavy and dense organ, that must be sectioned into smaller pieces according to its lobes in order to facilitate penetration of the fixative solution. Good fixative penetration is related to the size and thickness of the sample which is recommended to be inferior to 6 mm.

6. The fixation of the collected tissues is essential for the preservation of biological samples. Good fixation is critical for inhibiting tissue autolysis, coagulation or tissue hardening, preservation of cell and tissue components, improvement in optical tissue differentiation, and resulting in good staining preparations [12]. The fixative solution should be able to penetrate the tissue quickly, cause less damage to the tissue structure during penetration and should have a preservative action. Few fixatives fix and preserve tissue. 10% Formalin is considered one of the best universal fixatives [13].
7. It is important to identify the cassettes with the samples using a pencil or printed paper, given that the processing procedure using ethanol, xylene, and paraffin can remove pen inks. Optionally, the samples can be stored individually in identified 8 mL glass flasks or 15–50 mL conical tubes containing the fixative solution.
8. The removal of formaldehyde residues is important because it can react with antigenic epitopes, masking them directly or it can also react with the surrounding amino acids of the epitope, changing its shape and masking it indirectly, thus making immunohistochemical analyses difficult [14]. Additionally, formaldehyde residues can disrupt tissue staining, Hematoxylin and Eosin (H&E), causing splotches of histological sections. Preferentially, the whole processing procedures must be carried out 48 h after fixation of mouse samples (from washing the fixative to paraffin impregnation), respecting the periods described for each procedure. If not possible, let the samples be in 70% ethanol. The more accurate the processing procedure, the better the quality of the obtained sample, but it is still possible to obtain good samples after varying times in 70% ethanol. The sample tends to become more friable over time.
9. Tissues such as the liver can become very dried up after inclusion. To prevent this, liver samples can remain in water for 24 h, after washing the fixative and before starting processing. Partially reducing the clearing time of liver samples can also favor morphological preservation of the tissue.

10. The samples must be completely covered by the ethanol solutions. Never let the samples dry out when changing the solutions. Ethanol in the correct concentrations can be stored in a processing battery, to be used more than once. If this is the case, routinely check the ethanol graduation of the solutions and, if necessary, filter them before using.
11. In hard tissues the dehydration protocol is longer. It is necessary that the samples pass through 30 min baths in ethanol 70% I, 70% II, 80% I, 80% II, 90%, 95%, 100% I, 100% II, and 100% III.
12. Xylene is a colorless, volatile, flammable organic solvent with significant toxic potential, so it is necessary that this processing phase be carried out in an exhaust hood with the proper use of personal protective equipment (gloves and masks). It is also necessary to dispose this reagent at an appropriate disposal.
13. Make sure the samples are in a glass container, xylene can melt some plastic pots.
14. The xylene can be stored for reuse, but the ideal is that the third bath is new xylene.
15. For embedding in paraffin, it is necessary that it reaches its melting point. Therefore, use a histological oven at ~60 °C. Very high temperatures can damage the tissue and lower temperatures leave the paraffin solid. Paraffin can be used more than once, but it needs to be filtered when it becomes dirty. The traditional impregnation method is performed with paraffin, but other options can be used, including histology waxes (blend of paraffin wax and plastic polymers) or inclusion resin.
16. Processing can be stopped at this stage if necessary. Then, the next day, return the sample to paraffin III for 20 min before starting the embedding step.
17. For hard tissues, it is necessary to keep the samples longer in paraffin. The first paraffin bath should last 30 minutes, while the second and third should last at least 40 min each.
18. To cut the blocks, we use a microtome. There are several types of microtome but the most used, and the one that is discussed here, is the rotary microtome. In this microtome, there is a fixed blade and the block approaches the microtome at a distance defined by the operator, which generates cuts in the required thickness.
19. It is normal that cuts do not come out in the first turns of the handle, as the knife was not completely touching the paraffin block.

Acknowledgments

We acknowledge the funding agencies Coordenação de Aperfeiçoamento de Pessoal de Nível Superior (CAPES/Brazil), Conselho Nacional de Desenvolvimento Científico e Tecnológico (CNPq/Brazil), Financiadora de Estudos e Projetos (FINEP/Brazil), and Fundação de Amparo à Pesquisa do Estado de Minas Gerais (FAPEMIG/Brazil) for the financial support.

Celso Queiroz-Junior has received a research fellowship under the FINEP-funded project (MCTI/FINEP -MS/SCTIE/DGITIS/CGITS, grant number 6205283BBB28-4F9C-AA65-808FE4450542).

References

1. Capa N (2009) Antony van Leeuwenhoek–inventor do microscópio. J Bras Patol Med Lab 45(2). https://doi.org/10.1590/S1676-24442009000200001
2. Masters BR (2008) History of the optical microscope in cell biology and medicine. In: Encyclopedia of life sciences (ELS). Wiley, Chichester. https://doi.org/10.1002/9780470015902.a0003082
3. Schultz M (2008) Rudolf virchow. Emerg Infect Dis 14(9):1480–1481. https://doi.org/10.3201/eid1409.080667
4. Musumeci G (2014) Past, present and future: overview on histology and histopathology. J Histol Histopathol 1(5):1–3. https://doi.org/10.7243/2055-091X-1-5
5. Bracegirdle B (1977) The history of histology: a brief survey of sources. Hist Sci 15(2):77–101. https://doi.org/10.1177/007327537701500201
6. Becerra SC, Roy DC, Sanchez CJ et al (2016) An optimized staining technique for the detection of gram positive and gram negative bacteria within tissue. BMC Res Notes 9(1):1–10. https://doi.org/10.1186/s13104-016-1902-0
7. Brito MVH, Araújo MD, Acácio GJS et al (2001) Lesão intestinal após isquemia-reperfusão: estudo comparativo usando sal tetrazólico (MTT) e histologia. Acta Cir Bras 16:26–31. https://doi.org/10.1590/S0102-86502001000100005
8. Azevedo LSS (2012) Segmentação e Reconstrução 3D de Tecidos Animais em Imagens Histológicas. Dissertation, Universidade do Porto
9. Monteiro L, De Mascarel A, Sarrasqueta AM et al (2001) Diagnosis of helicobacter pylori infection: noninvasive methods compared to invasive methods and evaluation of two new tests. Am J Gastroenterol 96(2):353–358. https://doi.org/10.1016/S0002-9270(00)02311-X
10. Alturkistani HA, Tashkandi FM, Mohammedsaleh ZM (2016) Histological stains: a literature review and case study. Global J Health Sci 8(3):72. https://doi.org/10.5539/gjhs.v8n3p72
11. Brasil (2018) Diretriz da Prática de Eutanásia do Conselho Nacional de Experimentação Animal. Avaliable via MCTI. https://antigo.mctic.gov.br/mctic/opencms/legislacao/outros_atos/resolucoes/Resolucao_CONCEA_n_37_de_15022018.html. Accessed 1 Fev 2023
12. Junqueira LC, Carneiro J, Abrahamsohn P (2008) Histologia Básica: Texto e Atlas. Guanabara, Rio de Janeiro
13. Frasquilho SG, Sanchez I, Yoo C et al (2021) Do tissues fixed in a non-crosslinking fixative require a dedicated formalin-free processor? J Histochem Cytochem 69(6):389–405. https://doi.org/10.1369/00221554211017859
14. Jones ML, Bancroft JD, Gamble M (2008) Connective tissues and stains. In: Bancroft JD, Gamble M (eds) Theory and practice of histological techniques, 6th edn. Churchill Livingstone, London, pp 135–160. https://doi.org/10.1016/B978-0-443-10279-0.50017-8

Chapter 8

Measurement of MPO, NAG, and EPO as an Indirect Quantification of Leukocyte Infiltration in Mouse Tissues

Carla Daiane Ferreira de Sousa, Bruno Silva Amaral, Clara Soares de Souza, and Danielle G. Souza

Abstract

Enzyme assays are used to measure the activity or concentration of enzymes in biochemical or cell-based systems. Most enzyme assays are based on the detection of fluorescent, luminescent, or spectrophotometric endpoint signals. In recent years, they have been developed and widely used for diagnostics, mechanisms of action, and inflammatory activities. An enzyme assay essentially works by the conversion of a substrate into a product by the enzyme of interest. In this case, it is extremely important to know the optimal conditions for enzyme activity, as these affect the specificity and efficacy of the assay. For optimal reaction conditions, temperature, pH, and the presence of ions should be considered. In this chapter, the enzymatic assays for the detection of the enzymes N-acetylglucosaminidase (NAG), myeloperoxidase (MPO), and eosinophil peroxidase (EPO) are addressed. These assays are used to assess inflammatory parameters, for example, at the peripheral level in models of viral disease. They are based on an index of neutrophil, macrophage, or eosinophil accumulation in inflammatory tissues from animals by measuring the specific activity of the marker enzymes. The enzyme activity assays discussed here are based on colorimetric reactions compatible with any experimental model in which the respective cells has an active role. The advantage of using these enzymatic assays in inflammation response models is that they are simpler and less expensive compared to techniques such as Western blot or quantitative PCR.

Key words Enzyme assay, Leukocyte, Inflammation, N-acetylglucosaminidase, Myeloperoxidase, Eosinophil Peroxidase

1 Introduction

Enzyme assays can be used to obtain information such as the presence of cells or inflammatory features. Enzymes such as N-acetylglucosaminidase (NAG), myeloperoxidase (MPO), and eosinophil peroxidase (EPO) are present mainly in macrophages, neutrophils, and eosinophils, respectively. They are commonly associated with inflammatory processes and are used as an indirect measure of the presence or activation of leukocytes, as shown previously [1, 2].

Lucas Secchim Ribeiro (ed.), *Models in Inflammation and Immunopharmacology*, Methods in Molecular Biology, vol. 2993,
https://doi.org/10.1007/978-1-0716-5019-6_8,
© The Author(s), under exclusive license to Springer Science+Business Media, LLC, part of Springer Nature 2026

MPO is a proinflammatory enzyme stored in the granules of neutrophils and synthesized during myeloid differentiation [3, 4]. The function of MPO is associated with catalyzing the formation of hypochlorous acid (HOCl) from hydrogen peroxide (H_2O_2) [5]. HOCl is extremely reactive with biological molecules, a potent antimicrobial agent, and can lead to tissue damage characteristic of inflammatory diseases. MPO also possesses peroxidase activity by catalyzing H_2O_2-mediated oxidation of various substrates. Here we describe the use of 3,3,5,5-tetramethylbenzidine (TMB), a safe substrate, to detect peroxidase activity because that can be catalyzed by MPO [6].

NAG is an enzyme found in lysosomes and produced mainly by macrophages. NAG has been used to detect macrophage accumulation and activation in several tissues. For the assay described here, we used p-nitrophenyl-N-acetyl-β-D-glucosaminide (pNP-β-d-GlcNAc) as a sensitive substrate for the activity of NAG, which is mainly used for urine diagnostics [7].

EPO is an enzyme present in eosinophil granulocytes and is often used for the detection of the cells in different organs. The main function of this enzyme is to catalyze the formation of hypohalous acids from H_2O_2 and halide ions, oxidants that are highly cytotoxic to bacteria. EPO activity is commonly used as a measure of the presence or activation of eosinophils in biological tissue. Here, we have described the use of OPD as a substrate to measure EPO activity, which has already been shown to be a viable and more specific option [8–10].

Although measurement of enzyme expression or protein content can provide information about the abundance of the molecule, enzymatic activity can vary considerably even when the amount of enzyme present is similar. For example, MPO can be inhibited by endogenous inhibitors, so MPO protein or RNA expression does not always correspond to its activity levels [11, 12]. Therefore, one must be cautious in interpreting the results. The assays described here are based on enzyme reactions using specific reagents for each enzyme of interest, and the results are on a colorimetric basis that can be read with a spectrophotometer. Protocols were developed for rodent tissue samples in inflammatory models where abundant leukocytes are present in the tissue.

2 Materials

1. Buffer I (pH 4.7): 0.1 M NaCl, 0.02 M $NaPO_4$, 0.015 M Na_2EDTA (*see* **Note 1**).
2. Buffer II (pH 5.4): 0.05 M Na_3PO_4, 0.5% (w/v) Hexadecyltrimethylammonium bromide (HTAB) (*see* **Note 2**).
3. NaCl solution: 0.2% NaCl (w/v) (*see* **Note 3**).

4. NaCl/glucose solution: 1.6% NaCl, 5% glucose (w/v) (*see* **Note 3**).
5. 3,3,5,5-tetramethylbenzidine (TMB) solution: 1.6 mM TMB dissolved in Dimethyl-sulfoxide (DMSO) (*see* **Note 4**).
6. Hydrogen peroxide (H_2O_2): 0.002% H_2O_2 diluted in buffer II (v/v).
7. Sulfuric acid (H_2SO_4): 1 M H_2SO_4.
8. Saline/Triton solution: 0.9% NaCl, 0.1% Triton X-100 (v/v).
9. Citrate/phosphate buffer (pH 4.5): 0.1 M Citric acid, 0.1 M Sodium phosphate dibasic (Na_2HPO_4).
10. p-nitrophenyl-N-acetyl-β-D-glucosaminide (pNP-β-d-GlcNAc): 2.24 nM diluted in citrate/phosphate buffer.
11. Glycine Buffer (pH 10.6): 0.8 M glycine, 0.8 M NaCl, 0.8 M NaOH.
12. Phosphate saline buffer (PBS).
13. HTAB working solution: 0.5% HTAB in PBS.
14. Enzyme substrate o-phenylenediamine dihydrochloride (OPD, working solution): 1.5 mM OPD in 0.075 mM Tris-HCl buffer (pH 8) with 6.6 mM H_2O_2.
15. 96-well flat-bottom plate.
16. Microcentrifuge.
17. Tissue homogenizer (*see* **Note 5**).
18. Biochemical Oxygen Demand Incubator with humidified atmosphere.
19. 2-mL, Single-use polypropylene tubes (*see* **Note 6**).
20. Spectrophotometer compatible with 96-well plates.
21. Liquid nitrogen.
22. Vortex.

3 Methods

1. Mouse tissue can be collected and immediately stored in a freezer at −80 °C. For the assay, the tissue should be homogenized with 1 mL of Buffer I for 100 mg of tissue. The solution must be kept cold during tissue homogenization (~4 °C), for example, it can be placed in a box with ice. A simplified scheme of MPO processing and assay is shown in Fig. 1.
2. Centrifuge the tissue homogenate at 4 °C (3,000 × *g*, 10 min).
3. Discard the supernatant.

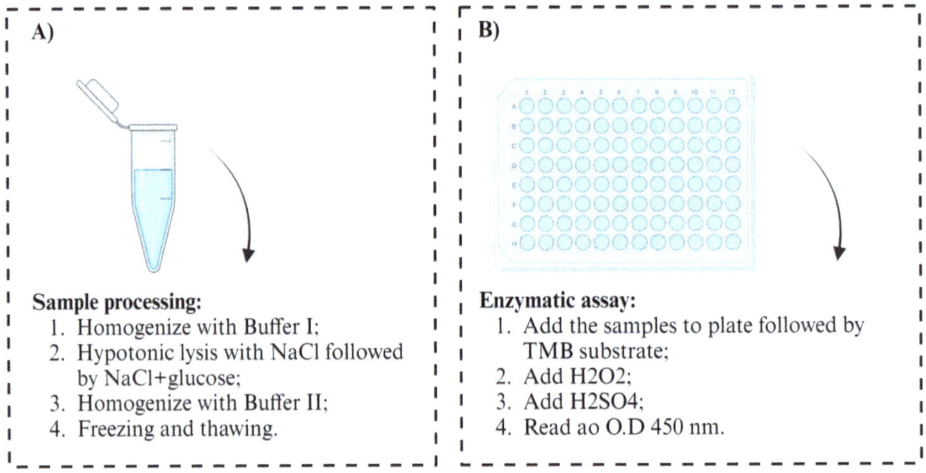

Fig. 1 Simplified scheme of MPO steps. Steps describing MPO procedure for (**a**) a processed animal tissue, followed by (**b**) the enzymatic assay

4. To promote hypotonic lysis, add cold NaCl solution (0.5 ml/100 mg tissue). After 30 seconds, add cold NaCl/glucose solution (0.5 ml/100 mg tissue).
5. Homogenize the samples. Until this step, the samples undergo the same procedure for all the enzymatic assays described here. After homogenization, the samples should be separated into three different 2 mL microcentrifuge tubes, each of which will then follow the specific steps required for its respective assay.

3.1 Myeloperoxidase

1. Centrifuge the sample homogenate at 4 °C (3,000 × *g*, 10 min).
2. Discard the supernatant.
3. Resuspend the pellet by adding 1 mL of Buffer II (at room temperature) per 100 mg of tissue.
4. Homogenize the sample quickly by vortexing. At this point, the tubes can be frozen for storage (−20 °C), or the protocol can be continued.
5. Freeze and thaw the samples three times with liquid nitrogen (*see* **Note 7**).
6. Centrifuge the thawed samples at 4 °C (3,000 × *g*, 15 min) and collect the supernatant for the enzyme assay.
7. Add 25 μl of the sample supernatant (diluted in assay Buffer II if necessary) to a 96-well plate (flat-bottom) in duplicates.
8. Add 25 μl TMB and incubate for 5 min at 37 °C in a Biochemical Oxygen Demand Incubator with a humidified atmosphere.

9. Add 100 μl H_2O_2 and incubate for 5 min at 37 °C in a Biochemical Oxygen Demand Incubator with humidified atmosphere.
10. Add 100 μl H_2SO_4 to stop the reaction (*see* **Note 8**).
11. Measure the absorbance of the plates in a spectrophotometer at 450 nm.
12. Values can be expressed as absorbance or in relative units of a standard curve (*see* **Note 9**).

3.2 N-Acetyl-glucosaminidase

1. Centrifuge the sample homogenate at 4 °C (3,000 × *g*, 10 min).
2. Discard the supernatant.
3. Resuspend the pellet by adding 1 mL of Saline/Triton solution per 100 mg of tissue.
4. Homogenize the sample quickly by vortexing. At this point, the tubes can be frozen (−20 °C) for storage, or the protocol can be continued.
5. Centrifuge the samples at 4 °C (1,500 × *g*, 15 min) and collect the supernatant for the enzyme assay.
6. Add 100 μl/well of samples at the appropriate dilution to a 96-well plate, flat bottom, in duplicate, using citrate/phosphate buffer as a blank (*see* **Note 10**).
7. Add 100 μl of the pNP-β-d-GlcNAc solution (*see* **Note 11**).
8. Incubate the plate for 10 min at 37 ° C in a Biochemical Oxygen Demand Incubator with a humidified atmosphere.
9. Add 100 μl glycine buffer.
10. Measure the absorbance of the plates in a spectrophotometer at 450 nm.
11. Values can be expressed in absorbance or in relative units of a standard curve (*see* **Note 9**).

3.3 Eosinophil Peroxidase

1. Centrifuge the sample homogenate at 4 °C (3,000 × *g*, 10 min).
2. Discard the supernatant and resuspend the pellet in HTAB working solution.
3. Homogenize the sample.
4. Freeze and thaw it 3 times in liquid nitrogen (*see* **Note** 7).
5. Centrifuge and save the supernatant for EPO assay.
6. Add 75 μl/well of the samples to a 96-well, flat-bottomed plate, in duplicate, using PBS as blank.
7. Add 75 μL of the substrate OPD solution.
8. Incubate the plate for approximately 30 min or until the reaction occurs at 37 °C in a Biochemical Oxygen Demand Incubator with a humidified atmosphere in the dark.

9. Stop the reaction with 50 μL of H_2SO_4.
10. Measure the absorbance of the plates in a spectrophotometer at 492 nm.
11. The values can be expressed as absorbance or as relative units from a standard curve (*see* **Note 9**).

4 Notes

1. The pH must be measured and adjusted, and Na_2EDTA should not be added until the pH is adjusted to 4.7. The solution can be stored in the refrigerator (~4 °C).
2. As mentioned in **Note 1**, it is important to measure and adjust pH. HTAB should be added as soon as the pH is adjusted to 5.4. The solution can be stored in the refrigerator (~4 °C).
3. The solution can be stored in the refrigerator (~4 °C).
4. TMB diluted in DMSO should be protected from light and handled in a fume hood. Use it immediately after preparation.
5. A benchtop homogenizer or a manual (portable) homogenizer can be used, which can be used to grind the biological material without damaging the subcellular structures.
6. It is necessary to use tubes in which the homogenizer reaches the bottom of the tube so that all the tissue can be properly homogenized. For this reason, 2 mL tubes are generally recommended.
7. The samples, properly stored in polypropylene tubes, are placed in a suitable container, such as a styrofoam box, and liquid nitrogen is poured over them. This procedure is repeated 3 times so that the samples undergo a series of rapid freezing and thawing in a short time. Use extra care with liquid nitrogen, as it can cause burns, cold burns, and frostbite. It is advisable to work in a well-ventilated area and pour the liquid out slowly to minimize the risk of boiling and splashing. When in doubt, consult your laboratory's regulations for specific biosafety guidelines.
8. The colorimetric reaction observed in the MPO assay consists of the oxidation of TMB by H_2O_2. TMB acts as a hydrogen donor. This gives the solution a blue coloration, as described in **step 15**. Interruption of the reaction by H_2SO_4 results in a yellow coloration, as described in **step 16** and shown in Fig. 2, which can be measured at 450 nm [13].
9. A standard curve is not described in this assay description, but can be constructed to estimate the enzyme concentrations present in the sample. The curve starts from known amounts of the enzyme at various concentrations.

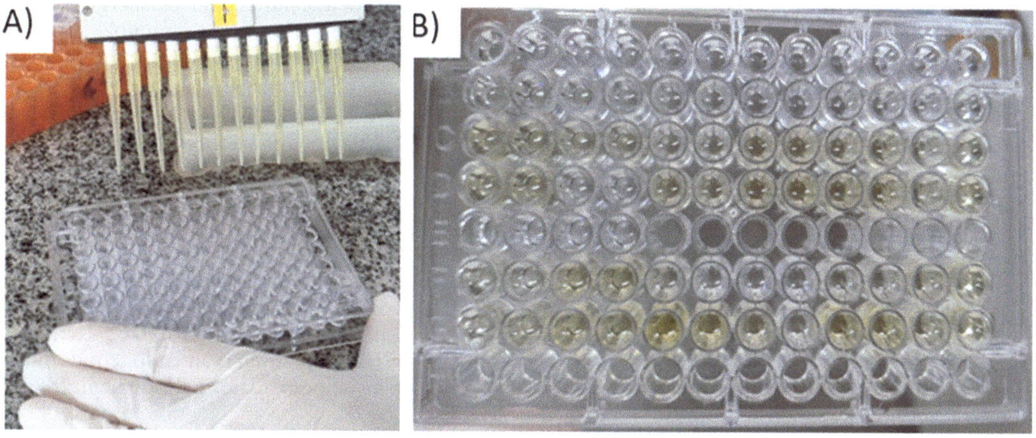

Fig. 2 Representative scheme of an enzymatic assay. A 96-well plate is used for the enzyme assay. (**a**) Control and samples are plated in duplicate, followed by the substrate. (**b**) After the addition of H2O2 and the stop solution, yellow coloration is observed in the samples where MPO is present

10. Depending on the tissue being analyzed and the stimulus protocol, samples might perform better at different dilutions, prepared in Citrate/Phosphate Buffer. For example, a dilution of 1:3 for samples such as paw tissue and joints shows better results in the analysis of alphavirus infections.
11. pNP-β-d-GlcNAc substrate is diluted in Citrate/Phosphate Buffer at the time of use.

Funding

This study was supported by Instituto Nacional de Ciência e Tecnologia (INCT) em dengue e interação microrganismo hospedeiro (Grant 465,425/2014–3), Coordenação de Aperfeiçoamento de Pessoal de Nível Superior (CAPES), Conselho Nacional de Desenvolvimento Científico e Tecnológico (CNPq), and FAPEMIG. The authors also indicate that they have no potential conflicts of interest.

References

1. de Araújo S, Costa VRM, Santos FM, Sousa CDF, Moreira TP, Gonçalves MR, Félix FB, Queiroz-Junior CM, Campolina-Silva GH, Nogueira ML, Sugimoto MA, Bonilha CS, Perretti M, Souza DG, Costa VV, Teixeira MM (2022) Annexin A1-FPR2/ALX signaling Axis regulates acute inflammation during chikungunya virus infection. Cells 11(17):2717
2. Moreira TP, Sousa CDF, Costa VRM, Queiroz-Junior CM, Santos FM, Bonilha CS, Ésper LM, Nogueira ML, Cunha TM, Teixeira MM, Costa VV, Souza DG (2023) Tumour necrosis factor plays a deleterious role in the pathogenesis of chikungunya virus infection. Immunology 168(3):444–458
3. Pulli B, Ali M, Forghani R, Schob S, Hsieh KLC, Wojtkiewicz G, Linnoila JL, Chen JW (2013) Measuring myeloperoxidase activity in biological samples. PLoS One 8(7):e67976

4. Soares AC, Pinho VS, Souza DG, Shimizu T, Ishii S, Nicoli JR, Teixeira MM (2002) Role of the platelet-activating factor (PAF) receptor duringpulmonary infection with gram negative bacteria. British J Pharmacol 137:621–628
5. Heinecke JW (1999) Mechanisms of oxidative damage by myeloperoxidase in atherosclerosis and other inflammatory disorders. J Lab Clin Med 133:321–325
6. Andrews PC, Krinsky NI (1982) Quantitative determination of myeloperoxidase using tetramethylbenzidine as substrate. Anal Biochem 127(2):346–350
7. Barcelos LS, Talvani A, Teixeira AS, Vieira LQ, Cassali GD, Andrade SP, Teixeira MM (2005) Impaired inflammatory angiogenesis, but not leukocyte influx, in mice lacking TNFR1. J Leukoc Biol 78(2):352–358
8. Adamko DJ, Wu Y, Gleich GJ, Lacy P, Moqbel R (2004) The induction of eosinophil peroxidase release: improved methods of measurement and stimulation. J Immunol Methods 291(1–2):101–108
9. Schneider T, Issekutz AC (1996) Quantitation of eosinophil and neutrophil infiltration into rat lung by specific assays for eosinophil peroxidase and myeloperoxidase Application in a Brown Norway rat model of allergic pulmonary inflammation. J Immunol Methods 198(1):1–14
10. Silveira MR, Nunes KP, Cara DC, Souza DG, Corrêa A, Teixeira MM, Deborah Negrão-Corrêa D (2002) Infection with Strongyloides venezuelensis induces transient airway eosinophilic inflammation, an increase in immunoglobulin E, and Hyperresponsiveness in rats. Infect Immun 70(11):6263–6272
11. Chapman ALP, Mocatta TJ, Shiva S, Seidel A, Chen B, Khalilova I, Paumann-Page ME, Jameson GNL, Winterbourn CC, Kettle AJ (2013) Ceruloplasmin is an endogenous inhibitor of myeloperoxidase. J Biol Chem 288(9): 6465–6477
12. Segelmark M, Persson B, Hellmark T, Wieslander J (1997) Binding and inhibition of myeloperoxidase (MPO): a major function of ceruloplasmin? Clin Exp Immunol 108(1): 167–174
13. Zhang X, Yang Q, Lang Y, Jiang X, Wu P (2020) Rationale of 3,3′,5,5′--Tetramethylbenzidine as the chromogenic substrate in colorimetric analysis. Anal Chem 92(18):12400–12406

Chapter 9

Bridging Immunity and Hemostasis: Technical Foundations of Platelet-Leukocyte Co-cultures

Lucas Secchim Ribeiro

Abstract

Platelets are immune cells with a great potential for the discovery of novel mechanisms regulating both innate and adaptive responses. Although it is known that the interaction between platelets and leukocytes plays a critical role in inflammation and thrombosis, the vast majority of experimental models focus on monotypic cell systems, ignoring the constant association of these two cell types in circulation and in the tissues. Co-culture systems that recapitulate this interplay are essential for understanding cellular processes and for evaluating therapeutic interventions. However, the technical requirements for isolating and handling these cells with such distinct physical and functional properties pose significant challenges. Here I provide a detailed guide for the isolation, characterization, and co-culture of human and murine platelets with leukocytes, while describing systematic procedures for obtaining high-purity platelet suspensions with minimal activation. Parallel instructions for isolating and differentiating leukocytes from human blood or mouse bone marrow are also provided. In addition, I outline flow cytometry-based methods to assess platelet purity and activation status using constitutive and activation-inducible markers. This comprehensive guide is aimed at experienced operators and newcomers alike, offering practical tips and troubleshooting advice to ensure reproducibility and biological relevance. By standardizing critical steps in co-culture protocols, this resource seeks to support high-quality experimental design in immunology and hemostasis research.

Key words Platelets, Leukocytes, Immune response, Co-culture, Inflammation

1 Introduction

Platelets, traditionally recognized for their essential role in hemostasis, are now acknowledged as dynamic participants in immune responses. Their direct effect and their interactions with leukocytes are critical in a wide range of physiological and pathological contexts, from vascular inflammation and infection to cancer and autoimmunity [1, 2].

As part of the pathogenesis of viral infections, platelet-monocyte aggregates have been implicated in exacerbating inflammation and coagulopathy. Studies have shown that in severe

Lucas Secchim Ribeiro (ed.), *Models in Inflammation and Immunopharmacology*, Methods in Molecular Biology, vol. 2993,
https://doi.org/10.1007/978-1-0716-5019-6_9,
© The Author(s), under exclusive license to Springer Science+Business Media, LLC, part of Springer Nature 2026

COVID-19 cases, these aggregates express high levels of tissue factor (TF), a key initiator of coagulation cascades. Mechanistically, platelet activation via SARS-CoV-2 spike protein enhances monocyte TF expression through P-selectin and integrin $\alpha_{IIb}\beta_3$ signaling pathways, contributing to the hypercoagulable state observed in patients [3]. Platelets also interact with neutrophils to modulate immune responses. This interaction is known to lead to the formation of neutrophil extracellular traps (NETs), which, while trapping pathogens, can also promote thrombosis and tissue damage. In COVID-19, increased platelet-neutrophil aggregates and NETosis have been associated with disease severity, including respiratory distress and multi-organ failure [4, 5]. Beyond innate immunity, platelets influence adaptive immune responses. They can present antigens via HLA class I molecules, modulating $CD8^+$ T cell responses. For instance, in sepsis, platelet antigen presentation has been linked to T cell suppression, whereas in malaria, it has been associated with T cell activation [6, 7].

The interaction between platelets and leukocytes also has an important value outside of the subject of infectious diseases, as an important component of non-communicable diseases. In vascular pathologies, platelet-leukocyte interactions contribute to disease progression. Activated platelets can adhere to leukocytes, promoting their recruitment to sites of endothelial injury, thereby exacerbating inflammation and atherosclerotic plaque formation [8, 9]. In oncology, platelets facilitate tumor progression by shielding circulating tumor cells, aiding in their evasion from immune surveillance, and promoting metastasis. They interact with leukocytes to modulate the tumor microenvironment, influencing processes like angiogenesis and immune cell recruitment [10–13].

Although the use of platelet-leukocyte aggregates is known in the scientific literature [14–20], understanding these cellular interactions requires in vitro models that preserve the biological properties of each cell type while allowing for controlled investigation of their crosstalk. However, the technical challenges involved in co-culturing platelets and leukocytes (given their different origins, activation sensitivities, and handling requirements) can significantly impact the validity and reproducibility of experimental findings. Moreover, platelets are extremely sensitive to external stimuli and small variations in temperature, pipetting speed, and centrifugation force can trigger pre-activation can confound downstream analyses. Thus, there is a clear need for description and discussion of these steps in a comprehensive format.

Therefore, this chapter brings a practical manual for researchers seeking to establish or refine co-culture systems involving platelets and leukocytes. It synthesizes detailed technical procedures with scientific rigor and usability. By emphasizing reproducibility and careful methodological design, the goal of this manuscript is to facilitate high-quality experimental research into the study of the immune roles of platelet-leukocyte interactions.

2 Materials

A good experiment starts with good planning. Check the **Note 1** for an initial checklist of important details.

2.1 Isolation of Human Platelets

1. Phlebotomy kit (*see* **Note 2**): butterfly-shaped catheter or vacuum tube adapter; elastic arm band; gloves; antiseptic spray solution; cotton pads; adhesive tape; bandages.
2. Citrate-coated blood collection tubes.
3. Serological pipettes and pipetting pump (manual or electronic).
4. 50-mL tubes.
5. Centrifuge at room temperature.
6. Prostaglandin E_1 (PGE_1, 200 μM aliquots).
7. Phosphate-buffered solution (PBS), pre-warmed.
8. RPMI without supplements, pre-warmed.
9. Cell-counting device (e.g., hemocytometer/Neubauer chamber with 0.4% Trypan Blue solution).
10. Pipettes and sterile tips (1000, 200, 10 μL).

2.2 Isolation of Mouse Platelets

1. Acid-Citrate-Dextrose (ACD) Buffer: 0.48% (w/v) Citric Acid; 1.32% (w/v) Sodium Citrate Tribasic; 1.47% (w/v) Glucose; adjust the pH (4.5 - 5.5) and filter the solution with a 0.2 μm membrane (*see* **Note 3**).
2. PIPES Saline Glucose (PSG) Buffer: 5 mM PIPES, 145 mM NaCl, 4 mM KCl, 50 μM Na_2HPO_4, 1 mM $MgCl_2$. $6H_2O$, 5.5 mM Glucose in ddH_2O. Adjust the pH to 6.8–7.2 and filter the solution with a 0.2 μm membrane.
3. 5-mL sterile tubes (e.g., flow cytometry tubes).
4. 28.2 mM PGE_1 (Stock solution, *see* **Note 4**).
5. Centrifuge at room temperature.
6. 23-25G needles.

1-. 2-mL syringes.

7. Pipettes and sterile tips (1000, 200, 10 μL).
8. Cell-counting device (e.g., hemocytometer/Neubauer chamber with 0.4% Trypan Blue solution).

2.3 Analysis of Platelet Activation and Purity (Human or Mouse)

1. Flow cytometer (*see* **Note 5**).
2. Antibody conjugates: anti-CD41a, anti-CD62p, anti-CD45 (associated with the available fluorophores) and their isotype controls (*see* **Note 6**).
3. Compensation beads.

4. Fc-Blocker.
5. Thrombin (aliquots at 100 U/mL).

2.4 Isolation of Monocytes and Macrophage Differentiation

2.4.1 Isolation of Monocytes from Whole Blood

1. Washing Buffer: PBS 1X + 0.5% BSA (w/v) + 2 mM EDTA, sterile and stored at 4 ° C.
2. Monocyte Medium: RPMI 1640 supplemented with 10% FCS, 1% antibiotics (from a stock of penicillin: 10.000 IU/ml and streptomycin: 10 mg/ml), 1% Glutamax, 1% sodium pyruvate, pre-warmed.
3. Sterile PBS, pre-warmed.
4. Ficoll.
5. CD14-tagged beads.
6. Separation columns with plunger and magnet.
7. Pre-separation filters (70 μm).

2.4.2 Differentiation of Monocytes into Macrophages

1. Counting device (e.g., Neubauer chamber and 0.4% Trypan Blue).
2. 17.8 μg/mL recombinant human GM-CSF or 50 μg/mL of recombinant human M-CSF (aliquoted in RPMI without supplements).
3. 6-well plate.
4. 96-well plate.
5. Cell scraper.

2.5 Isolation and Differentiation of Mouse Bone Marrow-Derived Macrophages

1. Sterile tweezers and scissors.
2. 70% ethanol.
3. Petri dish (10 cm).
4. Paper towels.
5. 25G needle.
6. 5- or 10-mL syringe.
7. 100-μm cell strainer.
8. Serum-free medium (DMEM).
9. Complete DMEM: DMEM +10% FBS + 1% antibiotics (*see* Subheading 2.4.1, item 2), pre-warmed.
10. L929 supernatant (containing murine M-CSF).
11. L929-DMEM (Complete DMEM supplemented with 20% v/v of supernatant of L929 cell culture, containing murine M-CSF).
12. Phosphate-buffered solution (PBS, 1X).
13. Cell scraper.
14. 175-cm^2 tissue culture flasks.

15. Incubator with controlled temperature and CO_2.
16. BMDM-detaching buffer: PBS + 2% FBS (v/v) + 5 mM EDTA, ice cold.
17. Cell-counting device (e.g., hemocytometer/Neubauer chamber with 0.4% Trypan Blue solution).
18. 6-well plate.
19. 96-well plate.

3 Methods

Since platelets cannot be frozen without drastic changes in their viability, structure, and function, they should only be used in co-cultures after fresh isolation. Therefore, plan your experiment in such a way that the leukocytes will be already isolated (differentiated, if needed) and seeded before the isolation of platelets starts (*see* Subheading 3.5.3).

3.1 Isolation of Human Platelets

1. With the assistance of a trained phlebotomist, collect ~24 mL of venous blood from a donor in citrate-coated tubes (*see* **Note 7**).
2. Carefully, transfer the blood from the citrate tubes to a 50-mL tube.
3. Centrifuge the blood for 15 minutes at 330 × *g* (slow acceleration, without brake, room temperature) to obtain platelet-rich plasma (PRP).
4. Add PGE_1 to a new 50-ml tube for a final concentration of 200 nM (1 μL of PGE_1 working aliquot per ml of plasma).
5. Transfer the PRP to the tube containing PGE_1 and dilute it 1: 2 with PBS to reduce leukocyte contamination. Centrifuge it at 240 × *g* for 10 minutes (slow acceleration, no brake).
6. Add PGE_1 to a new 50-ml tube for a final concentration of 200 nM (1 μL of PGE_1 working aliquot per ml of diluted plasma).
7. Transfer the platelet-rich supernatant to the tube containing PGE_1 and centrifuge it for 15 minutes at 430 × *g* (slow acceleration, no brake, room temperature).
8. Remove the supernatant without disturbing the platelet pellet. Add the proper volume of PGE_1 (for a final concentration of 200 nM of PGE_1) and carefully resuspend the pellet with 20 mL of PBS, followed by a new centrifugation (430 × *g*, 15 minutes, slow acceleration, no brake, room temperature).

9. Carefully remove the supernatant and even more carefully resuspend the pellet with 5 mL of your medium of choice (e.g., regular, non-supplemented RPMI, pre-warmed).
10. Count the platelets using the Neubauer chamber (suggested dilution = 1:50, *see* **Note 8**) or an automated device.
11. Dilute the platelets to the desired density.

3.2 Isolation of Mouse Platelets

3.2.1 Blood Withdrawal

1. It is critical that the blood from the mice is not exposed to any tissue factors that can start the coagulation cascade (*see* **Note 9**).
2. Add 500 μL of pre-warmed ACD buffer into a sterile 5-mL tube (cytometry tube).
3. Prepare a syringe/needle and flush it by aspirating ~2 mL of ACD buffer and dispensing it, several times. Leave a small volume (~ 100 μL) of ACD buffer in the syringe/needle.
4. Slowly take the blood from the mouse using the method approved by your ethics committee. Speeding up this step will lead to spontaneous platelet aggregation.
5. Carefully dispense the blood into the tube containing the ACD buffer. Homogenize it by a very gentle inversion of the tube, only once.
6. Pool the blood of genotype/sex/age-matched mice into one tube (*see* **Note 10**) and centrifuge it for 15 minutes at 340 × *g* (slow acceleration, no brake, room temperature).

3.2.2 Platelet Isolation from Fresh Mouse Blood

1. While the blood is in the centrifuge, prepare a fresh aliquot of PGE_1 solution by diluting 1.77 μL of stock PGE_1 (stock concentration = 28.2 mM) into 250 μL of PSG buffer, for a final concentration of 200 μM.
2. Prepare a 2X working PGE_1 solution (10 μM) by adding 50 μL of the previous solution (200 μM) into 950 μL of PSG Buffer. Depending on the initial volume of blood, prepare equivalent volumes of working PGE_1 solution.
3. Check the volume of plasma obtained after centrifugation. Add the same volume of 2X PGE_1 solution (10 μM) into a 15-mL tube. Next, slowly pipette the plasma on top of the 2X working PGE_1 solution (the final PGE_1 concentration will be 5 μM).
4. Centrifuge the tube for 10 minutes at 240 × g (slow acceleration, no brake, room temperature), to pull down leukocytes and red blood cells into a pellet. The supernatant constitutes the Platelet-Rich Plasma (PRP).
5. To a new 15 mL tube, add 25 μL of the 200 μM PGE_1 solution (from **step 1**) for every 1000 μL of PRP.

6. Slowly transfer the PRP to the new tube containing the PGE_1. However, don't take all of the PRP: leave ~10% of the total PRP volume to avoid leukocyte contamination.
7. Centrifuge the tube for 5 minutes at 1000 × g (slow acceleration, no brake, room temperature), to pellet down the platelets.
8. Next, remove the supernatant. Carefully, wash the platelet pellet with minimum 2 mL of a 5 μM PGE_1 solution (dilute 25 μL of the 200 μM PGE_1 into 1975 μL of PSG buffer). Use a 200 μL pipette tip for this step.
9. Centrifuge the tube for 5 minutes at 1000 × g (slow acceleration, no brake, room temperature), to pellet down the platelets.
10. Carefully resuspend the pellet in 1 mL of incomplete DMEM. Count the platelet suspension in a hemocytometer (1:20 dilution on Trypan-Blue) and adjust the platelet density to twice (2X) your final experimental conditions (*see* Subheading 3.5.3).

3.3 Analysis of Platelet Activation and Purity (Human or Mouse)

The purification of platelets from whole blood is achieved by a series of centrifugation steps. However, it is possible that interfering leukocytes remain in the final platelet suspension and an exceeding proportion of leukocytes can mask the actual effect exerted by the platelets. To assure the efficiency of the isolation, a flow cytometry experiment is performed, including an evaluation of platelet activation potential. These validation steps are essential for ensuring the interpretability of downstream assays and are often omitted in brief methodological descriptions elsewhere in the literature.

1. Turn on the flow cytometer at least 30 minutes before the measurements, so the lasers are warmed up. Ensure proper calibration and compensation status before starting the acquisition (*see* **Note 5**).
2. For the flow cytometry analysis of purity, take two aliquots of 100 μL of platelet suspension and transfer them to two separate microtubes.
3. Add 1 U of thrombin (1:100 dilution from stock—100 U/mL) to one of them and incubate both for 30 minutes at 37 °C, 5% CO_2.
4. After incubation, split each platelet suspension into two wells in a 96-well, U-shaped plate (or 0.5-mL microtubes). Add 10 μL of Fc-Blocker to each well and incubate it at room temperature for 15 minutes.
5. Add to one of the wells 2 μL of each antibody (*see* **Note 11**), or their respective isotype controls, like the example below.

Unstimulated platelets (50 μL)	Thrombin-activated platelets (50 μL)
Anti-CD41a*FITC	Anti-CD41a*FITC
Anti-CD62p*APC	Anti-CD62p*APC
Anti-CD45*PE	Anti-CD45*PE
Unstimulated platelets (50 μL)	**Thrombin-activated platelets (50 μL)**
*Host- and subclass-matched*FITC/APC/PE isotypes*	*Host- and subclass-matched*FITC/APC/PE isotypes*

6. Unstimulated platelets are CD41-positive and a low level of activation (< 5% of $CD62^+$ events amongst the total $CD41^+$ events) is acceptable. In the thrombin-activated samples, there should be a sizeable up-regulation of the CD62p expression, leading to an overrepresentation of $CD41^+CD62^+$ (double-positive) events. In none of the samples there should be more than 0.5% of $CD45^+$ events, which would indicate leukocyte contamination (*see* **Note 12**).

3.4 Isolation of Human Monocytes for Differentiation into Macrophages

Isolating and differentiating leukocytes is a critical step for many in vitro experiments. Ficoll gradient is used to separate leukocytes from whole blood and CD14-tagged beads are used to isolate monocytes, our focus in this chapter. With the addition of the growth factors GM-CSF or M-CSF, the monocytes will be differentiated into macrophages and become ready to activation and stimulation. These cells were chosen to be described in this chapter because, unlike neutrophils and other innate immune cell types, human- and mouse-derived macrophages are more resilient and long-lasting, allowing a broader option of experimental protocols. However, co-cultures of platelets with granulocytes or other cells of myeloid and lymphoid origin can also be performed.

For a higher leukocyte yield, the protocol below is based on blood fractions obtained from concentrated pouches (around 70–80 mL), available in most blood banks as a by-product of plasmapheresis. Nevertheless, the volumes of reagents can be adjusted proportionally for a smaller initial volume of blood.

3.4.1 Isolation of Human Monocytes from Concentrated Blood Pouch

1. Make sure the PBS, buffers and media are pre-warmed at 37 ° C.
2. Inside of a sterile laminar flow hood, add 15 mL of Ficoll into 4 sterile 50-mL Falcon tubes.
3. Decontaminate the external surface of the blood bag and with sterile scissors, open the pouch.
4. With a sterile pipette, distribute at least 20 mL of blood into sterile 50-mL tubes (4 tubes are enough).

5. Dilute the blood with PBS in equal parts (v/v).
6. Carefully pipette 35 mL of PBS-diluted blood on top of Ficoll, making sure to keep two separate phases. Tilting the tube during pipetting helps to prevent phase mixing. It is not recommended to proceed if the blood and Ficoll phases are mixed.
7. Set the centrifuge for 50% of maximal acceleration, without break, room temperature. Spin the samples for 20 minutes, 700 × *g*.
8. Remove most of the plasma without disturbing the buffy coat stratum. Then, transfer the leukocyte layer for new 50-mL tubes.
9. Complete the volume of the tubes up to 50 mL with PBS.
10. Spin the samples for 10 minutes, 340 × *g* (regular acceleration and break, room temperature).
11. Carefully discard the supernatant and combine the pellets into a new 50 mL tube.
12. Fill the tube up to 50 mL with washing buffer and spin it for 5 minutes, 340 × *g* (regular acceleration and break, room temperature).
13. Wash the cells once again with washing buffer (repeat **steps 11** and **12**).
14. Discard the supernatant and resuspend the cell pellet with 1 mL of washing buffer.
15. Add 250 μL of CD14-tagged beads into the 1 mL of cell suspension.
16. Incubate it for 15 minutes at 2–8 ° C.
17. After incubation, fill it up to 50 mL with washing buffer.
18. Spin it for 5 minutes, 340 × *g* (regular acceleration and break, room temperature).
19. Meanwhile, prepare the columns for monocyte enrichment. Attach the columns on the magnet and put the plunger aside. Put the pre-separation filters on top of the LS Columns (one column per donor, if multiple donors are being evaluated at the same time). Wash each column with 3 mL of washing buffer.
20. After centrifugation, discard the supernatant and resuspend the cells in 3 mL of washing buffer.
21. Add the cell suspension in the column, through a pre-separation filter.
22. Wash column with 3 mL of washing buffer, for 3 times.
23. Add 17 mL of washing buffer to a new 50-mL tube.
24. Remove the pre-separation filter, take the column out from the magnet and add 3 mL of washing buffer in it.

25. Use the plunger to press the cells throughout the column, right into the tube with 17 mL of washing buffer. Centrifuge it for 5 minutes, 340× *g* (regular acceleration and break, room temperature).
26. Discard the supernatant and resuspend the pellet in 10 mL of monocyte medium.
27. Take an aliquot of the cell suspension for a 1:10 dilution on Trypan Blue, count them on the Neubauer chamber (or automated device) and take note of the absolute number of cells.

3.4.2 Differentiation of Monocytes into Macrophages

1. Adjust the density for 2×10^6 cells/mL with Monocyte Medium.
2. Add 3.1 μL of rhGM-CSF (stock at 55 ng/mL) per mL of cell suspension, or 1.0 μL of rhM-CSF (stock at 50 ng/mL).
3. Seed 5 mL of cells (10^7 cells) per well in a 6-well plate and incubate for 3 days at 37 °C, 5% CO_2.
4. After the 3-day incubation, add 5.0 μL of rhGM-CSF (stock at 55 ng/mL) per well, or 1.0 μL of rhM-CSF (stock at 50 ng/mL) and incubate for 24 h.
5. After the final incubation, gently use the cell scraper to detach the cells. For a better harvest, wash the wells with warm medium or PBS 1X.
6. Count the cells using your preferred manual or automated technique and adjust the density with warm medium.

3.5 Isolation and Differentiation of Mouse Bone Marrow-Derived Macrophages

3.5.1 Bone Marrow Flushing and Cultivation

1. Have a 10-cm Petri dish filled with 5 mL of cold 1X PBS ready on ice (*see* **Note 13**).
2. Prepare syringes (5 or 10 mL) with cold, serum-free medium and place them on ice. Make sure to have an extra container (50-ml tube) with cold medium on ice as well.
3. Euthanize mice using the method approved by your local committee on animal experimentation.
4. Apply 70% Ethanol over the mouse fur, to avoid having it floating around, contaminating instruments and samples.
5. Make a small incision in the lower third of central abdominal region of the mouse with the scissors, making sure not to puncture the peritoneum.
6. Bring the skin over the hind legs and down towards the paws, always careful not to break any bones.
7. Remove the hind legs from the hip joints with scissors: find the head of the femur and cut around it.
8. Remove the paws by dislocating the ankle joint.

9. Cut the tendons connecting the knee and the ankle.
10. Dislocate the knee joint to separate tibia and femur.
11. Carefully remove any excess of tissue and muscles from the bones, carefully rubbing the bones with paper tissues.
12. Place intact, clean bones into the dish with PBS on ice (*see* **Note 14**).
13. Once all the bones are clean, the next steps must be done in a laminar flow hood under sterile conditions.
14. Add 70% ethanol in a Petri dish and PBS 1X in a second dish, both in enough volume to cover the bones (*see* **Note 15**).
15. With tweezers (sterile, if possible), transfer one bone into the dish/well with ethanol for 30 seconds, then place it in the new dish/well with PBS.
16. Use the tweezers to fix the bone and with sterile, sharp scissors, cut 2–3 mm at both ends of the bone, exposing the bone marrow cavities.
17. While still holding the bone with the tweezers, take one of the syringes pre-filled with cold medium, place the needle inside the bone marrow cavity and flush it with 5–10 mL of serum-free medium into a new 50-mL tube (*see* **Note 16**).
18. The bone marrow is usually flushed in big chunks of cells and matrix. Make sure to resuspend it with a 1000 μL tip to obtain a homogeneous cell suspension.
19. Place a cell strainer over a new 50-ml tube and transfer the cell suspension through it. Ensure that there are no large pieces of bone marrow on the cell strainer.
20. The hind bones from one mouse (2 femurs +2 tibias) can be pooled into a single tube.
21. Fill the tube up to 50 mL with cold, serum-free medium and centrifuge it for 5 minutes at 500 × *g* at 4 °C.
22. During the centrifugation, add 20 mL of complete DMEM + 20% of L929 supernatant to 4 culture flasks (175 cm^2) and leave the flasks and the sterile hood.
23. Once the centrifugation is over, discard the supernatant and resuspend the cell pellet in 20 mL of complete DMEM + 20% L929.
24. Add 5 mL of the bone marrow cell suspension to each of the flasks (*see* **Note 17**), for a final volume of 25 mL.
25. Incubate the flasks with the cells at 37 °C, 5% CO_2 for 3 days, checking daily for possible contamination.
26. On the third day of incubation, add 2.5 mL of L929 supernatant in each flask and incubate for another 3 days. After that, the cells are ready to be harvested.

3.5.2 Harvesting and Plating the BMDMs

1. Prepare the BMDM-detaching buffer in advance and leave it on ice. Ensure that enough complete DMEM and PBS 1X are warmed up to 37 °C as well.
2. After 6 days (in total) of differentiation in L929-loaded medium, the bone marrow-derived macrophages should form a monolayer of 80–90% confluence.
3. In a sterile hood, remove the culture supernatant with a pipette and carefully wash the cell layer with 5 mL of warm, sterile PBS.
4. Discard the PBS and add 10 ml of cold BMDM-detaching buffer.
5. Incubate the flasks on ice or 4 °C for 10 minutes.
6. Detach the BMDMs by hardly tapping the sides of the flask. Next, gently scrape the area and collect the cells into a 50-ml tube.
7. Use an additional 10 ml of PBS to wash out any remaining cells in the flasks and combine them into the 50-ml tube.
8. Centrifuge the cells for 5 minutes, 500 × *g* at 4 °C.
9. Discard the supernatants and resuspend the cell pellet in 5 mL of complete DMEM, making sure to create an homogeneous cell suspension by repetitive pipetting (with a 1000-μL tip).
10. Count the cells using your preferred manual or automated technique and adjust the density with warm medium.

3.5.3 Seeding the Cells and Preparing the Co-culture Experiment

1. The next steps are valid for both human- or mouse-derived cells.
2. In your experimental planning, remember to always include experimental groups composed only of leukocytes, as well as a group made exclusively of isolated platelets in culture (*see* **Note 18**).
3. Count the leukocytes and adjust the cell density to 1.0×10^6 cells/mL using complete medium. Next, seed 10^5 leukocytes/well (100 μl) in a 96-well plate, up to 3.0×10^6 leukocytes (3 mL) for 6-well plate (*see* **Note 19**).
4. Using gentle, circular movements, ensure the entire well is covered by the leukocyte suspension, without empty pockets.
5. Incubate the plates overnight at 37 ° C, 5% CO_2 (*see* **Note 20**).
6. After overnight incubation, warm up the non-supplemented medium and gather the freshly isolated platelet suspension. Then, gently remove the medium covering the leukocytes and replace it by half of the desired well volume of medium or platelet suspension (*see* **Note 21**). For example, if a final volume of the well is planned to be 100 μL, add 50 μL of non-supplemented medium or 50 μL of platelet suspension.

7. Complete the volume up the final volume with non-supplemented medium, immune stimuli (e.g., LPS), or any further pharmacological interventions.
8. For short-term experiments (up to 6 h), perform stimulation in serum-free medium. Longer experiments might require supplementation of the medium with 1–2% of FBS.
9. At the end of the experiment, centrifuge the plate (5 minutes at 500 × *g*, room temperature) and collect the supernatant into tubes or clean plates. Lysis buffer can be added to the wells to evaluate the cellular response, depending on the planned evaluation (e.g., qPCR, RNA-Seq, or Western Blotting). The supernatant can be used for cytokine quantification or soluble high-throughput assays (e.g., lipidomics or proteomics). Supernatants and cell lysated should be immediately frozen at −80 °C after collection.
10. At the conclusion of the experiment, it is also suggested to run a viability assay to account for cell survival as a confounder. Cell-based techniques (e.g., CTB or MTT) or supernatant-based (e.g., LDH) can be used. Keep in mind that platelets have functional mitochondria and that can also affect the output.
11. Compare the leukocyte-only group with those stimulated in co-culture with platelets using your preferred output assay. Remember to consider the effect seen in the platelet-only condition.

4 Notes

1. Before an experiment with human blood, contact the voluntary donor and the phlebotomist and confirm date and time for the collection. Print a copy of the Informed Consent with the identification of the approved Ethics Protocol and ask the donor to read and sign it. Make sure all the required material for the collection is reserved. Schedule the laminar flow hood and cytometer and make sure there are enough reagents and buffers.
2. The isolation of human and mouse platelets, as well as any procedure for isolation of leukocytes from human blood must be performed in a suitable laminar flow hood. For the isolation of mouse bone marrow-derived cells, the manipulation of the animal until the cleaning of the bones can be performed at a regular bench. After that, the protocol must be carried in a sterile manner.
3. Do not use cold reagents for any reason during platelet preparation, as they favor platelet activation!

4. Prepare PGE_1 in a volume compatible with your routine, since the re-freezing/thawing used PGE_1 aliquots is discouraged.
5. Check if the instrument is properly calibrated and perform a compensation if needed. Confirm if the device is set for gentle mixing with the proper nozzle size, as well as a setting for a low forward scatter trigger/threshold.
6. For the evaluation of platelet purity and activation (in human or mouse platelets), three antibodies can be used: CD41 (constitutive expression in platelets), CD62 (expression increased in activated platelets), and CD45 (negative in platelets, but positive for contaminating leukocytes). A suggestion of antibody panel would be CD41a-FITC, CD62p-APC, and CD45-PE, with their respective isotype controls: Mouse IgG1k-FITC/APC/PE. However, other combinations of conjugates are also acceptable.
7. In a healthy adult volunteer, there should be ca. $5–6 . 10^9$ platelets in 24 mL of blood. However, due to losses during the isolation procedure, an experienced operator can recover around $3 . 10^9$ platelets from that initial blood volume.
8. Since the platelet size is smaller than the depth of a hematocytometer chamber, remember to move the fine adjustment knob of the microscope constantly in both directions, to account for cells in various depths. Platelets can be identified for being refractive, unlike cellular debris of similar size.
9. Make sure to have any blood-collecting procedures approved by your local ethics committee in animal experimentation. The suggested methods for blood collection are the vena cava and heart puncture.
10. Most mice lineages have significantly more platelets per volume unit compared to humans. One can find up to $1 . 10^9$ platelets in 1 mL of murine blood and usually half of this amount can be successfully isolated for experimentation.
11. Adjust the antibody titer according to manufacturer- and lot-specific conditions. Remember to vortex and spin/centrifuge the antibody vial before use.
12. Isotype controls and a F-M_O protocol can be used to define the gating strategy.
13. If executed properly, **steps 2** to **11** can be carried out at a clean bench, outside of the sterile flow hood.
14. It is not recommended to use bone marrow content from broken bones, as external contaminants can access that compartment and activate the cells.
15. A 6-well plate can also be used in this step, as it facilitates the separation of bones from different mice lineages or conditions.

16. It is possible to see the marrow through the bone, in a dark red color. A successful marrow flow will leave the bone with a translucent shade of white. If there is still marrow in the cavity, try to flush the bone from the other end.
17. The number of flasks to be used depends directly on the size of the mouse and the efficiency of bone marrow flushing. This method is based on 8 to 10-week-old mice, ~ 20 grams. If the mice are smaller or the collection of the bone marrow was not fully executed, resuspend the pellet with less medium (e.g., 15 mL). Keep in mind that the cells need direct contact with each other for a proper differentiation.
18. To consolidate the regulatory effect of the platelets over leukocytes (i.e., over-activation of the evaluated biological pathway), it is important to confirm that the effect does not come simply due to an additive effect on the chosen output assay.
19. The initial cell density for seeding of human or murine macrophages is a suggestion for a confluence of 90% at the beginning of the experiment. The starting number of the cells or their optimal confluence might vary according to your own experimental setup. Moreover, I suggest 2 or 3 technical replicates for each condition in all experiments executed in 96-well plates.
20. An overnight incubation is strongly recommended to allow the cells to return to a basal state after so much handling and shear stress.
21. It is suggested to use at least three different fractions of leukocytes:platelets to properly observe a putative density-dependent effect. 3-fold and 10-fold proportions are recommended.

Funding

Lucas Secchim Ribeiro was supported by the Innovative Minds Program, DZNE Foundation, Project T0531/41807/2022/kg.

References

1. Tian Y, Zong Y, Pang Y, Zheng Z, Ma Y, Zhang C, Gao J (2025) Platelets and diseases: signal transduction and advances in targeted therapy. Signal Transduct Target Ther 10: 159. https://doi.org/10.1038/s41392-025-02198-8
2. Yan M, Wang Z, Qiu Z, Cui Y, Xiang Q (2024) Platelet signaling in immune landscape: comprehensive mechanism and clinical therapy. Biomark Res 12:164. https://doi.org/10.1186/s40364-024-00700-y
3. Hottz ED, Azevedo-Quintanilha IG, Palhinha L, Teixeira L, Barreto EA, Pão CRR, Righy C, Franco S, Souza TML, Kurtz P, Bozza FA, Bozza PT (2020) Platelet activation and platelet-monocyte aggregate formation trigger tissue factor expression in patients with severe COVID-19. Blood 136:1330–1341. https://doi.org/10.1182/blood.2020007252
4. Skendros P, Mitsios A, Chrysanthopoulou A, Mastellos DC, Metallidis S, Rafailidis P,

Ntinopoulou M, Sertaridou E, Tsironidou V, Tsigalou C, Tektonidou M, Konstantinidis T, Papagoras C, Mitroulis I, Germanidis G, Lambris JD, Ritis K (2020) Complement and tissue factor-enriched neutrophil extracellular traps are key drivers in COVID-19 immunothrombosis. J Clin Invest 130:6151–6157. https://doi.org/10.1172/JCI141374

5. Petito E, Falcinelli E, Paliani U, Cesari E, Vaudo G, Sebastiano M, Cerotto V, Guglielmini G, Gori F, Malvestiti M, Becattini C, Paciullo F, De Robertis E, Bury L, Lazzarini T, Gresele P, COVIR study investigators (2021) Association of neutrophil activation, more than platelet activation, with thrombotic complications in coronavirus disease 2019. J Infect Dis 223:933–944. https://doi.org/10.1093/infdis/jiaa756
6. Amoafo EB, Entsie P, Kang Y, Canobbio I, Liverani E (2023) Platelet P2Y12 signaling pathway in the dysregulated immune response during sepsis. Br J Pharmacol. https://doi.org/10.1111/bph.16207
7. Chapman LM, Aggrey AA, Field DJ, Srivastava K, Ture S, Yui K, Topham DJ, Baldwin WM, Morrell CN (2012) Platelets present antigen in the context of MHC class I. J Immunol 189:916–923. https://doi.org/10.4049/jimmunol.1200580
8. Schrottmaier WC, Mussbacher M, Salzmann M, Assinger A (2020) Platelet-leukocyte interplay during vascular disease. Atherosclerosis 307:109–120. https://doi.org/10.1016/j.atherosclerosis.2020.04.018
9. Vilahur G, Fuster V (2025) Interplay between platelets and coagulation: from protective haemostasis to pathological arterial thrombosis. Eur Heart J 46:413–423. https://doi.org/10.1093/eurheartj/ehae776
10. Zhou L, Zhang Z, Tian Y, Li Z, Liu Z, Zhu S (2023) The critical role of platelet in cancer progression and metastasis. Eur J Med Res 28:385. https://doi.org/10.1186/s40001-023-01342-w
11. Li J, Liu J, Yang S, Xia Y, Meng Q, Sun B, Liu Y, Zhao B, Jin J, Xu H, Wang L, Zhang P, Cheng Z (2025) PD-L1 positive platelets mediate resistance to immune checkpoint inhibitors in patients with colorectal cancer. Cell Commun Signal 23:29. https://doi.org/10.1186/s12964-025-02034-0
12. Kuratani A, Okamoto M, Kishida K, Okuzaki D, Sasai M, Sakaguchi S, Arase H, Yamamoto M (2024) Platelet factor 4-induced TH1-Treg polarization suppresses antitumor immunity. Science 386:eadn8608. https://doi.org/10.1126/science.adn8608
13. Yang J, Xu P, Zhang G, Wang D, Ye B, Wu L (2025) Advances and potentials in platelet-circulating tumor cell crosstalk. Am J Cancer Res 15:407–425. https://doi.org/10.62347/JAYK5667
14. Carestia A, Mena HA, Olexen CM, Ortiz Wilczyñski JM, Negrotto S, Errasti AE, Gómez RM, Jenne CN, Carrera Silva EA, Schattner M (2019) Platelets promote macrophage polarization toward pro-inflammatory phenotype and increase survival of septic mice. Cell Rep 28:896–908.e5. https://doi.org/10.1016/j.celrep.2019.06.062
15. Hally KE, Bird GK, La Flamme AC, Harding SA, Larsen PD (2019) Platelets modulate multiple markers of neutrophil function in response to in vitro toll-like receptor stimulation. PLoS One 14:e0223444. https://doi.org/10.1371/journal.pone.0223444
16. Tylek T, Schilling T, Schlegelmilch K, Ries M, Rudert M, Jakob F, Groll J (2019) Platelet lysate outperforms FCS and human serum for co-culture of primary human macrophages and hMSCs. Sci Rep 9:3533. https://doi.org/10.1038/s41598-019-40190-9
17. Rolfes V, Ribeiro LS, Hawwari I, Böttcher L, Rosero N, Maasewerd S, Santos MLS, Próchnicki T, de S Silva CM, de S Wanderley CW, Rothe M, Schmidt SV, Stunden HJ, Bertheloot D, Rivas MN, Fontes CJ, Carvalho LH, Cunha FQ, Latz E, Arditi M, Franklin BS (2020) Platelets fuel the inflammasome activation of innate immune cells. Cell Rep 31:107615. https://doi.org/10.1016/j.celrep.2020.107615
18. Albayati S, Li N, Unsworth AJ, Liverani E (2022) Platelet-lymphocyte co-culture serves as an ex vivo platform of dynamic heterotypic cross-talk. J Cell Commun Signal. https://doi.org/10.1007/s12079-022-00676-0
19. Winskel-Wood B, Padula MP, Marks DC, Johnson L (2023) The phenotype of cryopreserved platelets influences the formation of platelet-leukocyte aggregates in an in vitro model. Platelets 34:2206916. https://doi.org/10.1080/09537104.2023.2206916
20. Hawwari I, Rossnagel L, Rosero N, Maasewerd S, Vasconcelos MB, Jentzsch M, Demczuk A, Teichmann LL, Meffert L, Bertheloot D, Ribeiro LS, Kallabis S, Meissner F, Arditi M, Atici AE, Noval Rivas M, Franklin BS (2024) Platelet transcription factors license the pro-inflammatory cytokine response of human monocytes. EMBO Mol Med 16:1901–1929. https://doi.org/10.1038/s44321-024-00093-3

Chapter 10

Quantitative In Vitro and In Vivo Assays to Assess the Efferocytic Efficiency of Macrophages

Eduardo Gutseit, Vinícius Amorim Beltrami, Débora Gonzaga Martins, Mauro Martins Teixeira, Lirlândia Pires de Sousa, Juliana Priscila Vago, Vanessa Pinho, and Franciel Batista Felix

Abstract

Efferocytosis is a process in which specialized phagocytes, mainly macrophages, are responsible for recognizing and removing apoptotic cells from the inflammatory site, leading to tissue resolution, and preventing chronic inflammation. Herein we describe in detail a robust efferocytosis assay based on morphological criteria and flow cytometric for the quantification of efferocytic events and characterization of the engulfing efficiency of apoptotic cells by macrophage in vitro and in vivo, which can be used in many applications, including the development of genetic and pharmacological therapeutic targeting inflammation resolution.

Key words Efferocytosis, Macrophages, Inflammation resolution, Apoptosis, Flow cytometry

1 Introduction

Efferocytosis is a finely regulated process responsible for the efficient clearance of apoptotic cells from tissues [1]. Derived from the Latin term "efferō" meaning "to carry out for burial," efferocytosis represents the engulfment and subsequent degradation of apoptotic cells by phagocytes. Efferocytosis is performed by macrophages and to a lesser extent by other "professional" phagocytes (such as monocytes and dendritic cells) and non-professional phagocytes, such as epithelial cells [1]. Efferocytosis plays a critical role in maintaining tissue homeostasis by eliminating dead cells and orchestrating immunomodulatory responses. Basic, translational, and clinical studies have provided insights into the pivotal role of efferocytosis in several physiological and pathological contexts. The successful recognition and internalization of apoptotic cells by phagocytes not only prevent the release of intracellular contents

Lucas Secchim Ribeiro (ed.), *Models in Inflammation and Immunopharmacology*, Methods in Molecular Biology, vol. 2993,
https://doi.org/10.1007/978-1-0716-5019-6_10,
© The Author(s), under exclusive license to Springer Science+Business Media, LLC, part of Springer Nature 2026

[2] but also actively modulate immune responses, resolve inflammation, and promote tissue repair. The impairment of efferocytosis has emerged as a significant factor in the pathogenesis of numerous disorders, including autoimmune diseases, chronic inflammation, cancer, and neurodegenerative conditions [3, 4]. Understanding the dynamics of the efferocytosis process may be useful to develop novel therapies focused on the regulation of this event and boosting resolution in different inflammatory settings. Here we present several methods that allow us to quantify clearly and concisely manner the acquisition of apoptotic cells by macrophages [5–9]. Additionally, these methods also enable us to assess the promising pro-resolving properties of several molecules. Our method is advantageous for being a practical, fast, automated, and effective protocol. In this regard, the methods presented here provide detailed in vitro and in vivo experimental models for measuring the process and the efficiency of efferocytosis by macrophages. This involves the morphological examination using light microscopy to determine the proportion of cells exhibiting efferocytic morphology (macrophage with apoptotic cells observed in their cytoplasm) or the visualization of cytoplasmic fluorescence of carboxyfluorescein succinimidyl ester (CFSE)-labeled apoptotic bodies inside macrophages detected by flow cytometry. Importantly, this protocol can be applied to the co-culture of primary or cell lineages of macrophages with apoptotic cells of different origins, whether of human or mouse origin. This flexibility enables the investigation of efferocytosis in diverse experimental configurations and facilitates the analysis of specific macrophage-apoptotic cell interactions.

2 Materials

2.1 Isolation and Culture of Murine Bone Marrow-Derived Macrophages (BMDMs)

1. Healthy mice (*see* **Note 1**).
2. Anesthetic solution (240 mg/kg Ketamine and 30 mg/kg Xylazine diluted in 1× PBS).
3. 1× Phosphate-Buffered Saline (PBS) (137 mM NaCl, 2.7 mM KCl, 10 mM, Na_2HPO_4, 1.8 mM KH_2PO_4 in distilled water, pH 7.4).
4. Surgical equipment.
5. Gauze Pads.
6. Needle (30G, 12 mm).
7. Syringe (10 mL).
8. Fetal Bovine Serum (FBS).
9. Petri dish.

10. Dulbecco's Modified Eagle Medium (DMEM) with 2 mM L-glutamine and sodium bicarbonate, serum-free.
11. Macrophage-Differentiating (MD) media (20% FBS, 30% L929-enriched medium, and 50% DMEM) (*see* **Note 2**).
12. Culture Flasks (175 cm^2 surface area).
13. Cell scraper.
14. 96-well flat bottomed tissue culture plate.

2.2 Preparation of Apoptotic Cells

2.2.1 Human Neutrophil Isolation, Labeling, and Apoptosis Induction

1. 10 mL of Human Blood, collected in ethylenediaminetetraacetic acid (EDTA) tubes (*see* **Note 3**).
2. Ammonium-Chloride-Potassium (ACK) Lysing Buffer: weigh 8.26 g of NH_4Cl, 1.0 g of $KHCO_3$ and 0.037 g of EDTA. Dilute the reagents in distilled water to a volume of 900 mL. Mix and adjust pH to 7.2. Top up to 1 L with distilled water. Store at 4 °C.
3. Fetal Bovine Serum (FBS).
4. 1× PBS.
5. Pasteur pipette.
6. DMEM containing 1% FBS.
7. Histopaque (density 1119 g/mL).
8. Histopaque (density 1077 g/mL).
9. CFSE working solution (5 μM CFSE diluted in 5 mL 1× PBS) (*see* **Note 4**).
10. 1 μM Staurosporine, in dimethyl-sulfoxide.
11. Annexin-V/Propidium Iodide (PI) Apoptotic Detection Kit.
12. 1× PBS containing 0.5% Bovine Serum Albumin (BSA) (solubilize 0.25 mg BSA in 50 mL of PBS).
13. Hemocytometer.

2.2.2 Thymocytes Isolation, Labeling, and Apoptosis Induction

1. Healthy mice (*see* **Note 1**).
2. Anesthestic solution (*see* **item 2** in Subheading 2.1).
3. Surgical equipment.
4. Petri dish.
5. Fetal Bovine Serum (FBS).
6. 1× PBS.
7. Dulbecco's Modified Eagle Medium (DMEM).
8. Cell Strainer (100 μm).
9. Syringe (10 mL).
10. Hemocytometer.

2.3 Macrophage Migration into the Peritoneal Cavity Induced by Zymosan

1. Healthy animals (*see* **Note 1**).
2. 1× PBS.
3. 0.1 mg/mice Zymosan (diluted in 200 μL of 1× PBS).
4. Syringe (1 mL).

2.4 Assessment of Efferocytosis by Flow Cytometry

1. 96-well conical plates.
2. 1× PBS.
3. 1× PBS containing 1% BSA (solubilize 5 mg of BSA in 50 mL of 1x PBS).
4. MD media (see **item 11** in Subheading 2.1).
5. FACS tubes.
6. Fetal Bovine Serum (FBS).
7. DMEM containing 1% FBS.
8. Macrophage-specific antibodies. Particularly, we suggest the F4/80 antibody.
9. 2% Paraformaldehyde (PFA) in 1× PBS.
10. Flow Cytometer.

2.5 Assessment of Efferocytosis by Morphological Criteria

1. 24-well flat plates.
2. 1.5 thickness, 12 mm circular glass coverslips.
3. Bone holding clamp.
4. Panoptic kit staining solution.
5. Cytospin centrifuge.
6. Fetal Bovine Serum (FBS).
7. 1× PBS.
8. DMEM containing 1% FBS.
9. CFSE solution (5 μM CFSE diluted in 5 mL 1x PBS) (*see* **Note 4**).
10. 1 μM Staurosporine, in dimethyl sulfoxide.
11. 1× PBS containing 3% BSA (solubilize 15 mg of BSA in 50 mL of PBS).
12. Light microscope with 100× objective lenses.

3 Methods

3.1 Isolation and Culture of Murine Bone Marrow-Derived Macrophages (BMDMs) (See the Experimental Setting in Fig. 2a)

1. Euthanize three mice according to the locally approved Ethics protocol (see **item 2** in Subheading 2.1) and dissect them to expose the leg muscles. Cut the musculature around the tibia and femur until the Condyle (head of the Femur) is visible.
2. Using a sterilized gauze pad, remove the residual muscles around the bones by friction until most of the tissue is

removed. Transfer the bones into a tissue culture Petri dish containing 3 mL of serum-free DMEM.

3. Cut the cartilage between both bones and cut the upper and lower parts to expose the Bone Marrow.
4. Using a syringe containing 10 mL of serum-free DMEM and a needle, purge the Bone Marrow into a 15 mL centrifuge tube (*see* **Note 6**).
5. Centrifuge at 400 × *g* for 5 minutes at room temperature.
6. Aspirate the supernatant and resuspend the cell pellet on 25 mL of Macrophage-Differentiation (MD) media. Pipet up and down to achieve a homogenous and non-aggregated suspension.
7. Transfer cell suspension into culture flasks and incubate for three days at 37 °C + 5% CO_2.
8. Add another 25 mL of MD media and incubate for an additional 3 days at 37 °C + 5% CO_2 to complete differentiation.
9. On the sixth day of culture, the macrophages must achieve the desired confluency of around ~90%. At this point, remove the supernatant and wash cells three times with 10 mL of fresh serum-free DMEM to remove the non-adherent cells.
10. Carefully scrape cells using a cell scraper into 10 mL of fresh serum-free DMEM.
11. Transfer the cell suspension to a 15 mL centrifuge tube.
12. Centrifugate at 400 × *g* for 5 min at room temperature.
13. Remove the supernatant and resuspend the cell pellet on 1 mL of MD media.
14. Count cells on a hemocytometer.
15. Plate 3×10^5 cells into each well in a 96-well flat plate and incubate the plate overnight at 37 °C + 5% CO_2 to allow cell adhesion.
16. Replace well media with fresh DMEM containing 1% FBS (*see* **Note 7**).

3.2 Preparation of Apoptotic Cells

3.2.1 Human Neutrophil Isolation, Labeling, and Apoptosis Induction (See the Experimental Setting in Fig. 1a)

1. Collect 10 mL of human blood in EDTA tubes (*see* **Note 3**).
2. On a 15 mL centrifuge tube, add 3 mL of Histopaque (density 1119 g/mL).
3. Slowly add 3 mL of Histopaque (density 1077 g/mL) over the Histopaque (density 1119 g/mL) (*see* **Note 8**).
4. Slowly add 9 mL of human blood over the Histopaque gradient (be careful not to mix volumes of Histopaque and blood).

5. Centrifugate at 700n× *g* for 30 min at room temperature using null acceleration and null break.
6. Remove the upper layer rich in PBMC using a Pasteur pipette and collect the neutrophil layer that should be between two phases of the volumes of Histopaque and transfer it to a 15 mL centrifuge tube (*see* **Note 8**).
7. Bring volume to 10 mL with 1× PBS containing 0.5% BSA.
8. Centrifugate at 400 × *g* for 5 min at room temperature.
9. Aspirate the supernatant and resuspend the cell pellet in 10 mL of ACK and incubate for 10 min at 37 °C + 5% CO_2. This step is crucial for lysing erythrocytes present in the pellet.
10. Add 5 mL of FBS to stop the lysis reaction and centrifuge at 400 × *g* for 5 min at room temperature.
11. Discard the supernatant and resuspend it on 5 mL of sterile 1× PBS.
12. Add 5 mL of 5 μM CFSE solution into the 5 mL of neutrophils suspension under light protection (the final concentration of CFSE must be 2.5 μM). Importantly, do not expose the cells to prolonged visible light for the remainder of the protocol.
13. Incubate for 10 min at 37 °C + 5% CO_2.
14. Add 5 mL of FBS to stop CFSE staining.
15. Centrifuge at 400 × *g* for 5 min at room temperature (*see* **Note 4**).
16. Aspirate supernatant and resuspend CFSE-labelling neutrophils on 5 mL of serum-free DMEM.
17. Count cells using a hemocytometer and isolate the desired number of cells (9 × 10^5 cells/mL). Aliquot 1 mL of neutrophils suspension into each well in the 96-well plate (*see* **Note 9**).
18. To induce apoptosis of neutrophils, add 1 μM of staurosporine into each well and incubate the plate for 4 h at 37 °C + 5% CO_2.
19. Transfer neutrophils into a 15 mL centrifuge tube and wash cells three times with 1× PBS to remove residual staurosporine. Importantly, apoptotic neutrophils should be used on the same preparation day. The percentage of apoptosis can be confirmed by flow cytometry using fluorescent labeling with Annexin-V/PI Apoptotic Detection Kit (*see* **Note 10**).

3.2.2 Thymocyte Preparation, Labeling, and Apoptosis Induction (see Experimental Setting in Fig. 1b)

1. Euthanize two mice (*see* **item 2** in Subheading 2.1) and dissect them to expose the chest cavity. With fine-tip forceps, hold and remove the thymus, which lies above the heart (*see* **Note 1**).
2. Transfer the thymus into a tissue culture petri dish, and wash it three times using sterile 1x PBS.

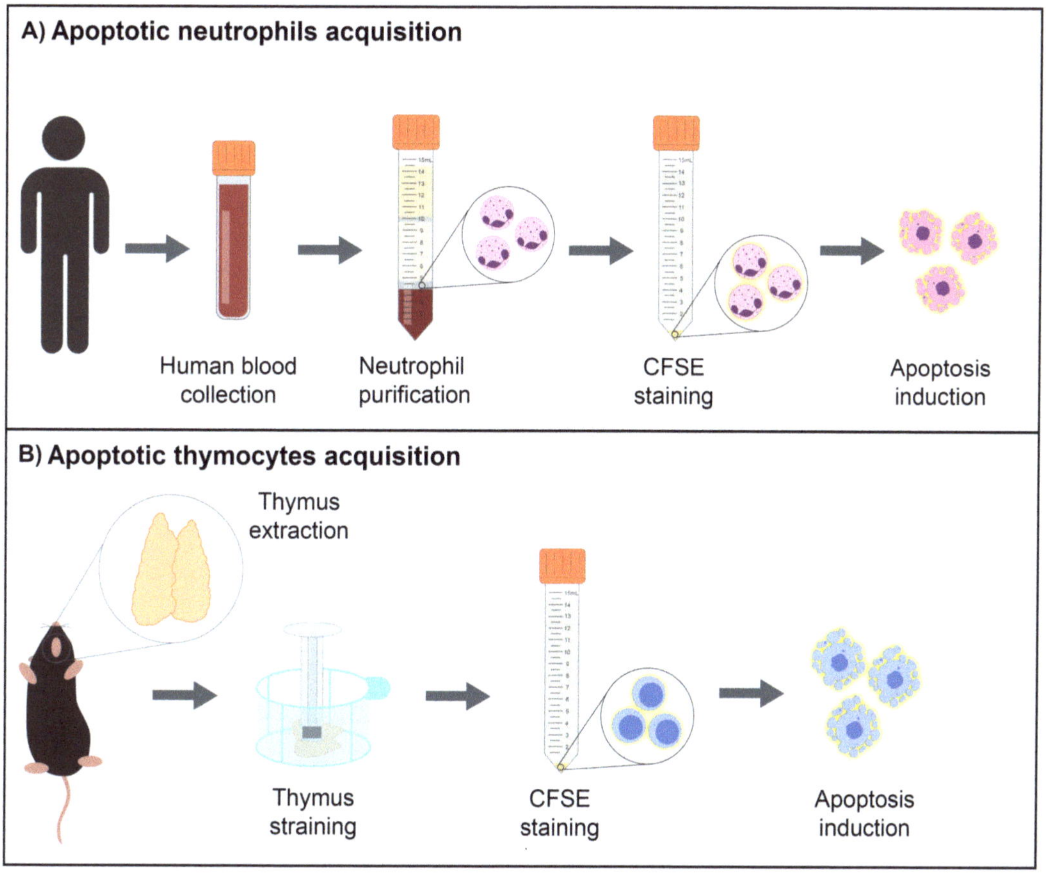

Fig. 1 Acquisition of apoptotic cells. (**a**) Isolation of human neutrophils and induction of apoptosis. Check this protocol in Subheading 3.2.1. (**b**) Isolation of thymocytes and induction of apoptosis. Check this protocol in Subheading 3.2.2

3. Grind the entire thymus using a syringe plunger until a cell suspension is obtained, and then filter the suspension using a cell strainer (100 μm).
4. Collect the cellular suspension and transfer it to a 15 mL centrifuge tube.
5. Centrifuge at 400× *g* for 5 min at room temperature.
6. Aspirate the supernatant and resuspend the cell pellet in 10 mL of sterile 1x PBS.
7. Perform the steps 12–19 from the protocol on preparation of apoptotic cells (Subheading 3.2.1).

3.3 Assessment of Efferocytosis by Flow Cytometry

3.3.1 In vitro Efferocytosis Assay (see Experimental Setting in Fig. 2b)

1. Remove the media from the macrophages incubated in **step 16** of the Isolation and Culture of BMDMs protocol (Subheading 3.1) and add 9×10^5 apoptotic human neutrophils (Subheading 3.2.1) or apoptotic thymocytes (Subheading 3.2.2) in 200 μl of DMEM containing 1% FBS, swirling the plate briefly to ensure even coverage (*see* **Note 9**).

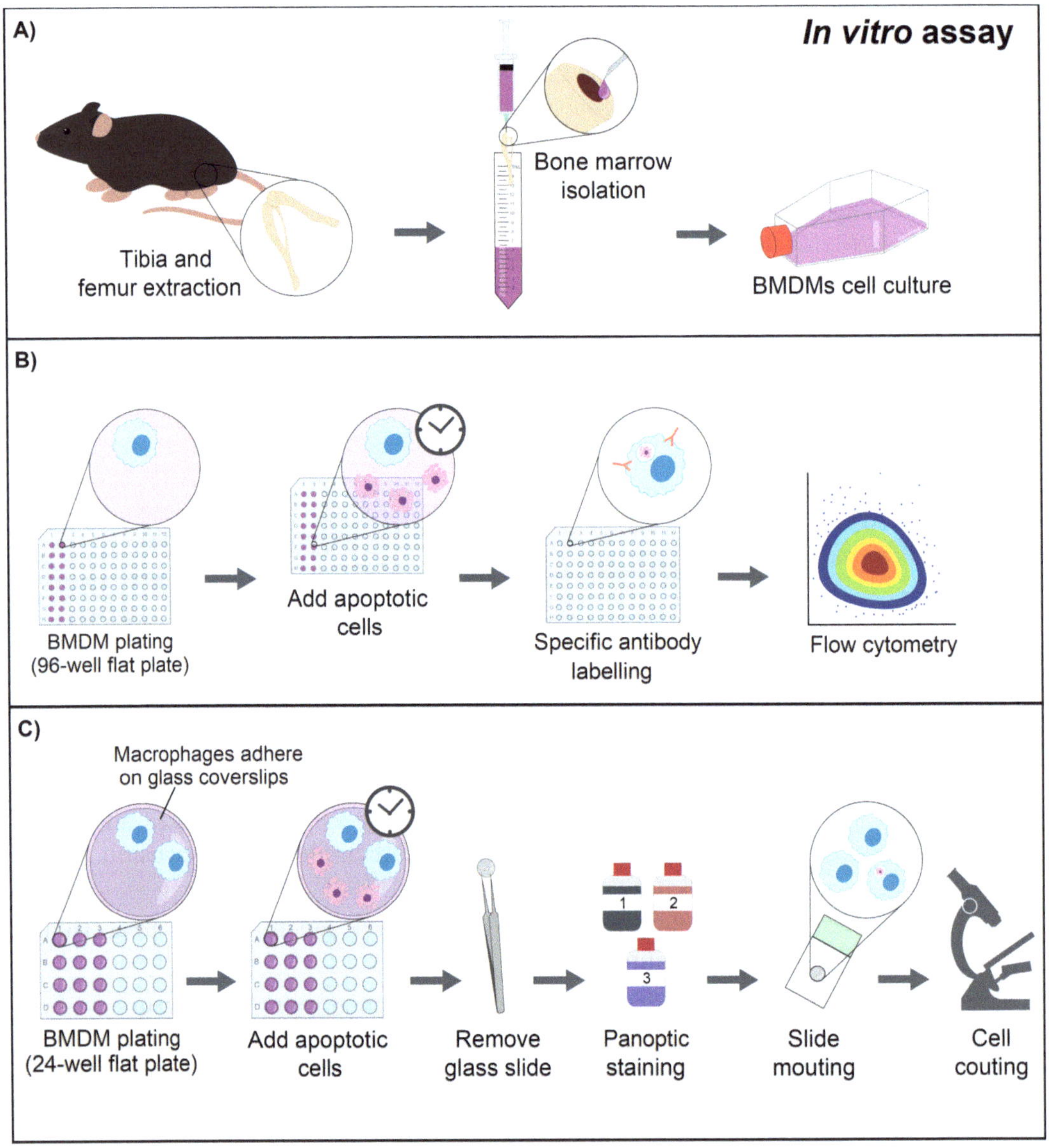

Fig. 2 In vitro efferocytosis assays. (**a**) Isolation of BMDM from the bone marrow of mice. Check this protocol in Subheading 3.1. (**b**) Investigation of efferocytosis by flow cytometry. Check this protocol in Subheading 3.3.1. (**c**) Investigation of efferocytosis using optical microscopy. In this experimental setting, glass coverslips were added to wells previous to the plating of cells. Check this protocol in Subheading 3.4.1

2. Incubate the plate for 60 min at 37 °C + 5% CO_2 (*see* **Note 11**).
3. Put the plate on ice, remove the media from wells (*see* **Note 5**), and wash cells three times with 1 mL of ice-cold 1x PBS. Be careful not to excessively rinse the macrophages away (*see* **Note 12**).
4. Add 1 mL of ice-cold 1x PBS and scrape the plate-bound macrophages using a tip of a 1 mL pipette.
5. Transfer the cell suspension into a 96-well conical-bottomed plate, maintaining it on the ice during the remainder of the process.
6. Centrifuge at 400 × *g* for 5 min at 4 °c.
7. Label with an appropriate dilution of macrophage surface-specific fluorescent antibodies (typically diluted 1:100 to 1: 200 in 1× PBS containing 1% BSA) for 45 min at 4°c.
8. Add 100 μL of 1× PBS containing 1% BSA and centrifugate at 400 × *g* for 5 min at 4 °c.
9. Turn the plate upside down at once, keeping cells on the bottom of the plate.
10. Fix cells using 100 μL of 2% PFA for 30 min at 4 °c. Repeat **steps 8** and **9**.
11. Resuspend the cell pellet into 100 μL of 1× PBS containing 1% BSA.
12. Transfer cell suspension into a FACS tube and proceed with flow cytometer analysis (*see* **Note 13**).

3.3.2 *In vivo* Efferocytosis Assay (see Experimental Setting in Fig. 3a-b)

1. Inject mice intraperitoneally with 0.1 mg of Zymosan diluted in 200 μL of 1x PBS to induce macrophage recruitment (transient peritonitis). There is a predominance of macrophages 96 h after zymosan injection. Optionally, mice can be treated with drugs or other treatments targeting the efferocytic signaling processes to investigate the direct impact on efferocytosis.
2. Perform an intraperitoneal injection of 9×10^5 apoptotic human neutrophils (Subheading 3.2.1) or apoptotic thymocytes (Subheading 3.2.2) in 200 μl of 1x PBS.
3. Euthanize mice 30 min after apoptotic cell injection by an overdose of anesthestic solution (*see* **item 2** in Subheading 2.1) and expose the peritoneal cavity.
4. Wash the peritoneal cavity three times with 3 mL of 1× PBS containing 3% BSA and transfer the cell suspension into a 15 mL centrifuge tube.
5. Centrifuge it at 400 × *g* for 5 min at room temperature.

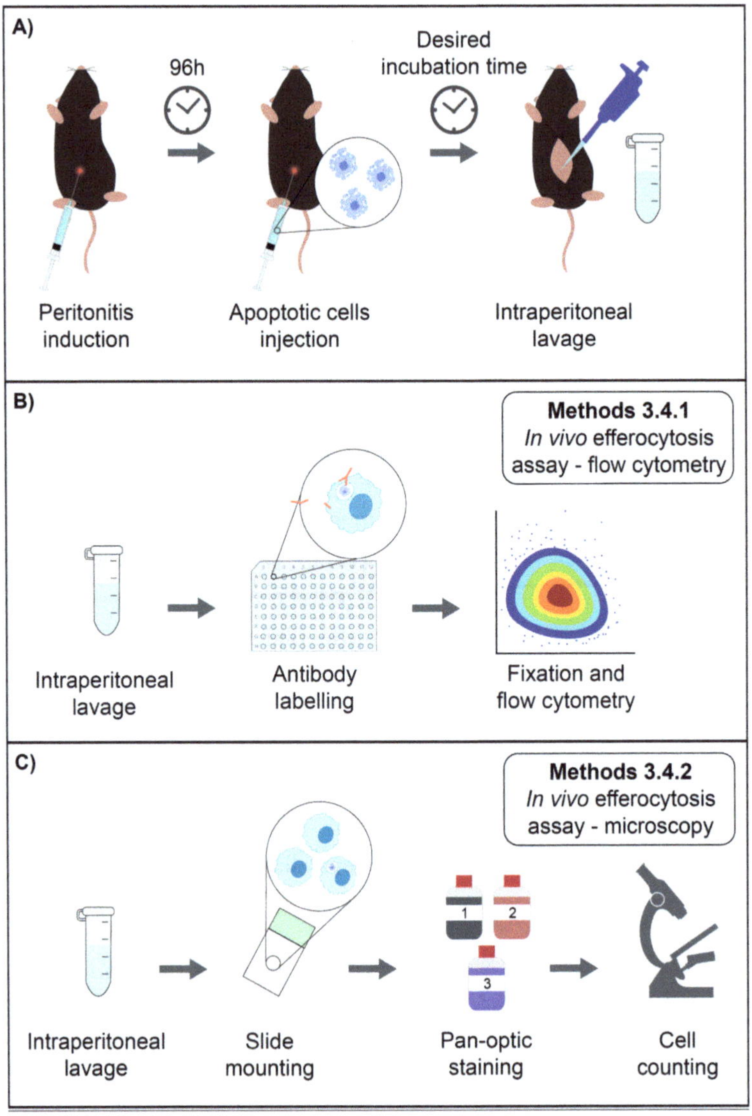

Fig. 3 In vivo efferocytosis assays. (**a**) Induction of peritonitis and apoptotic cell injection. Check this protocol in Subheading 3.3.2. (**b**) Investigation of efferocytosis by flow cytometry. Check this protocol in Subheading 3.4.1. (**c**) Investigation of efferocytosis using optical microscopy. Check this protocol in Subheading 3.4.2

6. Aspirate supernatant and resuspend the cell pellet in 100 μL of 1× PBS containing 1% BSA and transfer cells from the tubes to a 96-well conical plate, maintaining it on the ice during the remainder of the process.
7. Perform **steps 6–12** of the assessment of efferocytosis by flow cytometry protocol (Subheading 3.3.1).

3.4 Assessment of Efferocytosis by Morphological Criteria

Besides flow cytometry, efferocytosis can also be evaluated through optical microscopy by morphological examination (*see* **Note 15**).

3.4.1 In Vitro Efferocytosis Assay (See the Experimental Setting in Fig. 2c)

1. Place sterile round-shaped glass coverslips into the 24-well flat plate (*see* **Note 14**).
2. Seed macrophages (5×10^5 cells/well) (Subheading 3.1) in 500 μL of MD media.
3. Incubate cells for 24 h at 37 °C + 5% CO_2.
4. Replace MD media for media containing apoptotic cells (*see* protocol 3.2) and incubate the plate for 60 min at 37 °C + 5% CO_2 (*see* **Note 11**).
5. After incubation, put the plate on ice, remove the media and wash each coverslip with 1 mL of ice-cold 1x PBS (*see* **Note 12**).
6. Remove each glass coverslip from the bottom of the wells (*see* **Note 14**).
7. Stain the coverslips using a panoptic solution.
8. Mount the coverslips on a glass slide.
9. Analyze slides on an optical microscope (*see* **Note 15**).

3.4.2 In Vivo Efferocytosis Assay (see Experimental Setting in Fig. 3c)

1. Perform **steps 1–5** of the in vivo efferocytosis assay protocol (Subheading 3.3.2).
2. Resuspend cells pellet on 1 mL of 1× PBS containing 3% BSA.
3. Use 100 μL of the cell suspension in a cytospin assay.
4. Let slides dry overnight and stain them using a panoptic solution.
5. Analyze slides on an optical microscope.

4 Notes

1. For this protocol, the sex of animals will not be significant. Approximately equal numbers of male and female mice can be used as the source of bone marrow-derived cells. We suggest that the mice used for these methods be eight to ten weeks old. Young mice can provide more cells. Importantly, the procedures must be carried out following the recommendations of the local Animal Ethics Committee.
2. L-929 is an adherent type of mouse fibroblast cell line that produces a macrophage colony-stimulating factor (M-CSF), which can promote bone marrow cell differentiation into

macrophages [10]. If desired, the DMEM can also be enriched with the commercial growth factor instead of using the L929 supernatant.

3. Human blood should be collected from healthy volunteers via venipuncture on EDTA tubes, in order to prevent blood coagulation. The cell separation protocol should be performed immediately after collection. All enrolled subjects must inform consent for inclusion in the study. Carry out all procedures in accordance with the local Ethics Committee.
4. Avoid freezing and defrosting the CFSE in order to avoid degradation. After CFSE staining, the cell pellet turns yellowish demonstrating successful staining. Avoid long exposures of CFSE to light, keeping solutions and stained cells always in the shelter. Prepare the solution only on the day of use. Preferably, analyze the cells by Flow Cytometer on the same day as staining.
5. Additionally, the media from the co-culture of macrophages with apoptotic cells can be stored at −80 °C for further analysis of efferocytosis-related cytokines, including interleukin (IL)-10 and TGF-β.
6. Hold the bones with a clamp and insert the needle into the cavity of the bone marrow. Make an up-and-down movement with the syringe while purging the bone marrow with the medium. The bones should look white when done.
7. If desired, BMDM can be cryopreserved and used further [11, 12].
8. Use a Pasteur pipette to gently place the Histopaque with lower density above the Histopaque with higher density, avoiding mixing the solutions. This is essential for good isolation of neutrophils. A thin layer of neutrophils should be formed between the two Histopaque volumes if the centrifugation was successful.
9. It is recommended to perform the efferocytosis assays at a 1:3 ratio of macrophage: apoptotic cells, but it can be varying according to experimental need.
10. Apoptotic cells of different origins can be used to perform the efferocytosis assay as all apoptotic cells express phosphatidylserine molecules on their membranes, which are used by macrophages to bind and engulf these cells. To assess apoptosis in neutrophils or thymocytes resuspend the apoptotic cell pellet in 100 μL of 1x PBS containing BSA 1% and label it with 2.5 μL of APC-conjugated Annexin V and PI for 5 min at room temperature. Immediately, perform the flow cytometry analysis (*see* Fig. 4).

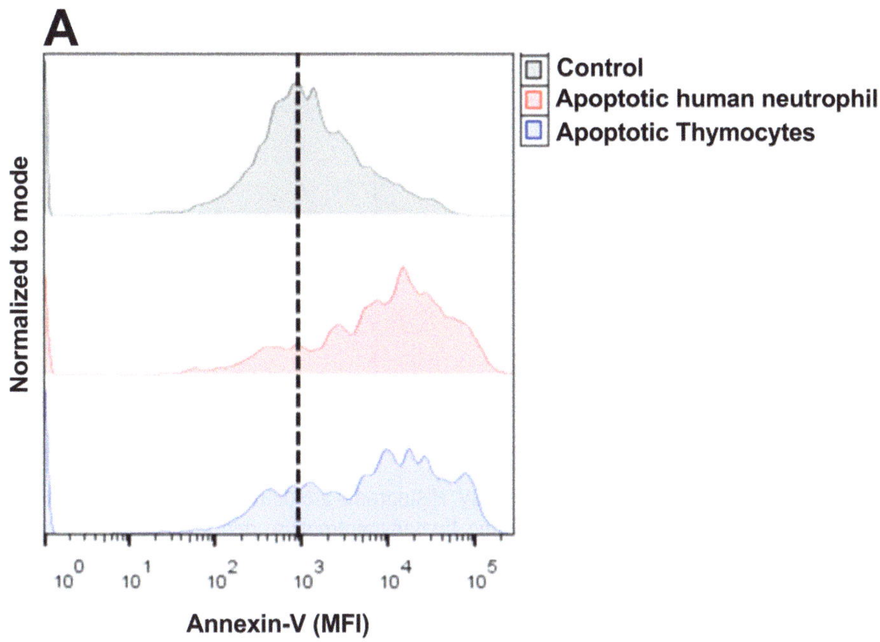

Fig. 4 AnnexinV staining of apoptotic cells. Annexin-V is a fluorescent molecule that binds to the phosphatidylserine molecules that are expressed on membranes of apoptotic cells and used as an apoptosis indicator for flow cytometry assays. Apoptotic human neutrophils and thymocytes were biochemically determined 4 h after treatment with staurosporine (1 M) and the median fluorescence intensity (MFI) of annexin-V+ cells was determined by flow cytometry. Representative histogram of annexin-V+ cells (A)

11. Efferocytosis is a rapid cellular event. The proper duration of efferocytosis assay can vary depending on the level of efferocytosis desired and the quality of BMDM differentiation. The establishment of a time-series experiment is helpful to understand times of exposure. Shorter times are recommended as dye decay may occur after a prolonged incubation time (for example, 180 min) or rapid digestion of engulfed apoptotic cells in the phagosome. In this protocol, we used 30–60 min. Due to experimental heterogeneity, it is important to carry out sufficient replications and repetitions of independent experiments with the appropriate statistical test to ensure statistically robust samples.
12. Once the time of exposure of macrophages to apoptotic cells is overdue, place the plate on ice, as the lower temperature will stagger the cell metabolism and interrupt the degradation of the efferosome.
13. For quantifying efferocytic efficiency by flow cytometry, it is recommended to use the mean fluorescence intensity (MFI)

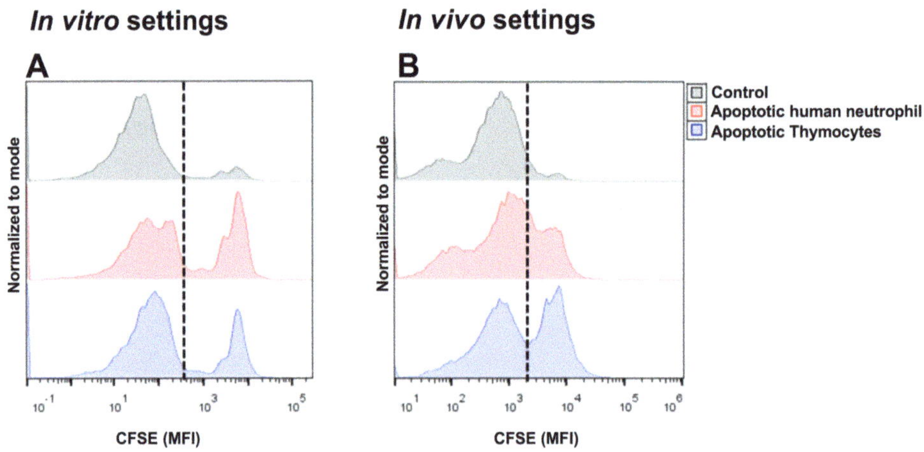

Fig. 5 Assessment of efferocytosis by flow cytometry. Histograms are demonstrating the median fluorescence intensity (MFI) of apoptotic thymocytes, apoptotic human neutrophils, and viable cells positive for CFSE detected inside macrophages (F4/80$^+$ cells) by flow cytometry (check protocol 3.3). Representative histograms are demonstrated for in vitro experimental settings (**a**) and in vivo experimental settings (**b**)

value of apoptotic cells (CFSE$^+$) inside macrophages (F4/80$^+$) (*see* Fig. 5).

14. Clean the glass coverslips with 70% ethanol before usage. Sterilization is also crucial for the experiment. A sterile serological pipette can be used to transfer slides from the Petri plate to the 24-well flat plate. Glass coverslips can be removed with the help of a needle. The inclination of the plate at 45° and the bending of the end of the needle are helpful tips.
15. For characterizing efferocytic morphology, use a 100× objective to perform light microscopy analysis of cytospin slides or glass coverslips (a counting 500 cells/slide) determining the proportion of macrophages with apoptotic bodies observed in their cytoplasm. So, divide the number of macrophages with engulfed apoptotic cells by the total number of macrophages x 100, therefore getting the % efferocytosis. Additionally, efferocytic events can be detected by imaging using a light microscope equipped with a camera (*see* Figs. 6 and 7).

Funding

This study was funded by Fundação de Amparo à Pesquisa do Estado de Minas Gerais (FAPEMIG, Brazil), grant number BPD-00484-22.

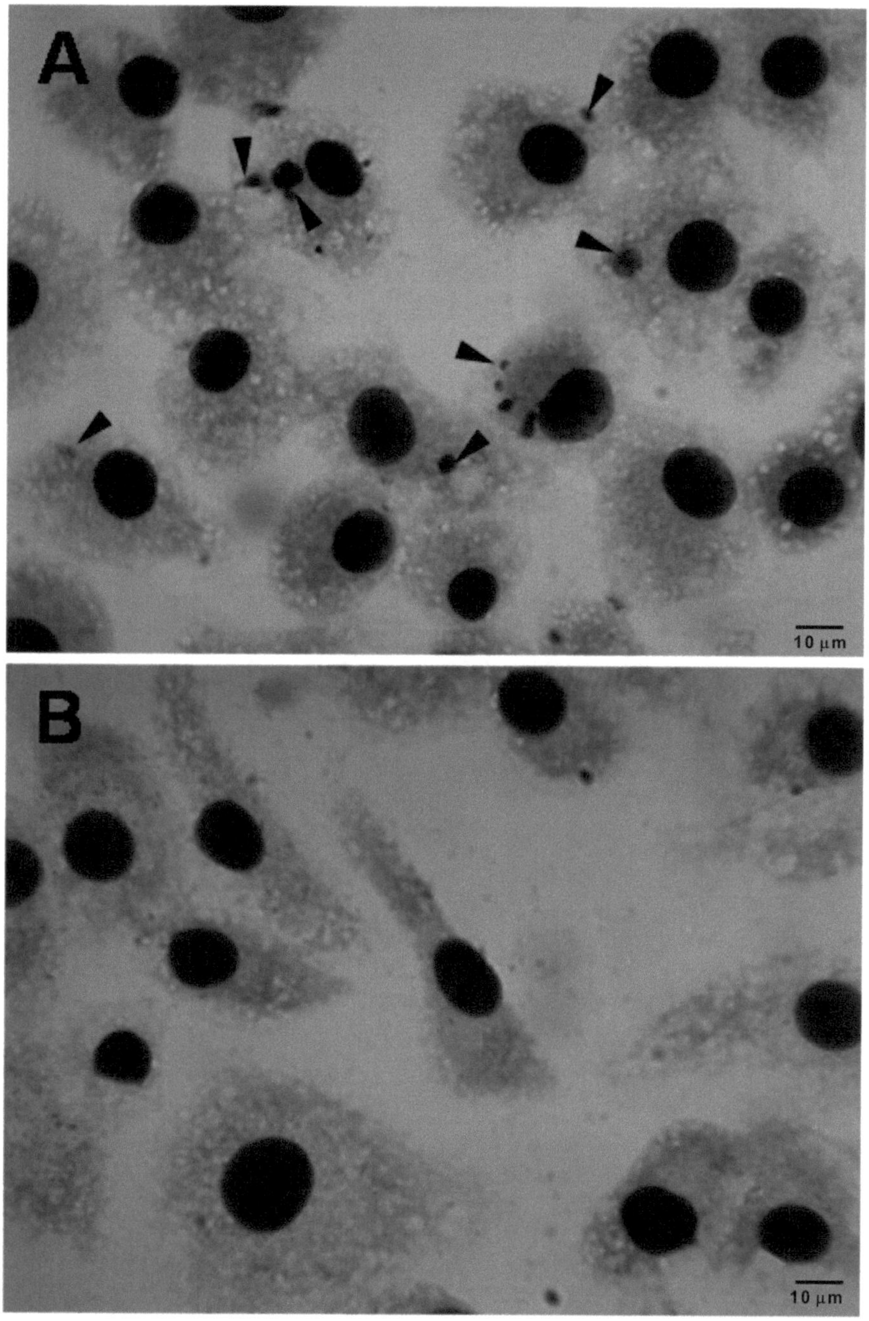

Fig. 6 Assessment of in vitro efferocytosis by morphological criteria. (**a**) Glass slide with adherent macrophages that were exposed to apoptotic cells. Arrows indicate engulfed apoptotic bodies by macrophages. Apoptotic cells can present different sizes inside efferosomes, due to their progressive degradation inside them. Generally, apoptotic cells have darker coloration than macrophage nuclei. (**b**) Glass slide with adherent macrophages that were not exposed to media viable cells. Check this protocol in Subheading 3.4.1

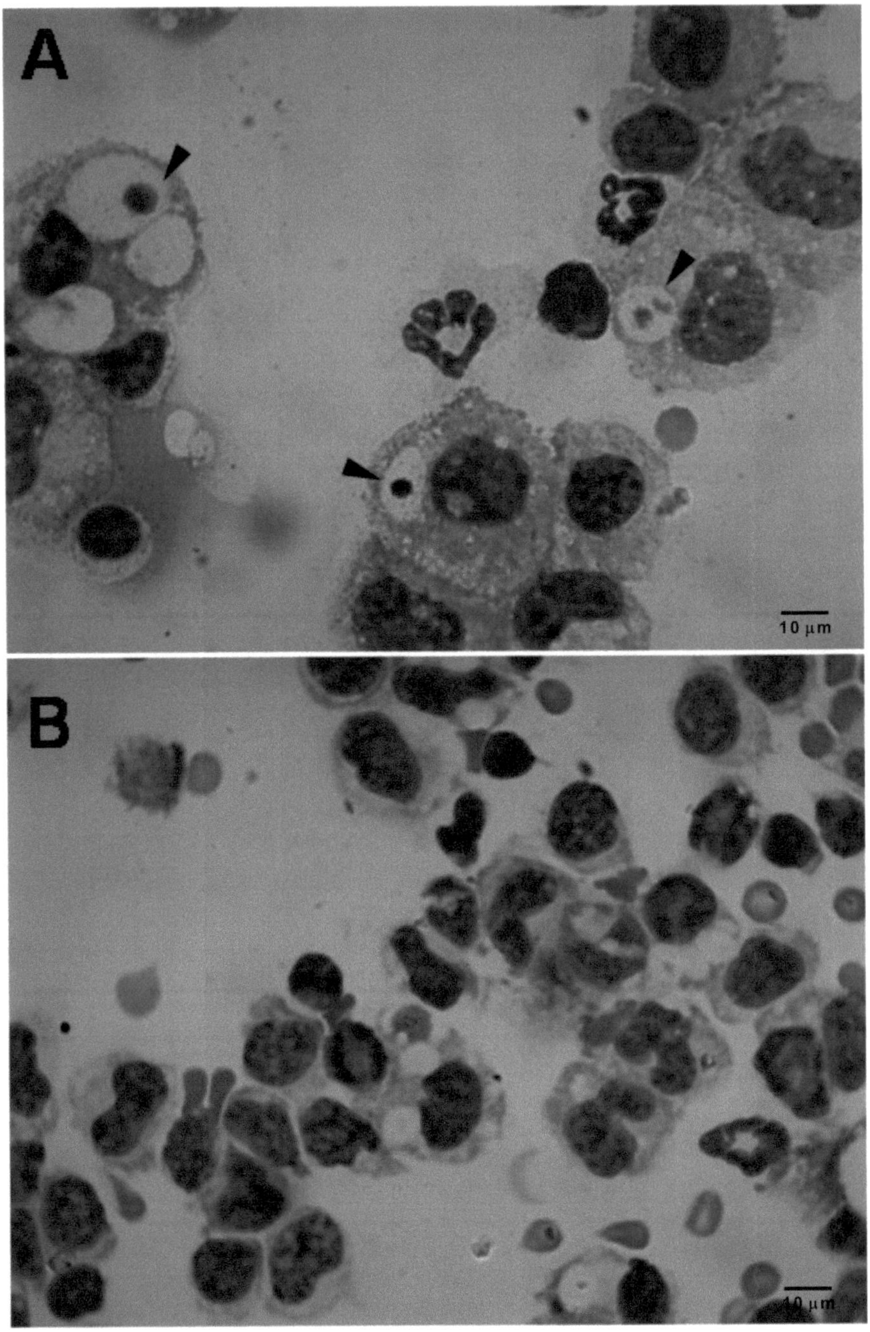

Fig. 7 Assessment of in vivo efferocytosis by morphological criteria. (**a**) Cytospin slide with peritoneal macrophages that were exposed to apoptotic cells. Arrows indicate engulfed apoptotic bodies by macrophages. Empty efferosomes indicate that the apoptotic cell has already been digested. (**b**) Cytospin slide with peritoneal macrophages that were exposed to viable cells. Check this protocol in Subheading 3.4.2

References

1. Morioka S, Maueröder C, Ravichandran KS (2019) Living on the edge: Efferocytosis at the Interface of homeostasis and pathology. Immunity 50(5):1149–1162. https://doi.org/10.1016/j.immuni.2019.04.018
2. Elliott MR, Ravichandran KS (2010) Clearance of apoptotic cells: implications in health and disease. J Cell Biol 189(7):1059–1070. https://doi.org/10.1083/jcb.201004096
3. Doran AC, Yurdagul A Jr, Tabas I (2020) Efferocytosis in health and disease. Nat Rev Immunol 20(4):254–267. https://doi.org/10.1038/s41577-019-0240-6
4. McCubbrey AL, Curtis JL (2013) Efferocytosis and lung disease. Chest 143(6):1750–1757. https://doi.org/10.1378/chest.12-2413
5. Zhen Y, Shao W-H (2019) Experimental analysis of apoptotic thymocyte engulfment by macrophages. JoVE 147:1–6. https://doi.org/10.3791/59731
6. Felix FB, Vago JP, Fernandes DO et al (2021) Biochanin a regulates key steps of inflammation resolution in a model of antigen-induced arthritis via GPR30/PKA-dependent mechanism. Front Pharmacol 12:662308. https://doi.org/10.3389/fphar.2021.662308
7. Felix FB, Dias J, Vago JP et al (2023) Blocking the HGF-MET pathway induces resolution of neutrophilic inflammation by promoting neutrophil apoptosis and efferocytosis. Pharmacol Res 188:106640. https://doi.org/10.1016/j.phrs.2022.106640
8. Negreiros-Lima GL, Lima KM, Moreira IZ et al (2020) Cyclic AMP regulates key features of macrophages via PKA: recruitment. Reprogramm Efferocyto Cells 9(1):128. https://doi.org/10.3390/cells9010128
9. Vago JP, Galvão I, Negreiros-Lima GL et al (2020) Glucocorticoid-induced leucine zipper modulates macrophage polarization and apoptotic cell clearance. Pharmacol Res 158: 104842. https://doi.org/10.1016/j.phrs.2020.104842
10. Heap RE, Marín-Rubio JL, Peltier J et al (2021) Proteomics characterisation of the L929 cell supernatant and its role in BMDM differentiation. Life Sci Alliance 4(6): e202000957. https://doi.org/10.26508/lsa.202000957
11. Marim FM, Silveira TN, Lima DS Jr, Zamboni DS (2010) A method for generation of bone marrow-derived macrophages from cryopreserved mouse bone marrow cells. PLoS One 5(12):e15263. https://doi.org/10.1371/journal.pone.0015263
12. Leung G, Petri B, Reyes JL, Wang A, Iannuzzi J, McKay DM (2016) Cryopreserved Interleukin-4-treated macrophages attenuate murine colitis in an integrin β7-dependent manner. Mol Med 21(1):924–936. https://doi.org/10.2119/molmed.2015.00193

Chapter 11

Physical Exercise as a Model for Investigation of the Inflammation Response in Mice

Albená Nunes-Silva, Antonio Felipe Souza-Gomes, Carolina Braga de Resende, Barbara Maximino Rezende, and William Antonio Gonçalves

Abstract

Animal models, particularly murine protocols on physical exercise, are widely used to investigate physiological adaptations and to study, prevent, and treat chronic non-communicable diseases. Controlled treadmill running enables precise manipulation of intensity and duration, overcoming limitations of voluntary wheel running for modeling inflammation. An incremental-speed treadmill test to fatigue disrupts homeostasis and acts as an acute inflammatory stimulus, eliciting local skeletal muscle and systemic immune responses. This chapter presents a standardized physical exercise protocol in mice and subsequent tissue collection, combined with intravital microscopy of skeletal muscle microcirculation and in vitro neutrophil migration assays, to quantify leukocyte recruitment and activation as a versatile model of exercise-induced inflammation.

Key words Physical exercise, Murine model, Inflammation, Intravital microscopy, Neutrophil migration

1 Introduction

Historically, animal model-based research plays a central role in physiological research [1]. In general, the murine models have been widely used by researchers to understand the functioning and adaptations of physiological mechanisms under many conditions, including acute and chronic response to physical exercise in many tissues [2–4]. In addition, acute and chronic physical exercise for rodents has been currently applied in studies as a non-pharmacological strategy, to study, treat, and prevent several chronic non-communicable diseases [5]. In this context, a relevant number of studies involving exercise have been conducted using different tools, protocols, and strategies for carrying out acute and chronic physical exercise protocols [6].

Lucas Secchim Ribeiro (ed.), *Models in Inflammation and Immunopharmacology*, Methods in Molecular Biology, vol. 2993,
https://doi.org/10.1007/978-1-0716-5019-6_11,
© The Author(s), under exclusive license to Springer Science+Business Media, LLC, part of Springer Nature 2026

In general, aerobic and anaerobic, voluntary, or involuntary protocols are used for scientific perspectives, and running on a treadmill model has been widely used for mice physical exercise training [2–4, 7–9]. However, to be used as a model for inflammation investigation, is crucial to guarantee, as much as possible, the control of the training overload, such as intensity and duration. This exercise protocol needs to break the body homeostasis and, in this sense, voluntary wheel running, in which the adaptations of the organism are restricted to the link to voluntary activity, without the training load manipulation is difficult to be measured.

Therefore, using the adequate intensity and duration of the overload, such as incremental-speed running test until fatigue could work as a canonical inflammation response model, which could be used to investigate inflammation response in many scenarios. Herein, we try to show the widely incremental physical exercise protocol until fatigue in a treadmill equipment could be helpful to study the inflammatory response in mice. It is possible, once, exercise works as an acute and chronic stressor of immune system during and after the end of the exercise session. This stress of the physiologic system can cause a local (skeletal muscle tissues and adipose tissue) and systemic (plasma and serum) inflammation response [2, 10].

In the skeletal muscle tissue, the fatiguing exercise protocol can induce local inflammation once it can increase the interaction of neutrophils with endothelial cells in quadriceps muscle by intravital microscopy after a fatiguing exercise protocol [2]. Some manuscripts were able to show clear rolling, adhesion, and subsequent transmigration of neutrophils induced by exercise [2]. This is consistent with other studies demonstrating leukocytes accumulation in tissue as determined by histology.

The fatiguing exercise protocol can also initiate a complex systemic cascade of inflammatory events, which depend on the type, intensity, duration, and familiarity of the exercise [11]. Plasma measurable immune parameters affected by exercise comprise changes in peripheral blood cell numbers, granulocyte activity of leukocytes, natural killer (NK) cell cytotoxic activity, lymphocyte proliferation, and alteration in the cytokine levels on plasma [11].

Here, we proposed a physical exercise acute protocol that induce an inflammatory response, as an inflammatory model, and in this sense could be useful for researchers' groups. To achieve this inflammatory response, a fatiguing exercise protocol test followed by extraction of many tissues that signaling this inflammation could be used. After the end of fatiguing exercise session, many protocols can be used to confirm the kinetic of inflammatory mediators. In this chapter we would like to show two techniques, such as the skeletal muscle microcirculation by intravital microscopy and in vitro neutrophil migration.

2 Materials

2.1 Incremental-Speed Running Test Until Fatigue

1. Electric treadmill (EP-132 – Insight).

2.2 Quadriceps Exposition Surgery

2.2.1 Surgery Equipment

1. *Syringe (1 mL) and needles (30G).*
2. *Tweezers.*
3. *Surgical scissors.*
4. *Shaving cream.*
5. *Cotton pads.*
6. *Pasteur pipette.*
7. *Modelling dough set.*

2.2.2 Solutions

1. *0.9% NaCl (w/v, sterile).*
2. *Anesthestic solution: ketamine and xylazine (diluted in sterile saline to a final dose of 100 mg/kg and 10 mg/kg, respectively).*
3. *Rhodamine 6G chloride (diluted in sterile saline for a final dose of 0.3 mg/kg).*

2.3 Skeletal Muscle Microcirculation by Intravital Microscopy

1. *Intravital microscope ECLIPSE 50i Nikon with a 20-objective lens (or similar equipment).*
2. *DS-Qi1MC Nikon digital camera for playback analysis with Nikon imaging software.*
3. *Rhodamine 6G chloride (diluted in sterile saline for a final dose of 0.3 mg/kg).*

2.4 Neutrophil Isolation and Migration

1. 15- and 50-mL tubes.
2. Sterile Pasteur Pipettes.
3. Laminar Flow hood.
4. 48-well Boyden chamber.
5. Neubauer chamber.
6. Polycarbonate filters.
7. Incubator with controlled temperature and atmosphere.
8. Histological slides.
9. Microtube centrifuge and Cytocentrifuge.
10. Microscope.
11. Tweezers.
12. Double distilled water.
13. Culture medium (RPMI with BSA 0.01%, w/v).

14. Histopaque 1119.
15. Histopaque 1077.
16. Turk Solution (2% glacial acetic acid, w/v; 1% Gentian violet in distilled water, w/v).
17. Panoptic Dye.
18. Trypan Blue.
19. N-Formyl-Met-Leu-Phe (fMLP, 0.1 uM).

3 Methods

A summary of the sequence to be followed in the next steps can be seen in Fig. 1.

3.1 Incremental-Speed Running Test Until Fatigue

1. Set the initial speed to 5 meters per minute (m/min), at a 5% grade.
2. The warm-up process is to establish a walking pace for 30 min to familiarize the mice with the apparatus and task.
3. Then, the speed is increased 1 meter/minute every 3 min, keeping the treadmill at a 5% grade, until the animal stops

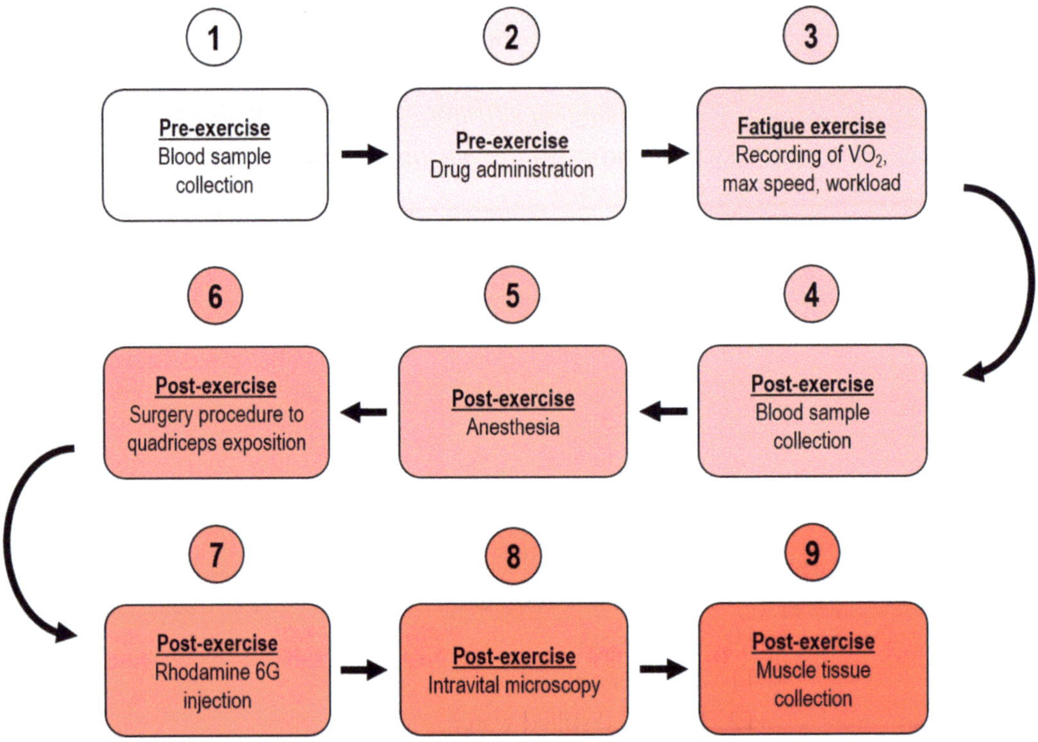

Fig. 1 Overview of the intravital microscopy steps of the procedure for the evaluation of leukocyte recruitment in the post-capillary venule of the rectus femoris

running and gets fatigued. The fatigue is judged by the refusal of the animal to continue moving on the treadmill belt more than 10 seconds.

4. A mild electrical stimulus (0.5 mA) is applied to mice that stepped off the treadmill to keep them exercising.

3.2 Quadriceps Exposition Surgery

1. Anesthetize the mice with a solution mix of Ketamine (100 mg/kg) and Xylazine (10 mg/kg).
2. Position the mice on a 37 °C thermal plate to prevent decrease in body temperature after anesthesia.
3. Shave the leg using a shaving cream to clean the skin in the surgery site.
4. The residual shaving cream must be removed from surgery site using a moistened cotton pad.
5. Next, remove the skin adjacent to anterior aspect of the leg. For this, hold the skin using a tweezer and cut it between knee join and hip (Fig. 2). During this step the contact of surgical materials with muscular tissue must be avoided.
6. The exposed quadriceps must be constantly moisturized with saline (NaCl 0.9%) to avoiding dehydration. This procedure could be realized using a Pasteur pipette for dripping the saline solution over the muscle prepared to evaluation.
7. Place a piece of play dough to set down the knee joint and mold it until quadriceps surface reach a position proximately to 90° related to laser direction (Fig. 2). During this step the total flexion of the knee joint stretching the quadriceps belly must be avoided. This step is essential to preserve normal flow in the muscle vessels.

3.3 Skeletal Muscle Microcirculation by Intravital Microscopy

1. Once the muscle is exposed, administer an intravenous (i.v.) injection (100 ul) of rhodamine 6G (0.3 mg/Kg) for fluorescent labeling of leukocytes.
2. Examine the muscle microcirculation using the intravital microscope ECLIPSE 50i Nikon with a 20-objective lens.
3. Choose a vessel with a diameter of 60–80 um.
4. Acquire the images and videos using the DS-Qi1MC Nikon digital camera for playback analysis with Nikon imaging software.
5. Quantify rolling and cell adhesion. Rolling leukocytes are quantified by the number of cells that pass by a determined point in the rectus femoris venule per minute. Cells are considered adherent when stay in the vessel for at least 30 s. Total cell adhesion is determined as the number of adherent cells in an area of 100 μm length of venule in 5 min.

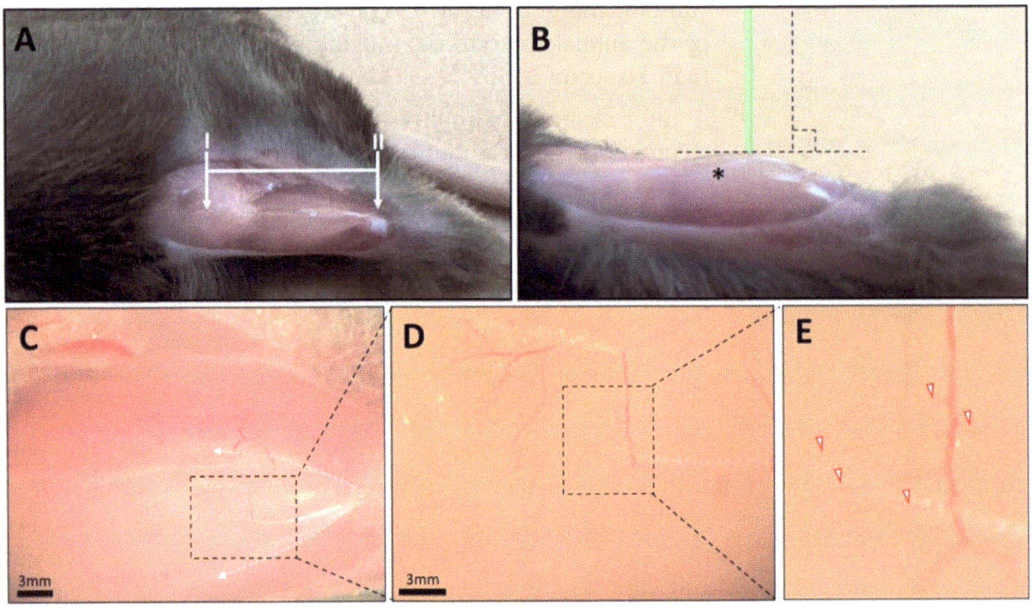

Fig. 2 Surgery to quadriceps exposure for subsequent evaluation by intravital microscopy. (**a**) Ventral aspect of leg after skin removal. (**b**) The surgical field must be opened between knee and hip joint (I and II). Lateral aspect of leg and situation of quadriceps surface for evaluation. It is expected that ventral part of distal muscle belly could be exposed to be explored by microscope light beam (represented by green line). This light beam must be parallel to muscular surface that will be analyzed. (**c**) The superficial vessels in ventral part of the quadriceps. The main vessels in antero-medial surface of the quadriceps could be observed in C, D, and E by 1X, 3X, and 4X in the Stereo microscope respectively. The red arrow points to the target vessels in E

3.4 Neutrophil Isolation and Migration Assays

3.4.1 Neutrophil Isolation

1. The collected blood sample should be added to a 50 mL tube and then the RPMI culture medium should be added in the same volume as the original volume of blood. The sample is then ready to be submitted to the process of cell separation through the density gradient technique.
2. Histopaque 1119 and histopaque 1077 are used to separate cells. Ensure there is enough volume for all samples.
3. In a 15-ml tube, pipette 3 mL of histopaque 1119 followed by 3 mL of histopaque 1077 to form 2 layers of solution with different densities. Make sure there are two layers of the solution.
4. The mixed solution of whole blood and RPMI medium should be carefully pipetted into the flask containing histopaque layers, forming then the third layer is formed in the tube.
5. Centrifuge the tube for 30 min, 700× *g*, with no break and at room temperature (25 °C).
6. After centrifugation, five layers of solution will be identified.

7. Remove the top layers of histopaque, serum and the PBMC, and discard them.
8. Remove the neutrophils with a Pasteur pipette and add them to a 15 mL Falcon tube and complete it with RPMI medium until it reaches a total volume of 15 mL.
9. Centrifuge the cell suspension for 15 min, 600 × *g*, at room temperature.
10. After that, remove the supernatant and resuspend the pellet with 1 mL of RPMI medium.
11. Homogenize the suspension and add 2 mL of double distilled water and keep the suspension in the refrigerator for 4 min. Afterward, fill up the tube with 10 mL of RPMI medium.
12. Centrifuge the suspension for 10 min, at 700 × *g* at 25 °C.
13. Remove the supernatant and resuspend it in 1 mL of RPMI medium.
14. Add 10 μL of these cells to a separate microtube and add 90 μL of Turk solution for counterstaining.
15. Count the cells in the Neubauer chamber and adjust the cell density to 10^6 cells/mL in RPMI + BSA 0.01%.

3.4.2 Neutrophil Migration

1. In the lower compartment of Boyden chamber, 28.6 μL of fMLP working solution (0.1 uM) is added and RPMI/BSA 0.01% was used to control "chance" migration.
2. Polycarbonate filter is placed on the lower compartment. A polyethylene membrane is placed on top of this, and then it is attached to the top of the chamber. After placing the membranes, the upper part of the Boyden chamber is attached, and screws are finally attached.
3. After attaching the parts, 50 μL of neutrophil suspension is added to the upper compartment of the chamber.
4. This chamber is taken to the incubator for 60 min, at 37 ° C with 5% CO_2.
5. After the incubation period, the polycarbonate filter should be removed and stained with the Panoptic staining kit.
6. After staining, the filter should be fixed on a glass slide to count cells that migrated and fixed to the membrane.
7. Neutrophil count should follow the arithmetic mean of cells counted in five random fields of the slide and all tests should be carried out twice for each sample tested to minimize experimental error. Results are expressed by the number of neutrophils per field.

4 Notes

1. It is very important to guarantee that the animals performed a fully fatiguing exercise protocol, once this type of physical exercise protocol is well known as a body homeostasis breaker.
2. When exposing the quadriceps muscle to count the number of rolling and adherent leukocytes, it is important to ensure that you are the same veins are being used for counting. It is possible to immobilize the leg of the animal exposing the same portion of quadriceps to observe the same veins.
3. Pipetting of histopaque solutions should be carefully carried out and it is suggested to incline the pipette so that histopaque 1077 may form a layer overlapping histopaque 1119.
4. Before carrying out this technique, it is recommended to mark the polycarbonate filter to identify the placement of the filter in the Boyden chamber.
5. It is recommended to develop a model of standardization of Boyden dish on paper and it should be done in the beginning of the procedure (Fig. 2).
6. Cell viability greater or equal to 97% is required for the experiment to help visualize cells by exclusion, using Trypan Blue.
7. It is important to make sure screws are tightly screwed in the chamber.
8. Neutrophil suspension should be quickly added to the upper compartment of the chamber and the pipette should be vertically positioned to avoid bubble formation. Bubbles prevent cell migration.
9. Membrane should be removed from the Boyden chamber carefully and with the help of tweezers.
10. In addition, muscle contraction releases a number of exercise-associated signaling molecules that have been identified as multiplied and are now called "exerkines". A multitude of organs, cells, and tissues release these factors, including skeletal muscle (myokines), the heart (cardiokines), liver (hepatokines), white adipose tissue (adipokines), brown adipose tissue (baptokines), and neurons (neurokines) [15].
11. Intense voluntary running protocol on treadmill can increase the number of circulating white blood cells and, in addition is also able to elevate the production and releasing of a several number of cytokines, such as interleukin-2 (IL-2), interleukin-4 (IL-4), interleukin-6 (IL-6), interleukin-10 (IL-10), tumor necrosis factor alpha (TNF-α), and interferons (IFN) from many cells and tissues [16] that can be analyzed by ELISA protocol by the researcher, in several tissues, such as adipose and skeletal muscle tissue.

Disclosure

The authors have no conflicts of interest to declare that are relevant to the content of this chapter.

References

1. Franco NH (2013) Animal experiments in biomedical research: a historical perspective. Animals (Basel) 3:238–273. https://doi.org/10.3390/ani3010238
2. Nunes-Silva A, Bernardes PT, Rezende BM, Lopes F, Gomes EC, Marques PE, Lima PMA, Coimbra CC, Menezes GB, Teixeira MM, Pinho V (2014) Treadmill exercise induces neutrophil recruitment into muscle tissue in a reactive oxygen species-dependent manner. Intra Microscop Stud PLoS One 9: e96464. https://doi.org/10.1371/journal.pone.0096464
3. Lacerda DR, Moraes MM, Nunes-Silva A, Costa KA, Rodrigues DF, Sabino JFL, Cordeiro LMS, Pinho V, Teixeira MM, Wanner SP, Soares DD, Ferreira AVM (2018) Aerobic training reduces immune cell recruitment and cytokine levels in adipose tissue in obese mice. Appl Physiol Nutr Metabol 1:1–9. https://doi.org/10.1139/apnm-2018-0523
4. Lacerda DR, Nunes-Silva A, Silveira ALM, Costa KA, Rodrigues DF, Moraes MM, Pinho V, Menezes GB, Teixeira MM, Wanner SP, Soares DD, Ferreira AVM (2023) Acute exercise modulates the inflammatory response in adipose tissue of lean and obese mice. Nutrition (Accepted) 115:112092
5. Fiuza-Luces C, Garatachea N, Berger NA, Lúcia A (2013) Exercise is the real polypill. Physiology (Bethesda) 28(5):330–358. https://doi.org/10.1152/physiol.00019.2013
6. de Souza Cordeiro LM, Mario EG, Moreira CCL, Rodrigues AH, Wanner SP, Soares DD, Boção LM, Ferreira AVM (2019) Aerobic training induces differential expression of genes involved in lipid metabolism in skeletal muscle and white adipose tissues. J Cell Biochem 120:18883–18893. https://doi.org/10.1002/jcb.29208
7. Allen DL, Harrison BC, Massa A, Bell ML, Byrnes WC, Leinwand LA (1985) Cardiac and skeletal muscle adaptations to voluntary wheel running in the mouse. J Appl Physiol 90(5):1900–1908. https://doi.org/10.1152/jappl.2001.90.5.1900
8. Bittencourt MA, Wanner SP, Kunstetter AC, Barbosa NHS, Walker PCL, Andrade PVR, Turnes T, Guglielmo LGA (2020) Comparative effects of two heat acclimation protocols consisting of high-intensity interval training in the heat on aerobic performance and thermoregulatory responses in exercising rats. PLoS One 15(2):e0229335. https://doi.org/10.1371/journal.pone.0229335
9. McKie GL, Medak KD, Knuth CM, Shamshoum H, Townsend LK, Peppler WT, Wright DC (2019) Housing temperature affects the acute and chronic metabolic adaptations to exercise in mice. J Physiol 597:4581–4600. https://doi.org/10.1113/JP278221
10. Nieman DC, Wentz LM (2019) The compelling link between physical activity and the body's defense system. J Sport Health Sci 8(3):201–217. https://doi.org/10.1016/j.jshs.2018.09.009
11. Peake JM, Neubauer O, Walsh NP, Simpson RJ (2017) Recovery of the immune system after exercise. J Appl Physiol 122:1077–1087. https://doi.org/10.1152/japplphysiol.00622.2016
12. Marques PE, Oliveira AG, Amaral SS, Nunes-Silva A, Almeida AFS, Reis AC, Braga AD, Perez DA, Alvarenga DM, Assis MHS, Procópio MS, Batista NV, Lima OCO, Bernardes PTT, Bilheiro RP, Menezes GB (2012) Intravital microscopy: taking a close look inside the living organisms. Afr J Microbiol Res 6(7): 1603–1614. https://doi.org/10.5897/AJMR11.1645
13. Tavares-Murta BM, Zaparoli M, Ferreira RB, Silva-Vergara ML, Oliveira CH, Murta EFC, Ferreira SH, Cunha FQ (2002) Failure of neutrophil chemotactic function in septic patients. Crit Care Med 30:1056–1061. https://doi.org/10.1097/00003246-200205000-00017
14. Rios-Santos F, Alves-Filho JC, Souto FO, Spiller F, Freitas A, Lotufo CM, Soares MB, Dos Santos RR, Teixeira MM, Cunha FQ (2007) Down-regulation of CXCR2 on neutrophils in severe sepsis is mediated by inducible nitric oxide synthase-derived nitric oxide. Am J Respir Crit Care Med 175:490–497.

https://doi.org/10.1164/rccm.200601-103OC

15. Chow LS, Gerszten RE, Taylor JM, Pedersen BK, van Praag H, Trappe S, Febbraio MA, Galis ZS, Gao Y, Haus JM, Lanza IR, Lavie CJ, Lee CH, Lúcia A, Moro C, Pandey A, Robbins JM, Stanford KI, Thackray AE, Villeda S, Watt MJ, Xia A, Zierath JR, Goodpaster BH, Snyder MP (2022) Exerkines in health, resilience, and disease. Nat Rev Endocrinol 18(5):273–289. https://doi.org/10.1038/s41574-022-00641-2
16. Suzuki K (2018) Cytokine response to exercise and its modulation. Antioxidants 7(1):17. https://doi.org/10.3390/antiox7010017

Chapter 12

Humanized Murine Model with Peripheral Blood Mononuclear Cells as a Strategy to Translate Experimental Biomedical Research

Matheus R. Gonçalves, Victor R. M. Costa, Jennifer R. Martins, Viviane L. Batista, Jéssica A. B. Pereira, Fernanda L. Tana, Walison Nunes Silva, Pedro P. G. Guimarães, Barbara Maximino Rezende, Vanessa Pinho, Mauro Martins Teixeira, Jordana G. A. Coelho-dos-Reis, Celso Martins Queiroz-Junior, and Vivian V. Costa

Abstract

One of the main limitations to developing new therapies for human diseases and understanding their pathogenicity mechanisms is difficulty to translate the results obtained in animal models to the clinic. Factors such as the existence of pathogens with unique tropism for human cells and the evolutionary differences between human and animal models often limit the effectiveness of the studies. On the other hand, animal models have been widely used to investigate disease mechanisms due to their shorter lifespans, small size, and easy maintenance. Many advances in the understanding of physiology and genetic architecture enable the development of new models and strategies to overcome species limitations. In this context, humanized mice have emerged as a powerful tool for the investigation of human diseases. Animal models without functional immune system allow the engraftment of human cells and tissues to study human infectious pathogens, tissue transplantation, as well as the human immune system function and ontogeny. Although this model can be limited by the onset of graft-versus-host disease (GVHD), it allows the investigation of this specific disease. Here in this chapter, we describe an experimental platform of peripheral blood mononuclear cell (PBMC) humanized mice and methods related to the model.

Key words Human diseases, Animal model, Humanized mice, PBMC

1 Introduction

Most of the investigations to understand the complex biological processes behind human diseases, as well as their treatment, require prototypes in vitro and in vivo [1]. Important advances in scientific research have been carried out using animal models, such as mice,

Lucas Secchim Ribeiro (ed.), *Models in Inflammation and Immunopharmacology*, Methods in Molecular Biology, vol. 2993, https://doi.org/10.1007/978-1-0716-5019-6_12,
© The Author(s), under exclusive license to Springer Science+Business Media, LLC, part of Springer Nature 2026

as valuable tools to understand human biology [2]. Nevertheless, some factors limit the use of animal models to study human diseases, such as: (i) evolutionary differences between humans and mice; (ii) differences in innate immune system molecules that are present in mice and not in humans [3], and (iii) some pathogens (i.e., denguevirus, HIV, Epstein-Barr virus, influenza virus, ebola virus, Mycobacterium, among others) that exhibit unique human tropism, not being able to infect mouse cells [4].

In this context, "humanized" mice have emerged as powerful tools in the investigation of human diseases [5–7]. Immunodeficient mice carrying mutations such as the common gamma chain of the IL2 receptor (IL-2rγ^{null}) lack adaptive immunity and exhibit severe innate immunity deficiencies [8]. There are three lines of immunodeficient mice most used for grafting cells of the human immune system: (i) NOD/Shi-*scid* IL-2Rγ^{null} (NOG) mice have a truncated cytoplasmic domain of the gamma chain that binds to cytokines but does not have the signaling domain, while (ii) NOD/LtSz-*scid* IL-2Rγc^{null} (NSG) and (iii) (BALB/c-Rag2$^{-/-}$/IL-2R$\gamma^{-/-}$ (BRG) mice do not fully possess the gamma chain [7, 9]. These mice lack functional B, T, and natural killer (NK) cells, and therefore, can be grafted with human cells.

One approach for mice humanization is the grafting of human PBMC [10–13] with or without previous irradiation to generate robust functional human CD4^{+} and CD8^{+} T cells [9, 14]. This model has been used to study human infectious agents, tissue transplantation, and immune functions of human T cells, as lymphocytes are the predominant cell in the graft [8, 14]. Another model very well-characterized, known as Hu-SRC-SCID, includes intravenous or intrafemoral injection of human CD34^{+} cells, derived from bone marrow, umbilical cord blood, or fetal liver. This model supports a complete graft of the human immune system [15].

Despite the benefit of studying human immune cells in mice, the humanized mouse models are limited by the relatively short time window for experimental observation, as all grafted mice will develop lethal GVHD within a few weeks. This can limit the study of infectious agents but allows the investigation of GVHD pathogenesis [9, 14, 16–18].

These humanized mouse models provide an opportunity to study species-specific agents that require human tissues for infection and replication and allow the study of human immune response development. In this chapter, we describe the humanization model of mice using PBMC as a strategy to translate experimental biomedical research. Table 1 presents the different mouse applications of humanized models, comparing their benefits and limitations.

Table 1
Application of humanized mouse models

Mouse strain I component Human	Applications I Benefits	Limitations	References
Hu-PBL-SCID injection of human peripheral blood leukocytes (PBMCs).	Rapid grafting of human $CD3^+$ T cells at the end of the first week. Good model for studying the function of human T cells in vivo.	Short experimental window due to the development of lethal xenogenous graft versus host disease (GVHD), usually within 4 to 8 weeks.	[14]
Hu-SRC-SCID | injection of intravenous human-derived $CD34^+$ cells from bone marrow (BM), umbilical cord blood (UBC), fetal liver, or peripheral blood mobilized with G-CSF.	Used for the study of the most complete human immune system, the host rarely produces immune rejection, thus preserving stabilization for up to 12 weeks.	T cells restricted to H2; long period of cell differentiation; indispensable sublethal radiation	[15, 21]
Hu-BLT-SCID | injection of cells of bone marrow, liver, and thymus.	More fully functional immune system, HLA-restricted T cells; development of the human mucosal immune system.	Human T cells are HLA-restricted. Greater possibility of GVHD; sophisticated technique and necessary material	[7]

2 Materials

The use of personal protective equipments (PPEs, e.g., lab coat, disposable gloves, safety glasses) is more than essential for the success of this protocol, given the increased sensitivity to collateral infections of the animals in this model.

2.1 Animal Work

1. NSG immunodeficient mice.
2. Irradiation device.
3. 70 mg/L Ciprofloxacin, in the drinking water.
4. Autoclaved microisolator cage and bedding.
5. Sterile laminar flow hood.

2.2 Human PBMC Isolation

1. Sterile laminar flow hood.
2. Centrifuge with conical plastic centrifuge tube rotor.
3. Cell counting device (e.g., Neubauer chamber).
4. Micropipettes and sterile tips.
5. Pasteur pipette.
6. 15 or 50 mL conical sterile plastic centrifuge tubes.

7. 0.5 mL centrifuge tubes.
8. Cryotubes.
9. K_2 or K_3 EDTA blood collection tubes.
10. 21G needle for human blood collection.
11. Density gradient media (e.g., Ficoll® Paque Plus).
12. Sterile phosphate-buffered saline 1X (1X PBS): 137 mM NaCl, 2.7 mM KCl, 10 mM. Na_2HPO_4, 1.8 mM KH_2PO_4 (pH 7.4).
13. Fetal bovine serum (FBS).
14. Dulbecco's Modified Eagle Medium (DMEM). Complete media can be obtained by supplementation with 10% FBS.
15. Dimethyl sulfoxide (DMSO).
16. Antibiotics (e.g., 90 μM gentamicin and 0.5 μM amphotericin B).
17. Cell culture flasks, petri dishes, or plates for cell culture.
18. 0.4% Trypan Blue diluted in 1X PBS (w/v).

2.3 Cell Inoculation in Mice

1. Laminar flow hood.
2. NSG mice.
3. Heat source (e.g., 250 W lamp).
4. Mouse restrainer for tail vein injection.
5. 1 mL syringes.
6. 30G/26G needle.
7. Previously obtained PBMC suspension.

2.4 Clinical Score (GVHD)

1. Laminar flow hood.
2. 30 × 30 cm delimited surface or mice microisolator cage.
3. Precision scale.
4. Sterile recipient for body weight measurements.
5. Score table (*see* Table 2).
6. Pen/pencil for notes.

2.5 Blood Collection from the Submandibular Vein

1. Sterile laminar flow hood.
2. 2 mL K_2 or K_3 EDTA blood collection tube or 1.5 mL microcentrifuge tube containing EDTA solution at a concentration of 1 g/mL of blood.
3. 16G or 18G needle or number 11 scalpel or lancet.
4. Gauze or cotton pad.
5. Craft paper to cover the sterile laminar flow hood surface.
6. 70% ethanol.

Table 2
Clinical parameters to be evaluated for the clinical score

Parameters to be evaluated	Characteristic of the parameter			
Stance (Score)	Absent (0 points)	Mild (0.5 points)	Moderate (1 point)	Severe/Inclined (2 points)
Ruffled fur	No change		Bristly	
Skin integrity	None	Moderate	Maximum peeling	
Lethargy	No change	Mild lethargy	Moderate lethargy	Maximum lethargy
Diarrhea	Not present	Mild/soft	Liquid	
Weight	Register every 2 days			
Δ weight	Up to 10%	Between 10–20%	Larger than 20%	

2.6 Flow Cytometry

1. Laminar flow hood.
2. 96-deep-well plates.
3. 15-mL centrifuge tubes.
4. Tube shaker.
5. Aluminum foil.
6. Refrigerated (4 °C) centrifuge with plate rotor.
7. Flow cytometer and analysis software.
8. 200–300 μL of whole mouse blood, freshly collected via the facial vein in EDTA-treated tubes.
9. FACS buffer: 1X Phosphate-Buffered Saline (PBS): 137 mM NaCl, 2.7 mM KCl, 10 mM Na_2HPO_4, 1.8 mM KH_2PO_4 (pH 7.4) + 2% (v/v) FBS.
10. Fluorochrome-labeled antibodies: antibodies to murine CD45 and human CD45.
11. Fc Blocker (anti-CD16/CD32). Dilution and usage should be verified according to the manufacturer's instructions.
12. 1X red blood cell–lysing buffer. Alternatively, Milli-Q® cold water and 10X Phosphate-Buffered Saline (PBS): 1.370 mM NaCl, 27 mM KCl, 100 mM. Na_2HPO_4, 18 mM KH_2PO_4 (pH 7.4).
13. 2% paraformaldehyde (PFA) diluted in 1X PBS (w/v).

2.7 Immunohistochemistry

1. Humid chamber (a sealed container that maintains high humidity and allows horizontal placement of slides to prevent reagent evaporation during antibody incubation).
2. Drying incubator.
3. Water bath.
4. Paper towel.
5. Hydrophobic pen (e.g., Pap Pen Hydrophobic Pen) or another hydrophobic material, such as paraffin wax.
6. 70%, 80%, 90%, 100% Ethanol.
7. Xylene.
8. Acid or basic antigen retrieval buffer (*see* **Note 1**).
9. Rinse solution: Tris-Buffered Saline (TBS) 1X: 200 mM Tris and 1500 mM NaCl (pH 7.6) or 1X Phosphate-Buffered Saline (PBS): 137 mM NaCl, 2.7 mM KCl, 10 mM. Na_2HPO_4, 1.8 mM KH_2PO_4 (pH 7.4) + Triton X-100 0.1%.
10. Endogenous peroxidase block solution: Hydrogen peroxide diluted in TBS 1X or rinse solution at a final concentration of 3%.
11. Protein block buffer: 1% BSA (w/v) in 1X PBS.
12. 0.1% BSA in 1X PBS (w/v).
13. Primary antibody: anti-hCD45 suitable for immunohistochemistry.
14. Biotinylated secondary antibody (against the host species of the primary antibody) (*see* **Note 2**).
15. Streptavidin-peroxidase (e.g., VECTASTAIN® ABC-HRP kit) (*see* **Note 2**).
16. Harris' hematoxylin.
17. 3,3′-Diaminobenzidine (DAB) chromogen substrate for immunohistochemistry staining.
18. Ready-to-use water-free mounting medium, e.g., Entellan™.

3 Methods

3.1 NSG Mice

1. Immunodeficient NOD/SCID/IL-2rγ^{null} (NSG) mice should be used to carry out the experiments, as they present complete null mutation of the common cytokine gamma chain receptor gene (IL-2Rγc) *(22)*. The protocol to work with these animals must be previously approved by the local Ethics committee.
2. NSG mice must be older than 4 weeks or young adults up to 8 weeks old, of either sex. To facilitate the grafting, mice must be irradiated using suitable devices, with 2 doses: 4.5 + 4.5 cGy

(source ^{60}Co), within an interval of 3 h. When irradiated, they are treated with ciprofloxacin (70 mg/L) diluted in their drinking water for 1 day before to 15 days after depletion by irradiation;

3. Due to immunodeficiency, mice are more likely to have a large number of infections. With this, it is necessary to be kept in a sterile microisolator cage, with sterile water, feed and bedding. The handling of these animals should be carried out in sterile biosafety laminar flow hood, with the use of sterile PPEs.

3.2 Human PBMC Isolation

1. 30 mL of human blood must be collected in EDTA collection tubes and can be kept at room temperature for up to 1 h. This procedure must be performed by a trained professional (*see* **Note 3**).
2. In a sterile laminar flow hood, add 10 mL of density gradient medium, e.g., Ficoll® Paque Plus in 50-mL conical plastic tubes. Alternatively, add 4 mL of the density gradient medium in 15-mL conical plastic tubes.
3. Transfer 30 mL of blood from the blood collection tube to 50-mL tube with Ficoll® Paque Plus. Alternatively, transfer 10 mL of blood from the blood collection tube to 15-mL tube. Using a Pasteur pipette, gently and slowly layer the blood on the top of the density gradient medium (Fig. 1a) (*see* **Note 4**).
4. Centrifuge the tube for 30 min at 200 × g at 15 ° C, with the brake and acceleration turned off. Next, carefully remove the tubes from the centrifuge. (*see* **Note 5**). Observe the formation of 4 layers (from top to bottom): plasma, PBMCs, Density Gradient Media, and RBCs (Fig. 1b).
5. Gently harvest the PBMCs by inserting a new Pasteur pipette directly through the upper plasma layer. Remove mononuclear cells at the interface without disturbing the other fractions. Alternatively, you can first remove the upper plasma layer and then collect the cells (*see* **Note 6**). Add the cells over 5 mL of PBS in a new tube.
6. Finally, wash the harvested cells. Add 40 mL of 1X PBS to the tube and centrifuge for 10 min at 400× *g* at 4 °C with maximum brake and acceleration.
7. Discard the supernatant and repeat **step 6**.
8. Discard the supernatant and resuspend the cells in 1 mL of DMEM medium supplemented with 10% FBS;
9. In a 0.5-mL centrifuge tube, dilute the cells in Trypan Blue in a 1:20 ratio. Count the cells in the Neubauer chamber (*see* **Note 7**).

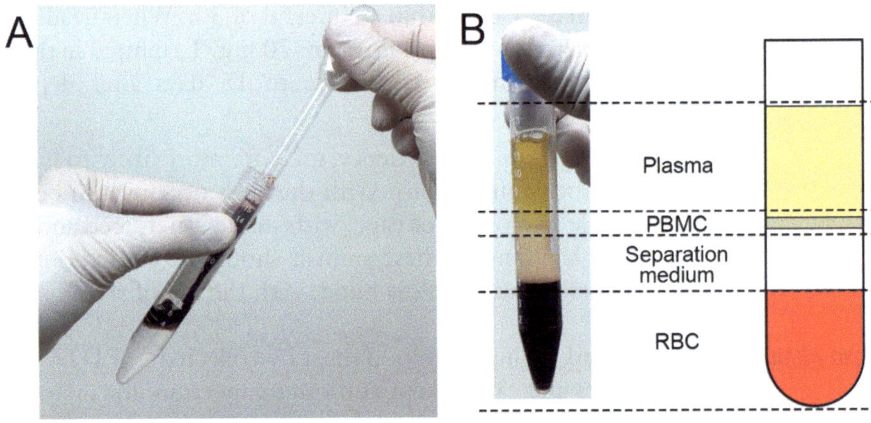

Fig. 1 Layers of density gradient media and whole blood prior to and after centrifugation showing the separation of layers in the Falcon-type tubes. (**a**) To avoid mixing the content of the tube, the Pasteur pipette must be kept as close as possible to the Density Gradient Media. It is recommended to keep tube at an angle of 45 degrees. (**b**) Observe the 4 layers (from top to bottom): plasma, PBMCs, Density Gradient Media, and RBC

10. After isolation, set the cell density for $5x10^6$ cells in 200 μL/ mice.
11. Store cells in ice for optimal conservation and use within a maximum of 2 h. For posterior use, freezing is needed (*see* **Note 8**).
12. For freezing, centrifuge the 10% FBS DMEM medium containing the cells for 10 min at 400 g at 4 °C with the maximum brake and acceleration. Discard the supernatant and resuspend the cells in 2 mL of freezing medium: 90% FBS + 10% DMSO and transfer it to cryotubes. Store the cryotubes at −80 °C overnight and transfer them to the liquid nitrogen for better cell preservation.

3.3 Cell Inoculation in Mice

Cells should be inoculated intravenously and the entire procedure should be performed in sterile laminar flow, with the use of PPEs.

1. Before starting to handle the mice, make sure that all the necessary material is separated (*see* **Note 9**).
2. Mice should be indirectly exposed to the heating source for 2 to 3 min for dilation of the blood vessels of the tail (Fig. 2a) (*see* **Note 10**).
3. While the animal is exposed to the heating source, load the syringe with the previously obtained cells.
4. After 2 min, place the mouse in the restrainer to expose the tail for injection (Fig. 2b) (*see* **Note 11**).
5. Observing the anatomy of the animal's tail, the veins are located laterally and dorsally. Observe whether veins are dilated (Fig. 3a, b) (*see* **Note 12**).

6. Ensure an aseptic procedure with a cotton moistened with 70% ethanol. Wait for the ethanol to dry (*see* **Note 13**).
7. Identify the cranial, middle, and caudal portion in the tail vein. Insert the needle in the middle portion of the tail (Fig. 2c) (*see* **Note 14**).
8. For inoculation, 1 mL syringes and 30G or 26G needles should be positioned with the bevel up and parallel to the tail plane, and the maximum inoculation volume must be 0.2 mL.
9. After needle insertion, gently press the plunger. If the needle bevel is indeed inside the vein, you will not find any resistance. However, if the needle is not inside the vein, the injection will not be possible. In this situation, remove the needle and insert it to the next possible anterior position of the previous site of injection (*see* **Note 15**).
10. After the injection, gently remove the needle and apply a cotton pad on the injection site.
11. Return the mice to the microisolator cage.

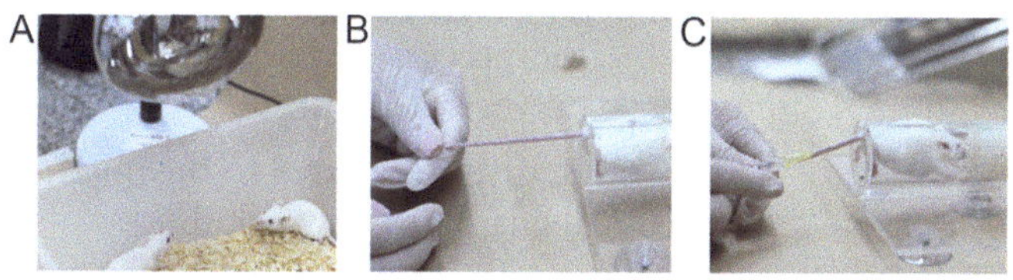

Fig. 2 Mice manipulation for intravenous injection. (**a**) Mice are exposed to a heating source for vein dilatation. (**b**) Exposure of the tail for caudal blood vessels location. (**c**) Intravenous injection is performed in the middle portion of the tail

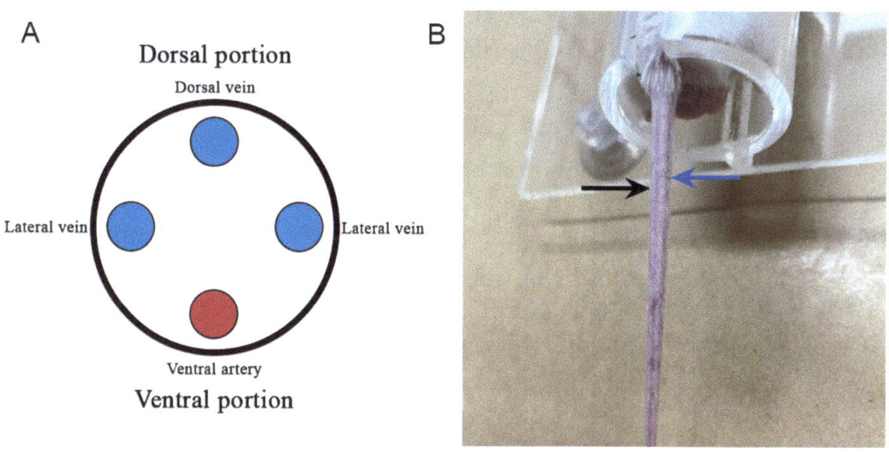

Fig. 3 Blood vessels in mice tail. (**a**) Schematic representation of the tail of mouse in transversal section. The animal has one ventral caudal artery, one dorsal caudal vein, and two lateral caudal veins. (**b**) A picture of a mouse tail showing the dorsal caudal vein (black arrow) and the right lateral caudal vein (blue arrow)

3.4 Clinical Score to Investigation of GVHD Development

The mice used for the humanized model study, such as NSG mice, are susceptible to the GVHD development [19], a condition characterized by a xenoreactivity of human cells to MHC molecules of the host, and consequently, may present altered parameters relevant to the determination of a clinical score according to Table 2. The animals are evaluated every day according to each parameter that receives scores as follows: a score of 0 (no alteration); 0.5 (mild/up to 9% of body weight loss); 1 (moderate/soft/between 10–19% of body weight loss); 2 (severe/inclined/maximum peeling/liquid/intense/greater or equal to 20% of body weight loss).

1. To determine the clinical score of the animals, the mice must be evaluated in a sterile laminar flow hood.
2. Prepare the necessary materials previously to handling the animal within the laminar flow, including the PPEs.
3. Open the microisolator cage and remove the mouse to start evaluating the score.
4. Visually evaluate each parameter of the first animal, according to the parameters described in Table 2 (*see* **Note 16**).
5. After that, weight the animal in a proper and sterile recipient on the precision scale. Write down any unexpected observations deemed relevant.
6. Return the animal to the microisolator cage and remove it from the laminar flow hood.
7. Repeat the **topics 3–6** for each animal.

3.5 Blood Collection from the Submandibular Vein

This procedure allows the obtention of a small amount of blood keeping the animal alive. This method is rapid and easy. Anesthesia is not required. The blood is obtained from the veins located in the mouse cheek [20] (*see* **Note 17**).

1. In a laminar flow hood, separate the needed materials.
2. Open the top of the microisolator cage and take an animal for blood collection.
3. Immobilize the mouse, inhibiting any head movement, exposing the animal's cheek (Fig. 4). With a gauze or cotton and 70% ethanol, clean the mouse's cheek and wait until it is dry.
4. Observing the mouse anatomy, localize a dark mark on the animal's cheek. The puncture site (submandibular vascular bundle) is located above the dark mark (Fig. 4) (*see* **Note 18**).
5. Puncture the vein with the needle or scalpel and quickly position a tube with anticoagulant for collection (*see* **Note 19**).
6. Collect the blood, close the tube, and apply pressure with the gauze or cotton on the animal's cheek to stop the bleeding (*see* **Note 20**).

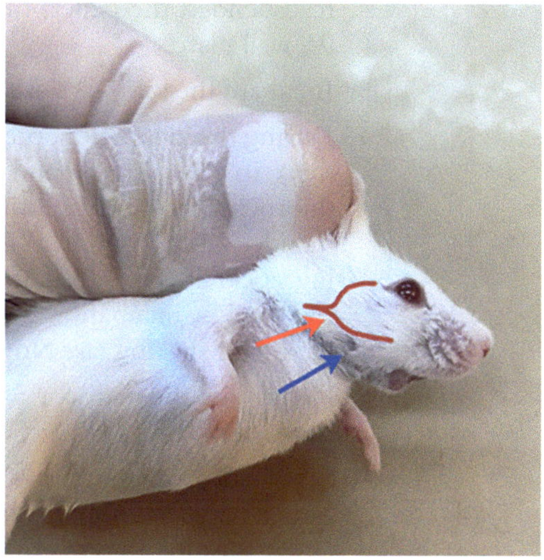

Fig. 4 Indication of the correct site to collect blood from a mouse cheek according to the veins in the animal's cheek, represented by the red lines. The blue arrow indicates the mark present on the mouse cheek and the red arrow indicates the correct site of puncture

7. Once the bleeding has stopped, return the animal to its previous conditions and continue the experiment.
8. Store the collected blood in ice.

3.6 Flow Cytometry

This step provides an important and detailed procedure to assess the levels of chimerism engraftment in immunodeficient mice implanted with PBMCs. The "humanized index" is the evaluation of the percentage of human cells in the total circulating leucocyte population of the mouse. This procedure must be performed once a week, initiating one week after the inoculation of the PBMC. Percent chimerism is calculated by flow cytometric evaluation of human pan-leukocyte marker (hCD45) levels in PBMC compared to total live cells, as follows.

PBMC chimerism (%) = (human $CD45^+$/murine $CD45^+$) × 100.

1. In a laminar flow hood, add 100 μL of whole mouse blood to a 15-mL conical centrifuge tube. Add 1.25 mL of 1X Red Blood Cell Lysis Buffer per sample. Incubate samples at room temperature in a tube shaker for 20 min, in the dark. Alternatively, the red blood cells can be lysed by adding 9 mL of freezing-cold Milli-Q® water and 1 mL of 10X PBS 30 s after the water (*see* **Note 21**);
2. Centrifuge the tubes for 10 min at 300× *g* at 4 °C and discard the supernatant by inversion. Add 100 μL of FACS buffer and

resuspended the pellet, then transfer the cell suspension to a 96-well plate.

3. Incubate the samples with Human Fc Block (*see* **Note 22**).
4. During the incubation, prepare a flow cytometry antibody stain mix in 2 mL round-bottom centrifuge tubes using fluorochrome-labeled antibodies at the manufacturer's recommended concentration in FACS buffer. Add the antibody staining cocktail (hCD45 + mCD45 to all whole blood samples, followed by incubation for 20 min at room temperature in the dark with shaking (*see* **Note 23**);
5. Centrifuge the plate for 5 min at 400× *g* at 4 °C, and discard the supernatant.
6. Wash off excess antibody buffer by adding 200 μL of FACS buffer per sample, centrifuging for 5 min at 400× *g*, at 4 °C, and discarding the supernatant. After the last wash, discard the supernatant and resuspend each cell pellet in a final total volume of 150 uL FACS Buffer or 2% PFA for posterior flow cytometry.
7. Acquire samples using a flow cytometer, ensuring that at least 20,000 cell events are collected.
8. Samples can be placed in plates or tubes, depending on the flow cytometer used.
9. Perform flow cytometry data analysis with flow cytometry analysis software (Fig. 5).
10. To monitor the humanization index, flow cytometry should be performed weekly until the humanization rate reaches 20–40%.

3.7 Immunohistochemistry and Immunofluorescence

Immunohistochemistry and immunofluorescence are ideal methods to evaluate cell chimerism. In this model, the localization of human leukocytes in the liver is an important approach to be analyzed. To that end, liver tissues can be obtained after mouse euthanasia and processed for routine histologic methods. Paraffinized sections of 5 μm thickness are used for immunohistochemistry procedures.

1. Before deparaffinization, slides must be incubated at 55 °C for 30 min (*see* **Note 24**).
2. Deparaffinize the slides in 3 rounds of xylene incubation: 30 min, 15 min, and 5 min in each solution, respectively.
3. The slides must be hydrated in 3 rounds of 100% ethanol, for 5 min each, and 90%, 80%, and 70% ethanol, respectively, for 5 min each.
4. Rinse the slides in deionized water for 10 min.
5. The antigen retrieval must be performed in acid or basic solution (*see* **Note 25**).

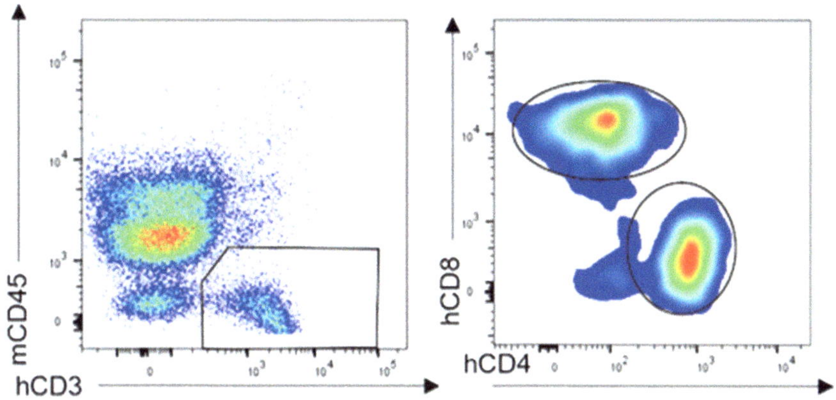

Fig. 5 Gating strategy for T-cell subset identification by flow cytometry. A representative flow cytometry gating strategy for analyzing T-cell subsets is shown. Cells were first gated on total live cells and then on mCD45$^+$ and hCD3$^+$ lymphocytes PBMC. Next, bivariate plots for hCD4$^+$hCD8$^+$ cells were generated from the hCD3$^+$ cell population

6. Arrange the slides in a container with the antigen retrieval buffer and keep the slides in it for 20 min at 96 °C. After that, the slides must be cooled to room temperature for 20 min.
7. Rinse the slides in 5 rounds of deionized water for 3 min each.
8. While the slides are rinsed, prepare the endogenous peroxidase block solution.
9. Submerge the slides in 2 rounds of the endogenous peroxidase block solution, 15 min each.
10. Rinse the slides in 3 rounds of TBS 1X or rinse solution, 5 min each.
11. Using a paper towel, dry the area around the tissue (Fig. 6A) (*see* **Note 26**).
12. Contour the slices using a hydrophobic pen or hydrophobic substance, such as paraffin (Fig. 6B).
13. From this step onwards the solutions are applied directly to the sections.
14. Block non-specific bindings with protein block buffer for 5 min. Alternatively, PBS 1X + BSA 1% for 1 h can be used (*see* **Note 27**). 20 to 50 μL is sufficient to cover the sections (*see* **Note 28**).
15. Drain the protein block solution and apply the primary antibody.
16. Dilute the primary antibody (anti-hCD45) in PBS 1X BSA 0,1%. Apply 30 to 50 μL of the primary antibody to the

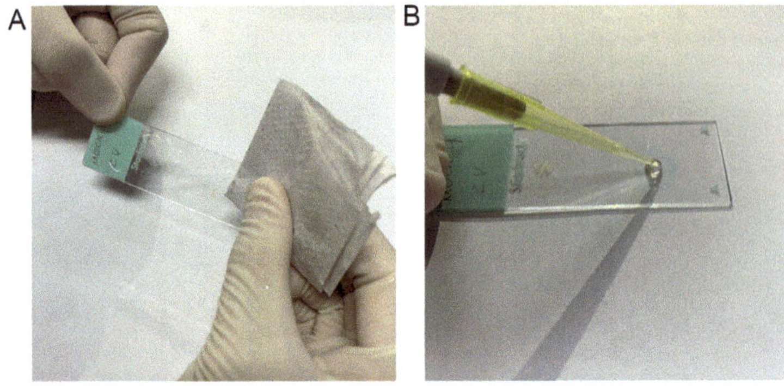

Fig. 6 Procedures for the histological mounting. (**a**) The correct form of drying the area around the sections with a paper towel. (**b**) The correct form of applying the reagents around the sections

sections. To the negative controls, apply only PBS 1X BSA 0.1% (*see* **Note 29**).

17. Incubate the slides in a humid chamber at 4°C overnight.
18. On the next day, rinse the slides in 3 rounds of TBS 1X or rinse solution, 5 min each.
19. Dry around the sections using a paper towel.
20. Dilute and apply 30 to 50 μL of the detection antibody on the sections and incubate the slides in a humid chamber for 30 min at room temperature (*see* **Note 30**).
21. Rinse the slides in 3 rounds of rinse solution, 5 min each.
22. Dry around the sections using a paper towel.
23. Apply 30 to 50 μL of streptavidin-peroxidase and incubate the slides in a humid chamber for 30 min at room temperature (*see* **Note 31**).
24. Rinse the slides in 3 rounds of TBS 1X or rinse solution, 5 min each.
25. Dry around the sections using a paper towel.
26. Dilute and apply the DAB chromogen solution. Observe the reaction. In the presence of peroxidase enzyme, DAB produces a brown color that is observed in the sections (*see* **Note 32**).
27. Stop the reaction by submerging the slides in distilled water. Rinse the slides in tap water for 10 min.
28. For counterstaining, rapidly submerging the slides in Harris' hematoxylin.
29. Rinse the slides in running water for 10 min.
30. Dehydrate the samples through 70%, 80%, and 90% ethanol for 1 min each and 3 rounds of 100% ethanol for 3 min.
31. Let the slides dry at room temperature.

32. The samples are cleared in 3 rounds of xylene for 5 min each (*see* **Note 33**).
33. Apply 3 drops of water-free mounting medium on the coverslips and put it on the slides gently.

4 Notes

1. Citrate buffer 0.1 M (acid buffer): Prepare two solutions: A and B. Solution A: 100 mL of citric acid (8 mM). Solution B: 50 mL of sodium citrate (0.01 M). Prepare a mixture of 9 mL of solution A, 41 mL of solution B and 450 mL of deionized water to a final volume of 500 mL (pH 6.0). Keep the buffer at 2 °C to 8 °C for up to 1 month; or EDTA buffer (basic buffer): prepare 300 mL of 0.4 mM of EDTA (pH to 8.0). Keep the buffer at 2 °C to 8 °C for up to 1 month. The antigen retrieval buffer should be selected according to the primary antibody datasheet.
2. The detection system must be compatible with the species in which the primary antibody was raised. In the ABC method, a preformed avidin–biotin–enzyme complex binds to the biotinylated secondary antibody previously applied to the tissue, enabling signal amplification through enzymatic conversion of the DAB chromogen. Alternative detection systems are also available; always refer to the manufacturer's instructions for appropriate use and reagent dilution.
3. Before starting human PBMC cell isolation procedures, ensure all reagents are at room temperature.
4. Take care not to mix the two layers. The Pasteur pipette must be kept as close as possible to the Ficoll® Paque Plus. To avoid mixing the content of the tube, it is recommended to add the blood on the tube wall with the tube at an angle of 45 degrees.
5. The centrifugation with brakes off and the careful removal of the tube from the centrifuge is needed to avoid mixing the layers.
6. To obtain the highest yield of PBMCs, it is recommended to include 1 mL above and 1 mL below the PBMC layer, which contains a small amount of plasma (above) or density gradient media (below). This step may yield 3–7 mL of fluid.
7. Cell viability is an essential parameter to ensure correct PBMC concentration. The manual counting of cells in Trypan Blue will show their characteristic biconcave feature. To measure live/dead cell concentration and viability using Trypan Blue, consider only cells that are not stained blue and remain translucent.

8. Ideally, the cells should be used immediately after the isolation. However, it can be stored for up to 6 months in liquid nitrogen when immediate use is not possible.
9. To avoid further stress to the animal, mice should be manipulated for the necessary and minimal time only.
10. Caution must be exerted since long expositions are harmful to the animal. The lamp must be positioned at a safe distance from the animals (25 cm distance). Inappropriate use of the heating source may lead the animal to death due to overheating. Alternatively, the animal's tail can be briefly soaked in warm water (45 °C).
11. To avoid further stress to the animal, it should be contained for the time necessary for inoculation only and then returned to the microisolator cage. To avoid accidents, the animal cannot move the tail during the inoculation procedure.
12. Vein dilation is characterized by an increase in its volume and by a characteristic pink color. Always compare the tail before and after heating source exposure to verify the difference in vein color and caliber.
13. This process also facilitates the veins visualization.
14. Inoculating in the caudal portion of the tail makes it possible for a new inoculation. In case of a missed inoculation, try right above the previous one.
15. The inoculation must be performed gently and slowly.
16. The clinical parameters must be evaluated by comparing the animals to the control group (non-injected cells group).
17. According to the ethical guidelines, the mouse must be euthanized if its body weight loss exceeds 20%. When needed, euthanasia is performed according to ethical guidelines. Anesthetic overdose (ketamine [80 mg/kg] + xylazine [10 mg/kg], i.p.) followed by cervical dislocation is the recommended method for ethical euthanasia.
18. Use a piece of paper to cover the surface of the laminar flow hood, since some blood may drop on its surface during the bleeding procedure.
19. To execute this procedure properly, the researcher needs experience. Punctures in the wrong position may lead to the submandibular salivary gland or ear canal perforation.
20. Due to anatomical variation, more than one attempt may be needed to collect the blood. It is safe to take up to 300 μL of blood using this method, refrain from taking more since it can lead to the animal's death.
21. The 10X PBS must be added exactly 30 s after the ice-cold Milli-Q® water, otherwise, leukocytes may be lysed.

22. Usually samples are incubated with 2.5 μL Human Fc Block +100 μL FACS Buffer per sample for 10 min at room temperature in the dark. Always check the manufacturer's instructions for correct use and dilution.
23. Antibody concentrations can be optimized through a titration study.
24. The paraffin temperature must be between 56 °C and 65 °C. At lower temperatures, the paraffin may be solid. This process melts the paraffin, making the deparaffinization easier. At higher temperatures, the tissue may be damaged by heat.
25. This procedure is important to expose the antigenic epitopes. The most commonly used antigen retrieval is citrate buffer 0,1 mol/L (pH 6.0), but EDTA buffer (pH 8.0) may also be used. Always check the antibody datasheet.
26. The area must be completely dry to apply the hydrophobic pen or hydrophobic substance around the section. The paper towel cannot touch the tissue and the tissue must remain humid.
27. This process avoids the presence of background caused by nonspecific bindings. The unspecific bindings blockage must be done for the correct time, otherwise, the primary antibody might bind to non-specific sites.
28. The amount of liquid must be sufficient to cover the entire section.
29. The concentration of the primary antibody may be found on its datasheet. However, previous test with different concentrations for standardization are highly recommended.
30. The detection antibody is a biotinylated anti-goat IgG, anti-rabbit IgG, or anti-rat IgG depending on the species in which the primary antibody was made. The ideal concentration of the secondary antibody must be checked in the antibody datasheet.
31. To allow the chemical reaction of the reagents to occur, the ABC kit compounds must be diluted 30 min before the application on the sections.
32. To achieve the correct concentration, DAB must be diluted according to the manufacturer's instructions.
33. The slides can stand in xylene as long as it takes for the mounting process.

Funding

This study was funded by the Fundação de Amparo à Pesquisa do Estado de Minas Gerais (FAPEMIG), grant numbers RED-00202–22, APQ-02281–18, APQ-04650-23, APQ-02618-

23, and APQ-04983-24. This work also received support from the National Institute of Science and Technology in Dengue and Host-Microorganism Interaction (INCT Dengue), funded by the Conselho Nacional de Desenvolvimento Científico e Tecnológico (CNPq, Brazil) [grant number 465425/2014–3], and by FAPEMIG [grant number 25036]. Additional support was provided by FINEP—Financiadora de Estudos e Projetos under MCTI/FINEP—MS/SCTIE/DGITIS/CGITS [grant number 6205283BBB28-4F9C-AA65-808FE4450542].

References

1. Justice MJ, Dhillon P (2016) Using the mouse to model human disease: increasing validity and reproducibility. Dis Model Mech 9:101–103. https://doi.org/10.1242/dmm.024547
2. Rosenthal N, Brown S (2007) The mouse ascending: perspectives for human-disease models. Nat Cell Biol 9:993–999. https://doi.org/10.1038/ncb437
3. Zschaler J, Schlorke D, Arnhold J (2014) Differences in innate immune response between man and mouse. Crit Rev Immunol 34:433–454
4. Douam F, Gaska JM, Winer BY et al (2015) Genetic dissection of the host tropism of human-tropic pathogens. Annu Rev Genet 49:21–45. https://doi.org/10.1146/annurev-genet-112414-054823
5. Legrand N, Weijer K, Spits H (2006) Experimental models to study development and function of the human immune system in vivo. J Immunol 176:2053–2058. https://doi.org/10.4049/jimmunol.176.4.2053
6. Theocharides APA, Rongvaux A, Fritsch K et al (2016) Humanized hemato-lymphoid system mice. Haematologica 101:5–19. https://doi.org/10.3324/haematol.2014.115212
7. Shultz LD, Lyons BL, Burzenski LM et al (2005) Human lymphoid and myeloid cell development in NOD/LtSz- scid IL2R γ null mice engrafted with mobilized human Hemopoietic stem cells. J Immunol 174:6477–6489. https://doi.org/10.4049/jimmunol.174.10.6477
8. Racki WJ, Covassin L, Brehm M et al (2010) NOD-scid IL2rγnull mouse model of human skin transplantation and allograft rejection. Transplantation 89:527–536. https://doi.org/10.1097/TP.0b013e3181c90242
9. Ito M, Hiramatsu H, Kobayashi K et al (2002) NOD/SCID/gamma(c)(null) mouse: an excellent recipient mouse model for engraftment of human cells. Blood 100:3175–3182. https://doi.org/10.1182/blood-2001-12-0207
10. Caimari A, Oliver P, Keijer J, Palou A (2010) Peripheral blood mononuclear cells as a model to study the response of energy homeostasis-related genes to acute changes in feeding conditions. Omi A J Integr Biol 14:129–141. https://doi.org/10.1089/omi.2009.0092
11. Riedhammer C, Halbritter D, Weissert R (2014) Peripheral blood mononuclear cells: isolation, freezing, thawing, and. Culture:53–61
12. Ye C, Yang H, Cheng M et al (2020) A rapid, sensitive, and reproducible in vivo PBMC humanized murine model for determining therapeutic-related cytokine release syndrome. FASEB J 34:12963–12975. https://doi.org/10.1096/fj.202001203R
13. Brehm MA, Kenney LL, Wiles MV et al (2019) Lack of acute xenogeneic graft- versus -host disease, but retention of T-cell function following engraftment of human peripheral blood mononuclear cells in NSG mice deficient in MHC class I and II expression. FASEB J 33:3137–3151. https://doi.org/10.1096/fj.201800636R
14. King MA, Covassin L, Brehm MA et al (2009) Human peripheral blood leucocyte non-obese diabetic-severe combined immunodeficiency interleukin-2 receptor gamma chain gene mouse model of xenogeneic graft- versus -host-like disease and the role of host major histocompatibility complex. Clin Exp Immunol 157:104–118. https://doi.org/10.1111/j.1365-2249.2009.03933.x
15. Watanabe Y, Takahashi T, Okajima A et al (2009) The analysis of the functions of human B and T cells in humanized NOD/-shi-scid/γcnull (NOG) mice (hu-HSC NOG mice). Int Immunol 21:843–858. https://doi.org/10.1093/intimm/dxp050

16. Ehx G, Somja J, Warnatz H-J et al (2018) Xenogeneic graft-versus-host disease in humanized NSG and NSG-HLA-A2/HHD mice. Front Immunol 9. https://doi.org/10.3389/fimmu.2018.01943
17. Widmer MB, Bach FH (1972) Allogeneic and xenogeneic response in mixed leukocyte cultures. J Exp Med 135:1204–1208. https://doi.org/10.1084/jem.135.5.1204
18. Freeman GJ, Borriello F, Hodes RJ et al (1993) Murine B7-2, an alternative CTLA4 counter-receptor that costimulates T cell proliferation and interleukin 2 production. J Exp Med 178:2185–2192. https://doi.org/10.1084/jem.178.6.2185
19. Akkina R (2013) New generation humanized mice for virus research: comparative aspects and future prospects. Virology 435(1):14–28. https://doi.org/10.1016/j.virol.2012.10.007
20. Golde WT, Gollobin P, Rodriguez LL (2005) A rapid, simple, and humane method for submandibular bleeding of mice using a lancet. Lab Anim 34(9):39–43. https://doi.org/10.1038/laban1005-39
21. Coelho-Dos-Reis JGA, Funakoshi R, Huang J, Pereira FV, Iketani S, Tsuji M (2020) Functional human CD141+ dendritic cells in human immune system mice. J Infect Dis 221(2):201–213. https://doi.org/10.1093/infdis/jiz432. PMID: 31647546; PMCID: PMC7457331

Chapter 13

Murine Models of Alphavirus Infection to Study the Immunopathogenesis of Pain

Victor R. M. Costa, Carla Daiane Ferreira de Sousa, Matheus R. Gonçalves, Gabrielly Mattos, William Antonio Gonçalves, Simone de Araújo, Mauro Martins Teixeira, Danielle G. Souza, Celso Martins Queiroz-Junior, and Vivian V. Costa

Abstract

Alphaviruses, such as Chikungunya and Mayaro, are emerging arthropod-borne pathogens that particularly affect developing countries. Alphavirus infection has great clinical relevance and must be identified and treated properly, as the disease caused by these viruses has a characteristic chronic phase with joint pain that can last for several months or even years, significantly impacting the quality of life of the infected subjects. Animal models of arthritis caused by Alphaviruses allow the study of inflammatory response, bone loss, and tissue damage, providing biological information similar to that obtained in humans. In this chapter, we detail protocols for the process of viral propagation and infection in a murine model, as well as monitoring of the clinical score, plantar edema, and hyper-nociception by methods such as Von Frey filaments and hot plate, in an attempt to better understand the clinical manifestations that occur in humans. This chapter also brings other techniques associated with the study of the disease such as viral titration, micro-CT processing to evaluate bone health parameters, as well as the extraction of neurological tissues associated with the disease and evaluation of antiviral antibodies by ELISA.

Key words Alphavirus, Chikungunya virus, Mayaro virus, Pain, Hyper-nociception, Viral propagation

1 Introduction

Alphaviruses are RNA viruses of the *Togaviridae* family. They are emergent viruses transmitted by arthropods [1] and cause diseases in humans associated with persistent arthritis and encephalitis, including the Venezuelan equine encephalitis virus (VEEV), Eastern equine encephalitis virus (EEEV), and Western equine encephalitis virus (WEEV) [2, 3]. Among the major Alphaviruses associated with musculoskeletal disease and persistent joint pain, which will be addressed in this chapter, we must mention the Chikungunya virus (CHIKV) and Mayaro virus (MAYV) [3–5].

Lucas Secchim Ribeiro (ed.), *Models in Inflammation and Immunopharmacology*, Methods in Molecular Biology, vol. 2993, https://doi.org/10.1007/978-1-0716-5019-6_13,
© The Author(s), under exclusive license to Springer Science+Business Media, LLC, part of Springer Nature 2026

The Chikungunya virus was first identified between 1952–53 in Africa and is now considered endemic in many countries in Africa, Asia, and the Americas [6]. The Mayaro virus was first isolated in Trinidad and Tobago in 1954 and has been responsible for outbreaks in Latin and South American countries ever since [7]. The disease caused by these viruses involves an acute phase, which includes symptoms common to many arboviruses, such as headache, fever, and joint pain. However, a common feature of these two viruses is a subacute or chronic phase of the disease, with debilitating joint pain that can last from months to years [8].

Several animal models have been developed to study the pathogenesis of the diseases triggered by Alphaviruses [9]. Our group has created a model that mimics the chronic pain caused by MAYV or CHIKV, as well as protocols for assays which help understanding the pathology of disease associated with these viruses [9–11]. Here we describe an infection model in young immunocompetent C57BL/6, SV129, or BALB/c mice, including analysis of paw edema and nociception assays. For techniques associated with the peripheral nervous system (PNS) and the study of pain, experiments with the dorsal root ganglion (DRG) will also be described. The DRG is an important component of the PNS, being a sensory organ where the pain neuron cell bodies are located [12]. In the study model described here, the lumbar sensory ganglia associated with sciatic nerves (L3, L4, and L5) were chosen as a target for investigation once they are directly related with knee joint pain and site of infection [13].

2 Materials

For the study of the immune pathogenesis of pain induced by alphavirus infections, cell lines and mice susceptible to infection must be used. In this chapter, we will describe methods using mouse strains (BALB/c, C57BL/6, SV129) and cell cultures (Vero, C6/36, primary neurons). The viruses include a clinical isolate of the Asian lineage (ECSA) of Chikungunya virus (strain BHI3762) and ATCC VR-66, strain TR 4675 of Mayaro virus.

2.1 General Cell Culture

1. Laminar flow hood.
2. Micropipettes and sterile tips.
3. 175-cm^2 cell culture flasks.
4. Biochemical Oxygen Demand Incubator with humidified atmosphere.
5. Electronic pipette controller and sterile serological pipette tips.
6. Inverted microscope.

7. Phosphate-buffered solution (PBS): 137 mM NaCl, 2.7 mM KCl, 10 mM Na_2HPO_4, 2 mM KH_2PO_4 (pH 7.4).
8. PBS supplemented with 0.25% Trypsin and 0.02% EDTA.
9. 0.4% Trypan Blue diluted in 0.9% NaCl solution (w/v).
10. Cell counter (e.g., Neubauer chamber).
11. Microcentrifuge.

2.1.1 Vero Cells

1. Dulbecco's modified Eagle's medium (DMEM) culture medium +10% FBS; 1% antibiotics (*see* **Note 1**).
2. Vero cells (ATCC #CCL-81).

2.1.2 C6/36 Cells

1. Leibovitz L15 culture medium with 1.5% of 4-(2-hydroxyethyl)-1-piperazineethanesulfonic acid (HEPES), 1% antibiotics (*see* **Note 1**); 1% L-glutamine; 2% Fetal Bovine Serum; 1% non-essential amino acids.
2. C6/36 clone cells (ATCC).

2.2 Viral Propagation

1. General cell culture equipment (*see* Subheading 2.1).
2. Supplemented L15 culture medium (*see* Subheading 2.1.2).
3. Cryotubes.
4. Alphavirus of choice (CHIKV or MAYV), with a known titer.

2.3 Infection

1. Tubes with previously aliquoted viruses, diluted at 2.10^7 PFU/ml.
2. Tubes with C6/36 supernatant (*see* **Note 2**).
3. 1-mL syringes with 29G–31G needles.
4. Micropipettes and sterile tips.

2.4 Clinical Score

1. 30 × 30 cm delimited paper surface.
2. Precision scale.

2.5 Viral Titration

1. General cell culture equipment (*see* Subheading 2.1).
2. Vero cells (ATCC #CCL-81).
3. Dulbecco's modified Eagle's medium (DMEM) +10% FBS.
4. DMEM culture medium, powder.
5. Flat-bottom RNAse/DNase-free, 12-well tissue culture plate.
6. Sterile mortar and pestle.
7. 0.22 μm filter.
8. 50-mL centrifuge tubes
9. Sterile solution of 3% carboxymethyl cellulose (CMC) (*see* **Note 3**) + 2× (concentrated) DMEM.

10. 10% formaldehyde solution, diluted in PBS (v/v).
11. 0.1% Methyl Violet solution, diluted in PBS (w/v).
12. Pre-weighed tissue samples on sterile microcentrifuge tubes.

2.6 Hyper-nociception

1. A quiet, temperature-controlled, dedicated room.
2. Digital Von Frey algesimeter (INSIGHT Instruments, Ribeirão Preto, SP, Brazil).
3. 0.5 mm^2 polypropylene blunt tip.
4. Elevated metal mesh stand for Von Frey use.

2.7 Hot Plate Test

1. Closed heat plate.
2. Chronometer.

2.8 Dorsal Root Ganglion Analysis

1. Surgical material (*see* **Note 4**).
2. Perfusion peristaltic pump.
3. PBS.
4. 22 G needle.
5. Stereomicroscope.
6. Anesthetic solution (80 mg/kg ketamine +10 mg/kg Xylazine in 0.9% NaCl sterile solution) (*see* **Note 5**).

2.9 Footpad Edema

1. Digital caliper.

2.10 Micro CT-Scan

1. Bone tissue samples.
2. 10% formalin solution, diluted in PBS (v/v).
3. 70% ethanol.
4. Micro-CT scanner.

2.11 Antiviral Immunoglobulin ELISA

1. Alphavirus stock with a known titer.
2. Ultraviolet light source.
3. 0.01% Tween 20 in PBS (v/v).
4. 1% BSA in PBS (w/v).
5. Microplate shaker incubator.
6. Biotinylated Anti-IgG or Anti-IgM antibody.
7. HRP-Conjugated Streptavidin.
8. o-phenylenediamine dihydrochloride (OPD) solution: 4 mg of OPD + 2 μL of H_2O_2 (30% v/v), diluted q.s.p. 10 mL with Citrate Buffer (38.5 mM Citric acid + 154.7 mM Na_2HPO_4, pH 5.0).
9. 1 M H_2SO_4 solution.
10. Spectrophotometer.
11. 96-well flat-bottom plates.

3 Methods

3.1 General Cell Culture

1. On a laminar flow hood, let the vial containing the frozen Vero or C6/36 cells thaw completely.
2. Meanwhile, prepare 175-cm^2 cell culture flasks and the supplemented culture media (DMEM for Vero, Leibovitz's L-15 for C6/36).
3. Once the cell vial is thawed, centrifuge the vial at 600 $\times g$ for 5 min at room temperature, discard the supernatant, resuspend the pellet in the supplemented culture medium for a final volume of 25 mL and transfer its contents to the cell culture flasks gently homogenizing its contents with the help of a pipette.
4. Close the flasks and check on an inverted microscope for cell clumps. If there are any, homogenize the contents with a pipette until no clumps are left.
5. Transfer the flasks to the incubator (Biochemical Oxygen Demand), with a humidified atmosphere (37 °C, 5% CO_2 for Vero; 25 °C, 0% CO_2 for C6/36) and check the flasks on the inverted microscope daily, checking for adherence and confluence.
6. Once the flasks are 80% confluent, discard its culture medium and, with a pipette, wash the cells with sterile 1× PBS to remove any remaining culture medium.
7. Discard the PBS and add the trypsin-EDTA solution. For a 175-cm^2 flask, 5 mL of the solution is recommended.
8. Make sure the solution completely covers the cell monolayer.
9. Return the flask to the incubator for 3 min or until the cells are no longer adhered to the wall of the flask. Gentle tapping the flask usually helps removing still adhered cells.
10. Once all the cells have detached, return the flask to the laminar flow hood and, collect them with a pipette into a 50-ml tube and add 15 mL of medium (DMEM +10% FBS for Vero, Leibovitz's L-15 + 10% FBS for C6/36).
11. Dilute 5 μL of cell suspension on a Trypan Blue solution at 1: 20 dilution, homogenizing the contents of the tube.
12. Count the stained sample on a Neubauer chamber to determine the amount of live cells on a single flask.
13. Once the number of cells and their concentration in the medium have been determined, they can be transferred to new cell culture flasks for colony expansion or transferred to plates for experimentation, according to the requirements and sample size of the experiment.

3.2 Viral Propagation

1. Once the number of cells and concentration is asserted, reserve in ice a frozen vial of virus suspension of known titer.
2. Calculate the MOI and transfer the desired viral load to the flasks (*see* **Note 6**).
3. Return the infected flasks to the incubator and check every 12 h for indirect signs of viral replication such as loss of adherence and formation of plaques.
4. Once the confluence of the cell monolayer has been reduced to around 20%, the viral inoculum is ready for collection.
5. Transfer the viral suspension to a transitory sterile recipient and properly discard the empty flasks.
6. Transfer the virus-rich culture medium to 50-ml centrifuge tubes for centrifugation at 600 ×*g* for 15 min at 15 °C.
7. Once clarified (free of cellular debris), aliquot the supernatant on cryogenic tubes for storage on a −80 °C freezer for later use (*see* **Note 7**).

3.3 Infection

1. Reserve the animals to be infected in groups and prepare the necessary materials (*see* **Note 8**).
2. Prepare the viral inoculum with the required biosafety level preparation for the desired pathogen manipulation. Dilute the viral content on sterile C6/36 supernatant until the desired titer is reached (*see* **Note 8**). Each animal should receive 10^6 PFU in 20 to 50 μL (*see* **Note 9**).
3. Load the diluted virus inoculum to a 1-mL syringe coupled to a 29-31G needle and reserve it.
4. Immobilize the animal and expose its right footpad (*see* **Note 10**).
5. Insert the needle subcutaneously and inject the desired volume on the footpad (*see* **Note 11**) (Fig. 1a, b).
6. Return the mouse to its cage and proceed with the experiment.

3.4 Clinical Score

1. Separate the groups of animals to be evaluated and prepare the necessary materials (*see* **Note 12**).
2. Evaluate the individual mouse behavior according to Table 1 (*see* **Note 12**).
3. Weight the mice on a precision scale.
4. Take notes of the weight and the clinical score.
5. Return the animal to their previous condition.

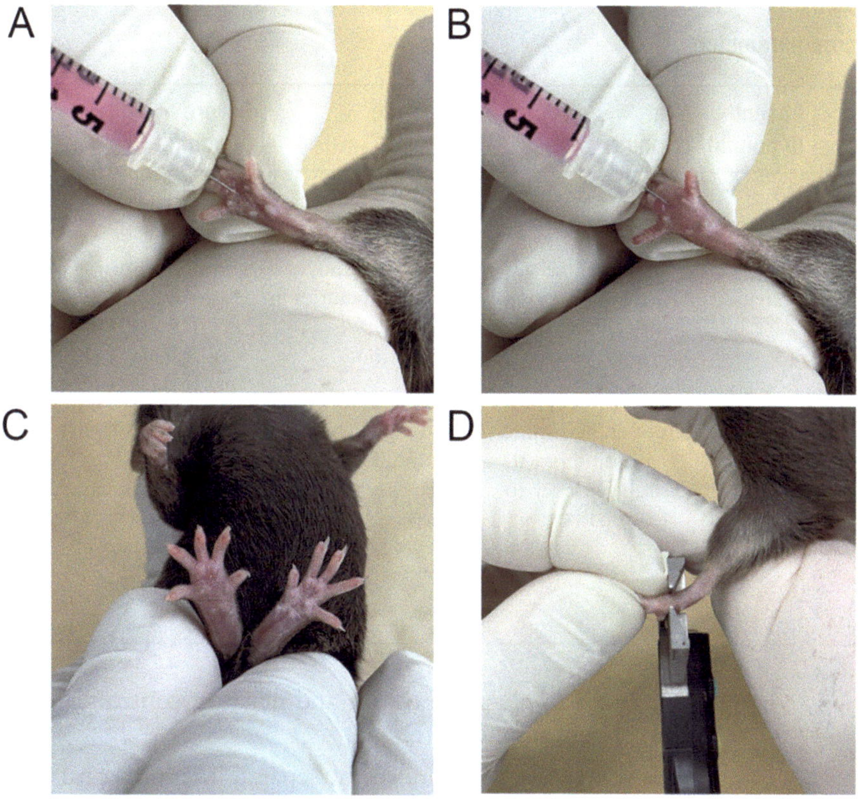

Fig. 1 Intraplantar inoculation and edema measurements. (**a**: Footpad right before inoculation with virus-rich cell culture medium in the correct position for inoculation. **b**: Footpad aspect after inoculation with 50 μL of virus-rich supernatant. **c**: Comparison between a footpad with an edema, left, and one without alteration, right. **d**: General position of the digital caliper for paw edema assessment)

3.5 Viral Titration

1. Perform the infection protocol described in Subheading 3.2, collect the required samples (e.g., blood and tissues) at the desired timepoints (usually at 1, 3, 5, 7, 10, 14, 21, and 28 days post infection) and store them at −80 °C or below. Solid samples (e.g., organs) and their fractions must be added to empty sterile tubes, previously weighed. The weight of the empty tube will be used later to control the weight of the tissue.
2. For viral titration of cell culture supernatants, perform the desired infection protocol, collect the samples by centrifuging the plate (600 ×*g*, 10 min, 4 °C) and transferring the supernatant to a collection tube (usually at 12, 24, 48, 72, and 96 h after infection) and store at −80 °C.
3. Perform the basic cell culture referenced in Subheading 3.1.1. Consider that a 175-cm^2 flask at 80% confluence is enough to seed 5 units of 12-well plates. Determine the number of plates to be used in the experiment by considering using 4 dilutions

Table 1
Clinical score parameters

	Absent (0 points)	Mild (1 point)	Moderate (2 Point)	Severe (3 points)
Ruffled fur	No change	Only face or cheeks	Small areas of the body	Over large areas of the body
Lethargy	No chance	The animal has a little restriction of movement	The animal moves with difficulties in one or two directions	The animal moves with difficulties after being touched in one or two directions. Usually, the animal also presents a severe rushed posture
Rushed posture	No change	Slight arched spine with hind paws slightly closer to front paws	Moderated arched spine. Hind paws are closer to front paws	Severe arched spine with hind paws are closer to front paws. Usually, the animal also presents severe bristling and severe lethargy
Δ weight loss	0–5%	6–10%	11–19%	Up to 20%

per sample plus positive and negative controls, i.e., 3 samples can fit in one plate.

4. Once the flasks are ready, separate them, discard their culture medium and, with a pipette, wash the cells to remove any remaining culture medium with sterile 1× PBS.
5. Discard the 1× PBS and add the trypsin-EDTA solution inside the flask. For a 175-cm^2 flask, 5 mL of the solution is recommended.
6. Return this flask to the incubator for 3 min or until the cells are no longer adhered to the flask's wall.
7. Once the cells are ready, pool the cells in a 50-mL microcentrifuge tube, centrifuge at $600\times g$ for 5 min at room temperature, discard the supernatant and resuspend the pellet in 1 mL of complete culture medium with a pipette. Dilute the resuspended cells on Trypan Blue solution at 1:20 dilution, homogenizing the contents of the tube with a pipette.
8. Count the stained sample on a Neubauer chamber to determine the amount of live cells available.
9. Reserve the desired amount of 12-well tissue culture plates.
10. Further dilute the cell suspension in DMEM until a concentration of 5×10^5 cells/mL.
11. Transfer 1 mL of the diluted cell suspension to each individual well of the desired number of plates to be used in the assay.

12. Transfer the plates to the incubator with humidified atmosphere (*see* **Note 13**).
13. The viral titration can be performed 1 or 2 days after the cell plating: check the plates on the inverted microscope for a confluence of 80%, proper adhesion, and possible contaminations.
14. Fill sterile 1.5-mL tubes with 450 μL of DMEM for later use in the serial dilution step. In general terms, 5 tubes per sample are enough, but the optimal dilution might vary according to the experimental conditions.
15. Take the tubes with the previously weighed samples (**step 1**) and place them on ice.
16. For solid tissues, completely macerate the tissue with sterile mortar and pestle. Add DMEM to the mortar in a proportion of 10% (weight/volume), for a minimum volume of 50 μL (*see* **Note 14**).
17. Perform the serial dilution in a 1:10 ratio by transferring 50 μL of the macerated suspension to the dilution tubes previously filled with 450 μL of DMEM. The number of dilutions needed depends on the expected viral titer in each tissue. For blood and tissues close to the infection site, such as the footpad, ankle, and popliteal lymph node, consider dilutions between 10^{-5} and 10^{-9} at early infection times (1–3 days after infection). For other samples, such as liver, spleen, and brain, among others, consider dilutions between 10^{-2} and 10^{-5} at early infection times. For long-term infection periods, dilutions of 10^{-1} to 10^{-4} should be considered since the infection is normally self-resolving around the seventh day after infection (*see* **Note 15**).
18. Bring the seeded plates into the laminar flow hood, identify the plate mapping for each well, and discard the cell culture medium from each well.
19. Using the tubes with the inoculum serial dilutions, infect the cells: for a 12-well plate, 300 μL is sufficient to cover the cell monolayer on a single well. The medium containing the diluted material must be added slowly on the well by its wall with the minimum amount of tissue debris possible.
20. Incubate the infected cell monolayer for 1 h at 37 °C, gently shaking the plate every 15 min to guarantee an even adsorption of the viral inoculum.
21. After adsorption, carefully discard the supernatant and add 2 mL/well of a sterile solution of DMEM +1.5% carboxymethylcellulose (CMC), made by mixing an autoclaved solution of 3% CMC with a 2× concentrated DMEM preparation (from powder). The CMC helps to stop the virus from infecting other cells in its surroundings.

22. Incubate the plate at 37 °C, 5% CO_2 for 3 days, checking every day on an inverted microscope for cytopathic signs such as plaque formation.
23. After this time, fix the cells by carefully adding, on top of the CMC solution, a solution of 10% formaldehyde (v/v, in PBS) to complete the volume of the well. Incubate the plates at room temperature for at least 3 h for fixation.
24. Next, discard the CMC + formalin and mix add a volume of 0.1% methyl violet solution to completely cover the well and incubate for at least 6 h.
25. Rinse the plates with distilled water and let them dry at room temperature.
26. Assure the negative control wells have no signs of contamination and count the plaques in each well and determine the viral titer by the following formula in Fig. 2.

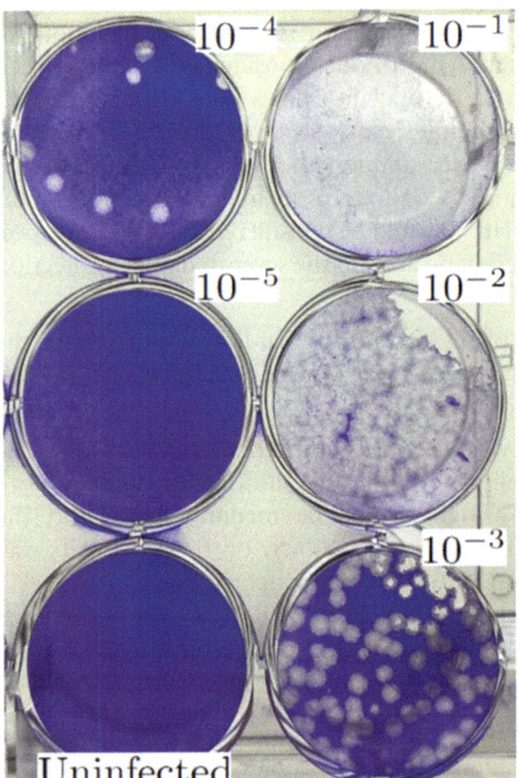

Viral Titration by Plaque Assay

The virus titer is calculated using the following formula:

$$T = \frac{1}{w} \sum_{i=1}^{w} \left(\frac{n_i \cdot d_i^{-1}}{v} \right)$$

where:

- T = virus titer (PFU/ml)
- n_i = number of counted plaques in well i
- d_i = dilution ratio of well i
- v = volume of the sample inoculated in each well (ml)
- w = number of countable wells used for the sample

Example Calculation of Viral Titer

Countable wells: 2
Countable dilution ratios: 10^{-3} and 10^{-4}

Thus,

$$T = \frac{1}{2}\left[\frac{95 \cdot (10^{-3})^{-1}}{0.3} + \frac{8 \cdot (10^{-4})^{-1}}{0.3}\right]$$

$$T = \frac{1}{2}\left[\frac{95000}{0.3} + \frac{80000}{0.3}\right]$$

$$T = \frac{1}{2}(316666.66 + 266666.66)$$

$$T = \frac{1}{2} \cdot 583333.32$$

$$T = 291666.66 \text{ PFU/ml}$$

$$\log_{10} T = 4.4648$$

Fig. 2 Example of a viral titration plate

3.6 Nociception Evaluation

Hypernociception can be assessed using the von Frey apparatus, which contains the animal in an individual confined space with a grid underneath, which allows the positioning of the apparatus's meter under the animal. The device records the amount of force exerted by the experimenter on the animal's paw in order to delimit a threshold for withdrawal of the animal's paw.

1. Bring the animals into the experiment room.
2. Identify each individual box with the ID of the animal to be placed inside.
3. Place the animals on their given boxes.
4. Make sure the experiment room is silent and inaccessible for any other person for the acclimatization period of 1 h (*see* **Note 16**).
5. Once the animals are properly acclimatized, they lose their regular exploratory behavior. Only then the operator may quietly return to the room for the use of the equipment (Fig. 3a).
6. Position the 0.5 mm^2 polypropylene blunt tip on the reading device and if necessary, according to the manual, perform the calibration protocols.
7. Position the apparatus parallel to the mesh with the open blunt tip perpendicular to it.
8. Search, with the aid of the inclined mirror, for the animals infected hind leg.
9. Without changing the general orientation of the apparatus, press with the blunt tip the center of the animal paw in such a way to cause flexion of the knee joint (Fig. 3b, c).
10. This movement should induce an aversive response of paw withdrawal precipitated by mechanical stimuli. If the reflex is present, take note of the measurement on the apparatus panel.

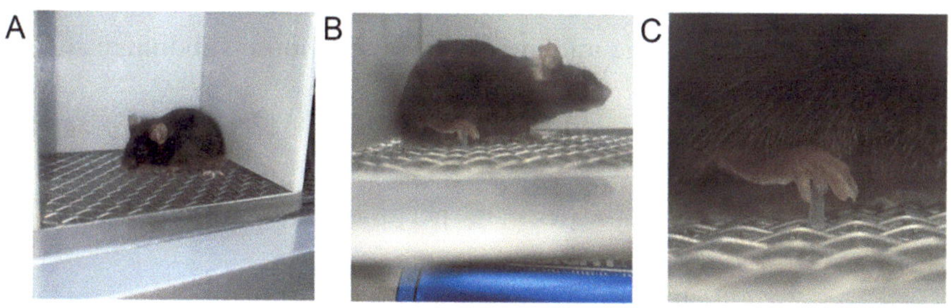

Fig. 3 Nociception assessment by Von Frey's apparatus. (**a**: Example of a position typical of correct acclimatization of the animal. **b**: Correct position of the apparatus on the grid and at the animal's paw. **c**: Zoomed in position of the tip on the animal's paw)

11. Repeat the measurement process at least 2 more times for the same animal and move on the next one until all animals are measured (*see* **Note 17**).
12. Once the values (recorded in grams) have been collected, transfer the animals back to their previous experimental conditions and clean all of the devices and transitory accommodations of the Von Frey equipment.

3.7 Hot Plate Test

1. Bring the animals into the experiment room.
2. Make sure the room in which they are being held is silent and inaccessible for any other person for the acclimatization period of 1 h (*see* **Note 16**).
3. Once the animals are properly acclimatized, quietly return to the room for the use of the equipment.
4. Warm up the metal plate to 50 ± 0.5 °C.
5. Prepare the chronometer for the reading process, the values are recorded as general time units.
6. Place the mouse on the heating plate and immediately start the time recording.
7. Observe the animal for abnormal behavior such as licking the infected hind paw, jumping or flicking. If any of these behaviors are observed, immediately stop the timer and record the reading (*see* **Note 18**).
8. Take the mouse back to its original containment and repeat the procedure to the rest of the animals.

3.8 Dorsal Root Ganglion (DRG) Removal

1. Reserve the animals to be infected in groups and prepare the necessary materials.
2. Prepare the perfusion solution buffer (PBS) and set the peristaltic pump.
3. Anesthetize the mouse and wait for the loss of response.
4. Place the mouse at the designed perfusion space.
5. With a midline cut, open the mouse's abdomen and expose the liver.
6. Remove the front-most part of its ribcage, exposing both lungs and heart.
7. Start the peristaltic perfusion pump with the needle attached to the tip.
8. Insert the needle to the left ventricular chamber (Fig. 4).
9. Make a deep incision in the right atrium, cutting it open.
10. At this point blood should be flowing out of the right atrium and the liver should be slowly losing its reddish color (*see* **Note 19**).

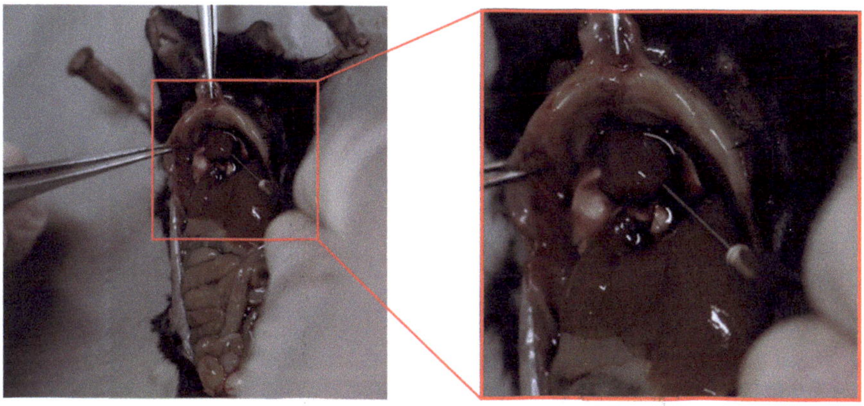

Fig. 4 Perfusion surgery for DRG removal and correct placement of the needle

11. After 5 min of a successful perfusion, take out the perfusion needle, clean any persistent blood clot that might be still present and transfer the mouse to the designed surgical space.
12. Remove all the mouse's skin for surgery (Fig. 5a, b).
13. Carefully remove the thigh muscles (i.e., gastrocnemius and hamstring muscles) to expose the sciatic nerve that appears as a thick pale cord (Fig. 5c, d) (*see* **Note 20**).
14. Remove the excess of muscles and soft tissue near the vertebral column being careful not to damage the nerve connection to the spine (Fig. 5e).
15. Once the bone of the vertebral column is exposed, remove the vertebral arch (laminectomy) of the lumbar segment to expose the spinal cord and DRG.
16. Follow the sciatic ramifications towards the spine and fish the DRGs on the exposed cord pulling them carefully by the nerve (Fig. 6).
17. Collect the DRGs and store them according to the experiment to be made (*see* **Note 21**).

3.9 Footpad Edema

1. Prior to infection, reserve the animals and the digital caliper.
2. Immobilize the animal and expose its right footpad (*see* **Note 10**).
3. Take 3 measurements of the paw's dimensions in the footpad region (Fig. 1c, d) and return the mice to their cage.
4. Repeat the measurement protocol at desired time points after infection.
5. Use the measurements to quantify the edema (*see* **Note 22**).

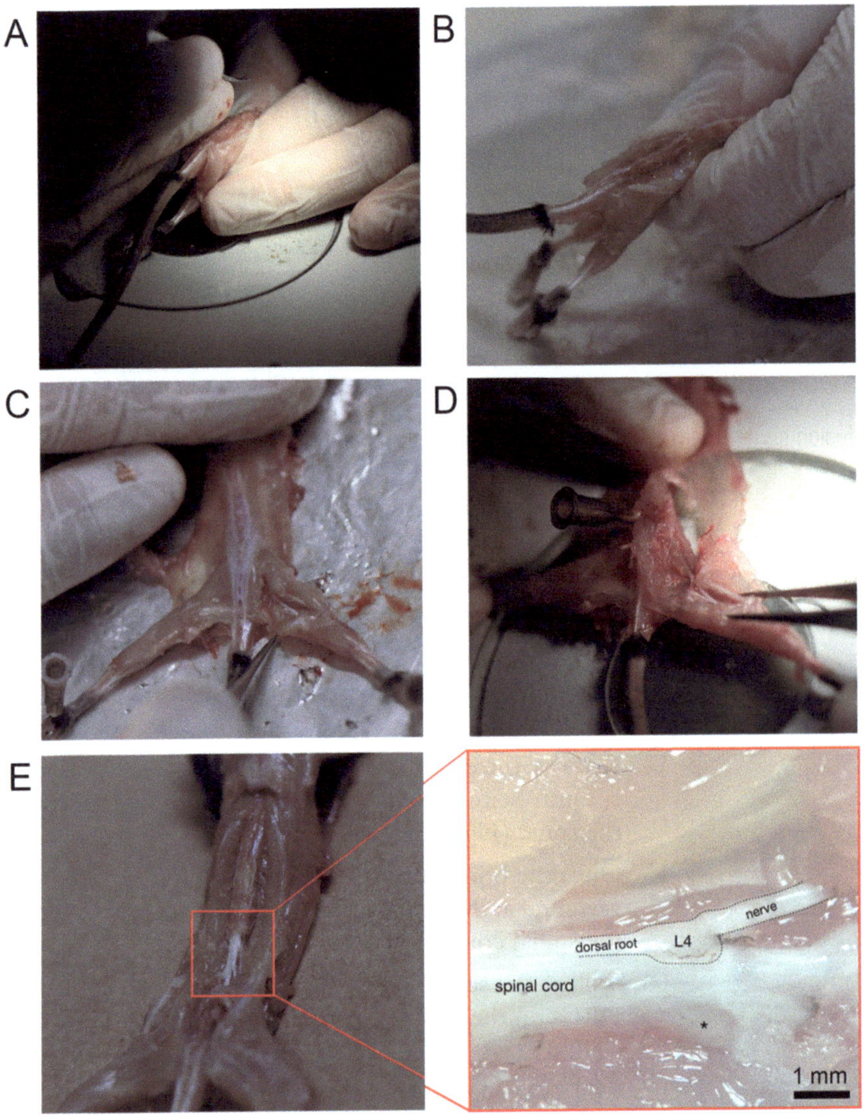

Fig. 5 Most critical steps for DRG removal surgery. (**a**, **b**: Removal of the animal's skin and excess muscle from the back. **c**, **d**: Removal of the paw's muscles to expose the sciatic nerve. **e**: Ramifications of the sciatic nerve to the spinal cord, zoomed in the L4 ganglia and its contralateral DRG signaled by *)

3.10 Computed Micro-tomography (Micro-CT)

Due to the osteoclastogenic profile observed in alphavirus infections (e.g., Mayaro virus), the evaluation of its impact on the loss of bone structure can be performed by computed micro-tomography analysis [14].

1. After euthanasia, collect the bones from the hind limb and transfer them to tubes containing a solution of 10% formaldehyde (v/v in PBS) for 24 h.

culture medium supplemented with 1.5% HEPES, 1% L-glutamine; 2% Fetal Bovine Serum, and 1% non-essential amino acids, for 72 h in maximum confluence. Next, centrifuge the sample at 600 × *g* for 10 min at room temperature and freeze it at −80 °C.

3. Carboxymethyl cellulose (CMC) dilution in distilled water at this concentration is often laborious. After initial mixing of the components and partial dilution, an autoclave cycle (121 °C, 30 min) must be performed to allow dilution into a homogeneous mass. Then, room-temperature CMC solution is ready to receive an equal volume of filtered DMEM 2×-concentrated solution (done by diluting powdered medium in half the required volume according to the instructions). It is also recommended that the solution be prepared a day before the assay so it can stay overnight in a heated magnetic shaker at 37 °C for better dissolution.
4. The surgery for DRG removal requires a set of surgical materials capable of bone cutting for both rib cage removal and spinal cord exposition and a set of instruments capable of fine and precise movements for nerve exposition, DRG removal, and heart manipulation.
5. A regular solution of ketamine (80 mg/kg) + xylazine (10 mg/kg) solution is usually administered, but other approved methods for loss of consciousness and analgesia may also be used. Use the euthanasia techniques approved in your institution.
6. For Chikungunya virus propagation a multiplicity of infection (MOI, expressed in PFU/cell) of 0.01 should be enough for sufficient recovery, for Mayaro Virus propagation a MOI of 0.001 should be enough for sufficient recovery, but several attempts might be needed to fine-tune the MOI for a better recovery on different lab conditions.
7. A titration assay must be performed to determine the viral aliquots' titers. If necessary, a concentration step can be added to the protocol using viral concentration columns or viral precipitation protocols.
8. Non-infected animal control groups should be inoculated with sterile C6/36 supernatant acquired from the collection of supernatant from confluent C6/36 cell culture flask after at least 24 h of contact from the culture media to the cell culture (*see* **Note 2**).
9. The volume of 20–50 μL is used to facilitate the injection of the correct inoculum to the animal. Volumes greater than 50 μL may cause discomfort in the animals.

10. For better and safer results, two people can execute this process: one for immobilization and the other to manipulate the caliper.
11. Needle insertion should be as superficial and as delicate as possible; the inoculum should be calculated for a maximum volume of 50 μL per animal and designed for better comfort and precision while administering.
12. The clinical score is evaluated for each animal, daily or at a desired time point after infection. Since the behavioral analysis is subjective, it is recommended that the operator run a blind analysis. If any animal reaches a score equal or above 16 at any given time point, euthanasia is recommended. If at any time point at the time of analysis an animal is found dead, the score is considered 20. According to the ethical guidelines, the animal must be euthanized if the body weight loss exceeds 20%. The animal must also be euthanized if it presents severe rushed posture associated with other clinical signs of severe disease.
13. The counting and plating steps should be done on the day before or maximum of 48 h after the assay.
14. In case of titration of cell culture supernatant or blood/plasma, the weighting and maceration steps are skipped straight to the dilution step.
15. Strongly homogenize the contents of the tube before and after pipetting.
16. Acclimatization times may vary according to different times of the day or year the experiments are being performed, for better results aim for a longer acclimatization time or wait for the natural loss of exploratory behavior inside the Von Frey box or hot plate.
17. A range of 3–5 measurements must be done per animal and the mean of these values should be considered for plotting. A good set of readings has low deviation between each individual reading. The values can be expressed as the absolute value of the paw withdrawal threshold (g) or as relative variation (delta, Δ), calculated from the subtraction of the measurements of the mean baseline value of the withdrawal threshold of each animal by the measurements of the mean after painful stimulus, for each time evaluated.
18. If there is no response within 30 s, the test is terminated and the mouse is removed from the hot plate to prevent heat-related injury. Mock-infected mice might also show abnormal behavior related to the heat at longer times than infected mice.
19. In case the liver does not lose its color, reposition the needle for a better perfusion and further wait for discoloration.

20. Nerves can be damaged from excessive stretching or pressure. Once exposed carefully expose its ramifications towards the spinal cord.
21. The collected DRG can be stored in a −80 ° C freezer for up to a month for biochemical analysis, such as ELISA. The tissue can also be immersed in 4% paraformaldehyde solution (pH 7.0–7.6) at room temperature for immune-histological analysis (immunohistochemistry). For cell culture, the DRG can be placed in cold DMEM and kept on ice until dissociation.
22. If only one dimension of the paw was measured (e.g., thickness), take the mean value of the measurements on a given time point for plotting; subtract this value from the baseline measured prior to the infection and plot as Δ. If all 3 dimensions of the pawn were measured, take the mean value of the measurements on a given time point, multiply them for the overall volume on the paw, subtract this value from the baseline measured prior to the infection and plot as Δ.
23. Operate the equipment at 50 kV of source voltage, 800 μA of source current, equivalent pixel size of 14.59 um and 0.5 mm Al filter.

Funding

This study was funded by the Fundação de Amparo à Pesquisa do Estado de Minas Gerais (FAPEMIG), grant numbers RED-00202-22, APQ-02281-18, APQ-04650-23, APQ-02618-23, and APQ-04983-24. This work also received support from the National Institute of Science and Technology in Dengue and Host-Microorganism Interaction (INCT Dengue), funded by the Conselho Nacional de Desenvolvimento Científico e Tecnológico (CNPq, Brazil) [grant number 465425/2014-3], and by FAPEMIG [grant number 25036]. Additional support was provided by FINEP—Financiadora de Estudos e Projetos under MCTI/FINEP—MS/SCTIE/DGITIS/CGITS [grant number 6205283BBB28-4F9C-AA65-808FE4450542].

References

1. Guzmán C, Calderón A, Mattar S, Tadeu-Figuereido L et al (2020) Ecoepidemiology of alphaviruses and flaviviruses. In: Emerging and reemerging viral pathogens, pp 101–125. https://doi.org/10.1016/B978-0-12-819400-3.00006-5
2. Guzmán-Terán C, Calderón-Rangel A, Rodriguez-Morales A, Mattar S (2020) Venezuelan equine encephalitis virus: the problem is not over for tropical America. Ann Clin Microbiol Antimicrob 19:19. https://doi.org/10.1186/s12941-020-00360-4
3. Santiago FW, Halsey ES, Siles C, Vilcarromero S et al (2015) Long-term arthralgia after Mayaro virus infection correlates with sustained pro-inflammatory cytokine response. PLoS Negl Trop Dis 9(10):e0004104. https://doi.org/10.1371/journal.pntd.0004104

4. Imad HA, Matsee W, Kludkleeb S, Asawapaithulsert P et al (2021) Post–chikungunya virus infection musculoskeletal disorders: syndromic sequelae after an outbreak. Trop Med Infect Dis 6(2):52. https://doi.org/10.3390/tropicalmed6020052
5. Hossain S, Choudhury MR, Ariful Islam A, Hassan M et al (2022) Post-chikungunya arthritis: a longitudinal study in a tertiary care hospital in Bangladesh. Trop Med Health:50–21. https://doi.org/10.1186/s41182-022-00412-9
6. Lumsden WH (1955) An epidemic of virus disease in Southern Province, Tanganyika Territory, in 1952–53. II. General description and epidemiology. Trans R Soc Trop Med Hyg:33–57. https://doi.org/10.1016/0035-9203(55)90081-x
7. Anderson CR, Downs WG, Wattley GH et al (1957) Mayaro virus: a new human disease agent. II. Isolation from blood of patients in Trinidad, B.W.I. Am J Trop Med Hyg:1012–1016. https://doi.org/10.4269/ajtmh.1957.6.1012
8. Suchowiecki K, Reid P, Simon GL, Firestein GS, Chang A (2021) Persistent joint pain following arthropod virus infections. Curr Rheumatol Rep 23(4):26. https://doi.org/10.1007/s11926-021-00987-y
9. Santos FM, Dias RS, Oliveira MD, Costa ICTA et al (2019) Animal model of arthritis and myositis induced by the Mayaro virus. PLoS Negl Trop Dis 13(5):e0007375. https://doi.org/10.1371/journal.pntd.0007375
10. Araújo S, Costa VRM, Santos FM, Sousa CDF et al (2022) Annexin A1-FPR2/ALX signaling axis regulates acute inflammation during chikungunya virus infection. Cells 11(17):2717. https://doi.org/10.3390/cells11172717
11. Moreira TP, Sousa CDF, Costa VRM, Queiroz-Junior CM et al (2023) Tumour necrosis factor plays a deleterious role in the pathogenesis of chikungunya virus infection. Immunology 168:444–458. https://doi.org/10.1111/imm.13583
12. Berta T, Qadri Y, Tan PH, Ji RR (2017) Targeting dorsal root ganglia and primary sensory neurons for the treatment of chronic pain. Expert Opin Ther Targets:695–703. https://doi.org/10.1080/14728222.2017.1328057
13. Martin SC, Macey AR, Raghu A, Edwards T et al (2020) Dorsal root ganglion stimulation for the treatment of chronic neuropathic knee pain. World Neurosurg 115(3):188–195. https://doi.org/10.1016/j.wneu.2020.07.102
14. Santos FM, Costa VRM, de Araújo S, de Sousa CDF, Moreira TP, Gonçalves MR, de Santos ACPM, Ferreira HAS, Costa PAC, Barrioni BR, Bargi-Souza P, de Magalhães Pereira M, Nogueira ML, da Glória Souza D, Guimarães PPG, Teixeira MM, Queiroz-Junior CM, Costa VV (2024) Essential role of the CCL2-CCR2 axis in Mayaro virus-induced disease. J Virol 98(1):e0110223. https://doi.org/10.1128/jvi.01102-23. Epub 2024 Jan 3. PMID: 38169294; PMCID: PMC10805060

Chapter 14

Platforms to Evaluate Flavivirus-Induced Immunoinflammatory Responses and Their Pathogenesis

Viviane L. Batista, Jennifer R. Martins, Danielle Cunha Teixeira, Jéssica A. B. Pereira, Victor R. M. Costa, Felipe Rocha da Silva Santos, Angélica Samer Lallo Dias, Mauro Martins Teixeira, Celso Martins Queiroz-Junior, and Vivian V. Costa

Abstract

Mosquito-borne diseases induced by viruses belonging to the *Flavivirus* genus such as Dengue, Zika, and Yellow fever have proven to cause widespread outbreaks over time. These viruses are prevalent mainly in tropical and subtropical regions and it is estimated that more than 100 countries are affected by one or more of these pathogens. The diseases may have mild clinical manifestations in common, such as fever, joint pain, headaches, and diarrhea; however, they can progress to serious reactions and, sometimes, trigger significant neurological malformations and post-infection sequelae, as in infection with Zika virus. To better understand their pathogenicity and tackle pharmacological development, animal and cell culture models have been developed. In the current chapter, we present in vivo platforms to study flavivirus-induced diseases, which allow us to mimic human infection to investigate clinical parameters, inflammatory, and viral mechanisms.

Key words Flavivirus, Dengue, Zika, Yellow fever

1 Introduction

Arboviruses belonging to the *Flaviviridae* family, *Flavivirus* genus are composed of single-stranded, positive-sense RNA enveloped virus species, transmitted to vertebrates by hematophagous arthropod vectors. Among the viruses belonging to this group, some of the most important ones can infect humans, such as the *Yellow fever virus* (YFV), *Dengue virus* (DENV), and *Zika virus* (ZIKV) [1, 2].

Flavivirus infections induce a broad spectrum of clinical manifestations, from asymptomatic occurrences to hemorrhagic fever,

Authors Viviane L. Batista and Jennifer R. Martins have equally contributed to this chapter.

Lucas Secchim Ribeiro (ed.), *Models in Inflammation and Immunopharmacology*, Methods in Molecular Biology, vol. 2993, https://doi.org/10.1007/978-1-0716-5019-6_14,
© The Author(s), under exclusive license to Springer Science+Business Media, LLC, part of Springer Nature 2026

biphasic fever, flaccid paralysis, and encephalitis, as seen among yellow fever, dengue, and zika diseases. These different manifestations allow the polarization of the phenotypes of the diseases triggered by flaviviruses, thus dividing them into hemorrhagic diseases and neurological diseases [2, 3]. This variability in disease presentation among flaviviruses infections can be explained due to the unique cellular and tissue tropism of each virus [2, 3].

ZIKV infection is well known for causing severe neurological damage, especially in pregnant women and their fetus. Zika virus has strong neurotropism for neural progenitor cells (NPCs) affecting gene expression, neurogenesis, differentiation of developing neurons, and causing apoptosis, leading to various neurological abnormalities including microcephaly in developing infants [4–6]. On the other hand, YFV replicates in the liver cells causing severe hepatitis, kidney failure, hemorrhagic shock, and even death [7, 8]. Under another perspective, dengue virus infects skin macrophages and resident dendritic cells (Langerhans cells), inducing a cascade of systemic events that result in vascular leakage, thrombocytopenia, mucosal bleeding, hemoconcentration, and hypotension [9–11].

One of the problems of studying these three Flaviviruses is the scarcity of animal models suitable for the study of the disease [12, 13]. Several models have been described in the literature, but many do not faithfully reproduce characteristics of the infections as seen in humans [12, 14, 15]. In addition to not mimicking the pathophysiological characteristics of human disease, some animal models use inoculums with high viral loads leading to a widespread infection inconsistent with what is observed during human infection [16–18]. For example, immunocompetent mice such as BALB/c and C57BL/6 are resistant to flavivirus infections, but being infected by high titers of dengue, some mice might develop neurological signals, which is not commonly observed in humans [14, 16].

On the other hand, partially immunocompromised mice with targeted disruption of the genes that encode the receptor of type I (A129 mice), II, or both IFNs (AG129 mice) are extremely susceptible to DENV infections [14, 16]. Thus, the use of these mice contributes to the study of several parameters of DENV and other flavivirus infections, as well as their interaction with the host, the investigation of cell and tissue tropism, and the screening of potential antiviral drugs specific for Flaviviruses. In this chapter, we will describe experimental platforms to study flavivirus infection in A129 mice, which allow a better understanding of the infections caused by flavivirus.

2 Materials

For the study of the diseases induced by flavivirus infections, the most used experimental models are performed in immunocompromised A129 mice, which are deficient for the IFN-α/β receptor (IFN-α/βR$^{-/-}$), both males and/or females, from 6 to 10 weeks old (*see* **Note 1**).

2.1 Cell Lines and Viruses

1. Vero cell line (African green monkey kidney cells—*Cercopithecus aethiops*).
2. C6/36 cell line (*Aedes albopictus* clone C6/36).
3. DENV2 (Strain 91 15 AM BR 2015).
4. DENV3 (Strain 3295: Genbank EU081177.1).
5. YFV (17DD strain, vaccine seed-lot 993FB013Z, lot 19PVFA059Z).
6. ZIKV (HS 2015 BA1 strain, isolated from a patient).

2.2 Culture of Monolayer Cells and Preparation of Virus Stock

1. Laminar flow hood.
2. Micropipettes and sterile tips.
3. Dulbecco's modified Eagle medium (DMEM) supplemented with fetal bovine serum (+ 2.5% or 10% FBS, depending on the required step).
4. Frozen vial of Vero cells (ATCC #CCL-81).
5. 37 °C, 5% CO_2 humidified incubator.
6. *Optional*: Frozen vial of C6/36 (ATCC #CRL-1660).
7. *Optional*: Leibowitz L-15 culture medium with 1.5% HEPES, 1% gentamicin, 1% L-glutamine (2 mM), 1% MEM non-essential amino acids (100× stock solution).
8. *Optional*: 28 °C biochemical oxygen demand humidified incubator.
9. Water bath at 37° C .
10. 75- and 175-cm^2 cell culture flasks.
11. 10-mL and 50-mL serological pipettes.
12. Inverted microscope.
13. Sterile PBS: 137 mM NaCl, 2.7 mM KCl, 10 mM. Na_2HPO_4, 2 mM KH_2PO_4 (pH 7.4).
14. Trypsin/EDTA solution: 0.25% (w/v) trypsin +0.02% (w/v) EDTA diluted in PBS, warmed up to 37 °C.
15. 15-mL and/or 50-mL centrifuge tubes.
16. Refrigerated centrifuge.

17. DENV, YFV or ZIKV strains.
18. 2.0-mL cryotubes with O-ring.

2.3 In Vivo Platform for Studying DENV

2.3.1 Intraplantar Infection

1. Laminar flow hood.
2. 1-mL syringe with fixed 30–26-gauge needle.
3. Tubes with previously aliquoted viruses.
4. Micropipettes and sterile tips.

2.3.2 Clinical Score

1. Precision Scale.
2. Pen and Printed table for annotations.

2.3.3 Platelet Count

1. 0.5-mL microtube for blood collection with K_2 EDTA.
2. Neubauer chamber.
3. 1% Ammonium oxalate, sterile and freshly prepared.
4. 2.0-mL sterile microcentrifuge tubes.
5. 0.22 μm pore size filter.
6. 10-mL syringe.

2.3.4 Degranulation of Mast Cells

1. Surgical scissors.
2. Surgical tweezers.
3. Histological cassettes.
4. 4% formalin, diluted in PBS.
5. 2-mm circular biopsy punch.

2.4 In Vivo Platform for Studying YFV

2.4.1 Intravenous Infection

1. 1-mL syringe with fixed 30–26-gauge needle.
2. Tubes with previously aliquoted viruses.
3. Container for Mouse: Internal Measurements in mm (H × W × L): 25 × 25 × 100. Made in 4 mm Crystal Acrylic, removable front, and rear, with locking system through handles, side holes for air flow and non-slip feet.
4. Laminar flow hood.
5. Sterile Vero supernatant acquired from cell culture.
6. Microisolator.
7. Infrared lamp.
8. Sterile gauze.

2.4.2 Indocyanine Clearance (IGC)

1. Aqueous solution of Indocyanine Green (for a final dose of 20 mg/kg of body weight).
2. Micropipettes and sterile tips.
3. 1-mL syringe with fixed 30–26-gauge needle.
4. Flat Bottom 96-well plates.

5. Sterile PBS: 137 mM NaCl, 2.7 mM KCl, 10 mM Na_2HPO_4, 2 mM KH_2PO_4 (pH 7.4).
6. Spectrophotometer.

2.5 In Vivo Platform for Studying ZIKV

2.5.1 Intravenous Infection

1. *See* Subheading 2.4.1.

2.5.2 Intraocular Pressure (IOP) Evaluation

1. Tono-Pen® Vet applanation tonometer and Tip Covers.
2. Topical anesthetic (1% tetracaine hydrochloride and 0.1% phenylephrine hydrochloride).
3. Parafilm.

2.6 Euthanasia (See Note 2)

1. Anesthetic: Ketamine (100 mg/kg), Xylazine (10 mg/kg) diluted in 0.9% sodium chloride.
2. 0.5 mL microcentrifuge for blood collection with K_2 EDTA.
3. 2-mL sterile Microcentrifuge tubes.
4. Ice box.
5. Sterile scissors and tweezers.
6. Laminar flow hood.

2.7 Viral Titration

1. 175-cm^2 cell culture flasks.
2. Vero cells, in culture.
3. DMEM culture medium.
4. Micropipettes and sterile tips.
5. Electronic pipette controller and sterile serological pipette tips.
6. Laminar flow hood.
7. Inverted microscope.
8. Microcentrifuge tube shaker.
9. 2% carboxymethyl cellulose + 2× DMEM diluted in sterile water to a final volume of 1 liter.

2.8 Viral Propagation and Dilution

See Subheading 2.2.

3 Methods

3.1 Viral Propagation and Dilution

1. Prepare a sterile environment in the flow hood using UV light and proper aseptic procedures before starting the experiment.

2. Rapidly place a frozen vial of Vero cells in a 37 °C water bath. Remove the vial from the bath immediately after complete thawing.
3. Disinfect the external part of the cell vial with 70% EtOH and, in the flow hood, transfer its content with a pipette into a 50-mL centrifuge tube. Add 10 mL of DMEM + 10% FBS. Centrifuge the tube for 5 min at 420 × *g*, at room temperature. Then, discard the supernatant. Resuspend the formed pellet in 1 mL of DMEM + 10% FBS and transfer it to a 75-cm^2 cell culture flask containing 20 mL of DMEM + 10% FBS. Gently tilt the flask to evenly distribute the cell suspension and incubate it overnight at 37 °C, 5% CO_2 and check daily for confluency.
4. When the cell monolayer is 90% confluent, carefully remove the culture medium. Gently wash the cell monolayer twice with PBS or DMEM to remove any remaining serum and then pipette it out.
5. Overlay cells with warm trypsin/EDTA solution using enough volume to cover the monolayer (e.g., 2 mL for a 75-cm^2 flask). Incubate up to 5 min in a 37 °C, 5% CO_2 humidified incubator. Tap the flask on the side to promote a better detachment of the cells.
6. Pipette the cell suspension up and down several times to break up any clumps. Transfer the cells to a tube, dilute them with Trypan for counting in a Neubauer chamber. Seed 1×10^5 viable cells in 1 mL of a 24-well plate or 2×10^5 viable cells in 2 mL of a 12-well plate (*see* **Note 3**).
7. The cells are maintained with a sub-cultivation schedule at each 3–4 days after seeding.
8. For viral infection, it is recommended that cells are in the growth phase, between 60% and 80% confluence.
9. For the infection, prepare 5 mL of inoculum for each 175-cm^2 tissue culture flasks. For that, thaw the cryotube of the virus to be propagated in an ice bath. Upon thawing, transfer the contents of the vial (1 mL) into a 50 mL tube containing 4 mL of DMEM culture medium.
10. Gently homogenize the solution using a micropipette or vortex.
11. Remove the Vero cell culture flasks to be used, identify it with the name of virus to be used for infection.
12. Decant or aspirate the culture supernatant and rinse the flask with 15 mL of PBS solution, and with light movements, wash the cell monolayer. Repeat this step another 2 times.

13. Add 5 mL of inoculum over the cell monolayer, ensuring that its distribution is homogeneous.
14. Incubate the infected cells at 37 °C, 5% CO_2 for 1 h, to promote the viral adsorption, gently moving the flask every 15 min.
15. At the end of the viral adsorption time, add 45 mL of DMEM +2% FBS to the flask.
16. Incubate the infected cells (preferably in a separate incubator, to avoid cross-contamination) and monitor the cytotoxic effects daily, until the cytopathic effects (CPE) reach 80% of the cells, as seen by an inverted microscope.
17. At this point, perform the collection of the virus suspension. In a laminar flow hood intended for viral manipulation, remove all cells still adhered with a cell scraper.
18. Transfer the contents of the flask into a 50 mL tube and centrifuge it at $2000 \times g$ for 8 min, at 4°.
19. Transfer the supernatant to another 50 mL tube.
20. Aliquot the collected stock in cryovials properly identified, with the date and viral isolate used.
21. Store virus aliquots in a − 80 °C freezer for later virus titration.

3.2 In Vivo Platform for Studying DENV (See Table 3)

3.2.1 Intraplantar Infection

1. Prepare the viral inoculum with one of the dengue serotypes (*see* **Note 1**). Use DENV-2 or DENV-3 strains with 2×10^4 PFU (plaque-forming units) in 30 μL of phosphate buffered saline (PBS). Keep diluted viral samples on ice until starting the inoculation (*see* **Note 4**).
2. Immobilize the mouse by holding it with the index finger and thumb between the interscapular, auricular, and occipital region of the animal. Keep it firmly in hands, so that the hind footpad is well-positioned and allows easy visualization for viral inoculation (Fig. 1a, b) (*see* **Note 5**).
3. For inoculation, the syringe needle should be positioned with the bevel up and parallel to the plane of the plantar cushion (Fig. 1b, c) (*see* **Note 5**). Insert the needle in the subcutaneous tissue and inject 30 μL of the viral inoculum.
4. Return the mouse to its chamber and proceed with the experiment by inoculating the other mice.

3.2.2 Clinical Score

1. Evaluate the clinical score of the mice visually, according to the criteria described in Table 1 (*see* **Note 6**).
2. Weigh the animals every day on the precision scale.
3. Return the animal to their previous condition.

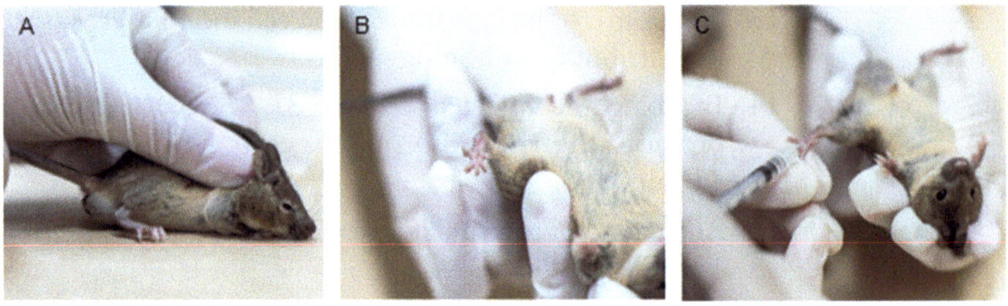

Fig. 1 Intraplantar infection. Two researchers are needed to conduct intraplantar inoculation. (**a**) The mouse must be restrained by its back by one of the researchers, preventing any unexpected movement. (**b**) After containing the mouse, a second operator locates the appropriate place for the inoculation, which is carried out in the center of the hind foot pad. (**c**) Inoculation is done with an insulin needle, the bevel must be facing upwards at an angle of 15°, the solution is injected slowly, it is possible to observe the formation of a kind of bubble when you inject it

Table 1
Parameters evaluated in the clinical score of mice

Points	Healthy animal	Mice infected with dengue	Mice infected with ZIKV	Mice infected with YFV	Procedure
Weight loss	Mice from 8 to 10 weeks weigh on average 20 g	Perform daily weight variation and score. 0% (0 points), 1–5% (0.5 points), 6–10% (1 point), 20% euthanasia (2 points)			Weigh the animals from day zero (before infection) until the day of euthanasia.
Diarrhea	Solid feces	Pasty feces with odor and secretion		X	Daily evaluation
Locomotion activity	Active, moves a lot around the box keeping its explorer habit	Tend to move less around the box, hide and avoid contact with each other		X	Daily evaluation
Ocular secretion	No secretion	May present ocular secretion from the third day onward		X	Daily evaluation

Score for each parameter of each mouse: 0 for absent; 0.5 for mild; 1.0 for moderate; and 2 for severe

3.2.3 Euthanasia and Sample Collection

1. The procedures described in this topic are suitable for all models proposed in this chapter.
2. All the steps must be performed in a Laminar flow Hood.
3. Anesthetize the animal intraperitoneally with a solution of 100 mg/kg of ketamine and 10 mg/kg of xylazine, in 200 microliters of sterile 0.9% NaCl saline (*see* **Note 7**).

Table 2
Sample collection of tissue specimens from flavivirus-infected animals

Viral infectionand suggested sample collection of tissue specimens	YFV	ZIKV	DENV
Brain		X	X
Spleen			X
Liver	X		X
Blood/plasma	X	X	X
Eye		X	
Optical nerve		X	
Hind footpad			X

4. Place the animal in the dorsal recumbent position. Execute a medial laparotomy and expose the vena cava of the animal that lies below the small intestine and stomach. Using a syringe, aspirate and plunge out 1 mL of EDTA, then insert the needle with the bevel facing up toward the blood flow, aspirate the blood and place it in a microtube with K_2 EDTA. (*see* **Note 8**). This procedure typically results in the euthanasia of the mouse due to exsanguination. Therefore, after blood collection, it is essential to confirm the absence of vital signs - such as heartbeat and respiration - to ensure that euthanasia has been completed before proceeding with tissue collection or disposal.
5. Collect the organs of interest (Table 2).
6. Dispose of the carcass in an appropriate place and disinfect all material used.

3.2.4 Platelet Count

1. Platelets should be counted manually using a Neubauer chamber.
2. Dilute the blood in the ratio of 1:100 in 1% ammonium oxalate solution. For example, add 990 μL of ammonium oxalate to 1% in a Microcentrifuge tube of 2 mL and right after the euthanasia of the animals take 10 μL of the total blood and mix with oxalate present in the tube (*see* **Note 9**).
3. Place 10 μL of this solution (blood + oxalate) in the Neubauer chamber and leave it resting for at least 20 min, so that the platelets are deposited in the chamber.
4. Count the platelets and express the final cell suspension as the total number of platelets/μL of blood.

3.2.5 Degranulation of Mast Cells

1. To assess mast cell degranulation, collect the footpad of the animals after euthanasia. Use surgical scissors to cut the entire paw of the mouse, at the height of the heel. The tips of the fingers must be removed so that the formalin penetrates well into the tissue. Store it in 50 mL tubes with 4% buffered formalin for 48 h.
2. After 48 h, transfer the tissue to a vial with water and leave it for 2 h. Next, transfer it to a recipient containing 70% ethanol solution until the tissue is processed.
3. Take the tissue out of the ethanol and transfer it to a bench for excision. Position the biopsy punch just above the cushion area and make circular movements to collect the skin tissue in that region. Use surgical tweezers to facilitate the collection.
4. After collecting the sample, place it in histological cassettes for standard histological processing. During embedding of the tissue in paraffin, position the sample perpendicularly to the bottom of the metal mold, so that epidermis and dermis can be visualized together in the slide. Assemble the paraffin blocks for cutting in a microtome and stain with toluidine blue, which consists of consecutive passages of the slide with the tissue through different solutions (xylene, ethanol water) for deparaffinization, hydration and staining, and then the same solutions are used for dehydration and assembly of the slides.
5. The microscopic analysis is done by evaluation of total and differential counts (intact and degranulated mast cells) performed using a 100× objective. Mast cells are found in the connective tissue of the dermis. Count the total toluidine blue-stained mast cells and also the cells with morphology showing degranulation (Fig. 2). Also record the total number of evaluated microscopic fields. The count result is expressed in cells/mm^2. Each field at 100x magnification is equivalent to = 0.01 mm^2 (*see* **Note 10**).

3.3 In Vivo Platform for Studying YFV (See Table 3)

3.3.1 Intravenous Infection

1. Separate the groups of animals to be infected and the necessary materials.
2. Inside the sterile flow hood, prepare the viral inoculum by diluting the viral content on sterile PBS solution until the desired titer is reached (lethal dose = 1×10^6 PFU).
3. Aspirate the diluted inoculum with the syringe and reserve it on ice.
4. Place the microisolator next to a warm light source for 2–3 min to allow dilation of the tail vein (Fig. 3a) (*see* **Notes 2** and **11**).
5. Place one mouse at a time in a suitable mechanical retainer, so that the tail is exposed (Fig. 3b) (*see* **Note 12**).

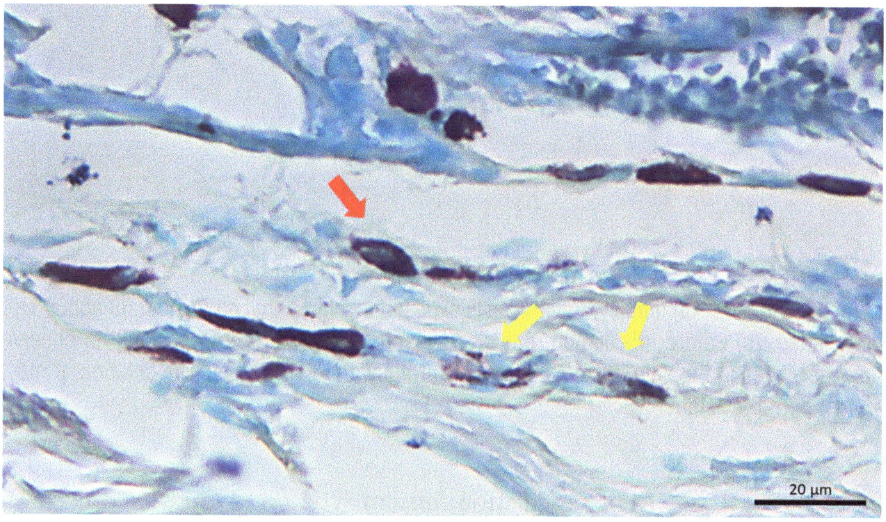

Fig. 2 Mast cell morphology. Visualized by optical microscope at 100× magnification. Intact mast cell (indicated by the red arrow) and degranulation mast cell (indicated by the yellow arrows)

Table 3
Frequently used techniques and assays and tissues of interest

Tissue specimens/assay methods	Brain	Spleen	Liver	Blood	Plasma	Hind footpad	Eye	Optic nerve
H&E staining	X	X	X			X		
Plaque/median tissue culture infectious dose					X		X	X
qPCR	X		X	X	X		X	X
ELISA	X		X		X			
IHC	X	X	X			X		
Mast cell degranulation						X		
Myeloperoxidase activity	X	X	X					
Transaminase activity	X				X			
Indocyanine green clearance			X					
Hemocytometer				X				

6. Locate the animal lateral vein of the mouse. Based on mouse vein anatomy descriptions, there are two small and easily accessible lateral veins directly under the skin. There is also a dorsal vein in the tail, however, it is more difficult to access and should not be used for injection (Fig. 3c) (*see* **Note 11**).

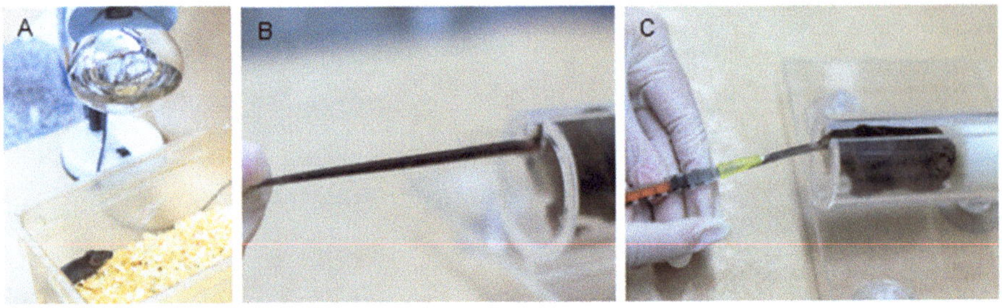

Fig. 3 Intravenous infection. (**a**) Initially, the mouse is exposed to a lamp that emits heat for about 3 min so that the tail vein dilated. (**b**) Before inoculation, the tail vein must be located. It is possible to observe it more laterally. (**c**) After locating the vein, insert the needle carefully and lightly inject the solution, it must be fluid and there must be no resistance

7. For inoculation, insulin syringes (1-mL, fixed needle) should be positioned with the bevel up and parallel to the plane of the tail at an angle of 25° (Fig. 3c) (*see* **Note 13**).
8. Inject the YFV inoculum intravenously (1×10^6 PFU in 200 μL of PBS) (*see* **Note 1**).
9. Remove the syringe from the vein, apply local pressure with gauze until the bleeding stops.
10. Return the animal to its microisolator and proceed with the experiment, by evaluating the clinical signs over time.

3.3.2 Clinical Signs

1. Measure the body mass on a daily basis, always at the same time, until the third day post-YFV infection or when 20% of the body mass is lost.
2. Determine the body mass change (%) over the time.
3. For assessing the score based on clinical parameters, follow the criteria described in Table 1.

3.3.3 Green Indocyanine Clearance (IGC) to Assess Liver Function

1. Shortly before euthanasia, dissolve the appropriate amount of green indocyanine crystals in sterile water at the concentration of 20 mg/kg of body mass. For example, considering a mouse weighing 20 g, resuspend 0.4 mg of green indocyanine crystal in 100 μL of sterile water. Adjust the amount for the required groups (*see* **Note 14**).
2. Inject green indocyanine (20 mg/kg) intravenously 20 min before anesthesia administration, following the steps of Subheading 3.3.1.
3. Anesthetize the mice by the injection of 100 μL via intraperitoneal of 100 mg/kg of ketamine and 10 mg/kg of xylazine solution (*see* **Note 15**).

4. Collect blood samples from Inferior vena cava (using a needle and syringe treated with EDTA) and protect them from light until plasma obtention (12,000 × *g*, room temperature for 15 min).
5. For IGC level measurement, dilute the plasma samples at least 1:2 in PBS.
6. Transfer 50 μL of green indocyanine (working concentration: 2000 μg/mL) solution to the 96-well plate and perform serial dilution (1:2) using sterile water to obtain a standard curve.
7. Add samples and the standard curve to a flat-bottom 96-well plate and read it in a spectrophotometer at 800 nm.
8. Interpolate samples concentration using the values from the standard curve, subtracting the blank. Data should be expressed in μg/mL.

3.4 In Vivo Platform for Studying ZIKV (See Table 3)

3.4.1 Viral Inoculum Preparation

1. Based on the viral titer of the ZIKV stock, calculate the concentration of the inoculum that will be used in the infection, 4×10^3 PFU per mouse (lethal dose).
2. Considering the total number of mice that will be infected, prepare the volume of viral inoculum to be used.
3. The diluent may vary according to the viral stock freezing solution used, which may be medium or saline.

3.4.2 Intravenous Infection

1. Infect animals with the adequate inoculum intravenously as described in Subheading 3.3.1 and return the animal to its chambers afterwards.

3.4.3 Clinical Signs

1. After viral inoculation (day 0), the clinical signs of each animal are evaluated daily until 5 d.p.i (days post-infection), being defined by a standard clinical score, described in Table 1.

3.4.4 Intraocular Pressure (IOP) Evaluation

1. In addition to neurological damage, ZIKAZika infection can trigger ophthalmic damage. Tono-Pen® Vet applanation tonometer is used (Fig. 4a). Set up the tonometer by placing the battery and the protective tip at the end of the equipment, right after calibrating the tonometer according to the equipment manual.
2. Place a drop of anesthetic in the animal's eye and wait for 1 min (*see* Subheading 2.5.2).
3. Approach the tonometer perpendicularly to the most apical region of the cornea, gently touching the animal's eye with the tonometer (Fig. 4b).
4. Perform and record three measurements of each animal considering only the measurements with a maximum error

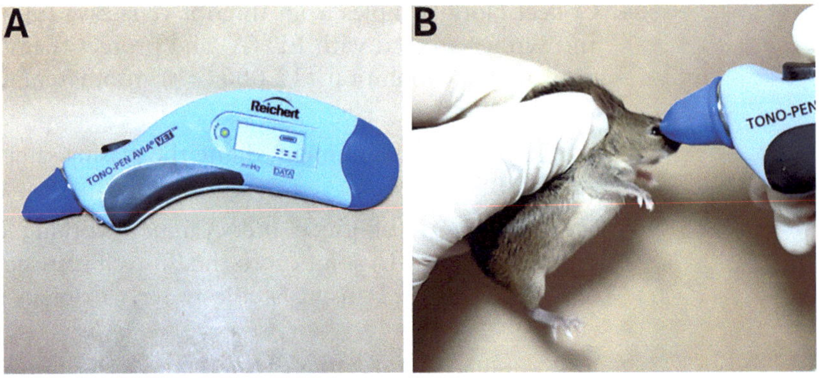

Fig. 4 Intraocular pressure (IOP) evaluation. Two researchers are needed to conduct evaluation of intraocular pressure. (**a**) Set up the tonometer by inserting the battery and the protective tip. (**b**) Immobilize the animal by the dorsal region of the neck in an upright position and apply ocular anesthetic. Soon after, approach the device gently touching the most apical region of the cornea, just for a second, and remove the tonometer

deviation of 5%. To avoid variations, standardize which animal's eye will be evaluated.

5. After measuring the IOP, average the three measurements to obtain a final value for data plotting. The measurement of the IOP must occur on the day of infection (day 0), on the third day after infection and on the day of euthanasia on the fifth day.

3.4.5 Viral Titration

1. One day before performing the viral titration, culture plates must be produced with Vero cells. To produce the plates, use 12-well adhesion culture plates, plated with Vero cells at a density of 1×10^6 cells/well, to form a homogeneous cell monolayer and a confluence of 80% must be observed.
2. For titration of plasma samples, a volume of 50 uL of sample will be used in serial dilution for viral titer quantification. For solid tissue samples, they must undergo tissue processing in a Tissue Homogenizer. The tissue must be weighed on a precision scale and consider 100 mg of tissue for 1 mL of DMEM.
3. Turn on the laminar flow hood intended for virus handling.
4. Check the cell viability of the culture plates to be used in the viral titration process. They must not have empty spaces nor the formation of lumps.
5. Fill sterile 1.5-mL microcentrifuge tubes with 450uL of DMEM culture medium. Five Microcentrifuge tubes will be used for each sample to be titrated.
6. For solid tissue samples, check the weight of the sample to be titrated, maintaining its sterility. For every 100 mg of tissue, add 900 ul of DMEM to the respective tube and perform tissue

homogenization, inside the sterile laminar flow hood. After homogenization, the sample should be considered to be at the dilution of 10^{-1}.

7. Collect 50 μL of the sample (plasma or tissue homogenate) at 10^{-1} and carry out the serial dilution in the microcentrifuge tubes previously filled with 450 μL of DMEM medium without supplementation, always passing 50 μL from one tube to the other, until reaching the fifth tube destined for that sample. Change the tips and vortex the solution formed before proceeding to the next tube.
8. Reserve the 5 tubes on ice or in the refrigerator.
9. Remove the previously produced Vero cell plates from the incubator. In the laminar flow hood, remove the culture supernatant from each well using a micropipette.
10. Add 300 μL of each diluted sample (10^{-2}, 10^{-3}, 10^{-4}, and 10^{-5}) in each well of the Vero cell culture plate, gently, so not to disturb the monolayer of cells.
11. Plate the 4 dilutions, preserving the rest of the first tube (10^{-1}) produced in the freezer −80 °C.
12. Properly identify the plate and each dilution used in each well.
13. Gently swirl the plate, ensuring that the diluted sample solution evenly covers the well of the culture plate.
14. Place the plate in an incubator with 5% CO_2 atmosphere, at 37 °C for 1 h to promote viral adsorption.
15. Gently swirl the plate every 15 min to ensure homogeneous adsorption throughout the well.
16. At the end of the viral adsorption period, remove the plates from the incubator and using a micropipette, gently remove all culture supernatant and discard.
17. Carefully add 2 mL of DMEM medium containing 1.6% carboxymethylcellulose (CMC), gentamicin (50 mg/mL) and 2% FBS.
18. Place the plate back in the culture incubator.
19. Take note of cytopathic effects over a period of 4 days (*see* Fig. 5).
20. After the end of this observation period, remove the plate from the incubator and in a laminar flow hood for virus handling, remove the medium (DMEM +2% FBS + CMC) with a vacuum pump or with a sterile pipette.
21. Initially, inside the laminar flow hood add 2 mL of 10% buffered formalin to each well and store overnight at room temperature to promote cell monolayer fixation. At this stage, the plates can be stored outside the hood and stove.

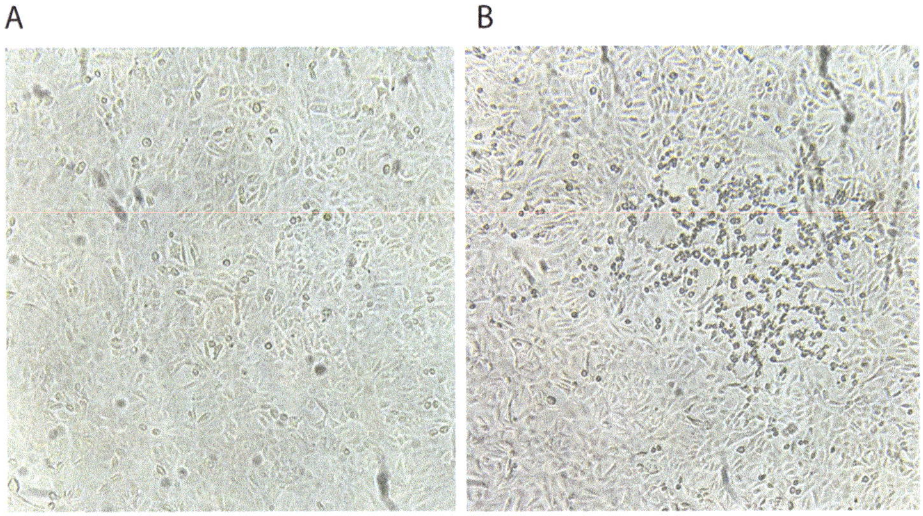

Fig. 5 Cytopathic effects on cells. (**a**) In uninfected Vero cells, the monolayer remains intact, with no signs of lysis or cellular detachment observed. (**b**) In Vero cells infected with the dengue virus, significant morphological alterations are observed, including cytoplasmic vacuolization, cellular detachment from the monolayer, and the formation of lysis plaques

22. Discard the formalin from the plates and add the dye (1% crystal violet in PBS, w/v) to the wells, leaving at room temperature for a minimum of 4 h, 24 h are sufficient for staining.
23. Remove the dye, and gently wash the plates to remove excess dye by submerging them in a bowl of water.
24. Allow the plate to dry at room temperature for further counting.
25. To quantify the titer of the samples, count the plaque-forming units (PFU). The number of plaques counted per well is then multiplied by the sample dilution factor (10^{-1}, 10^{-2}, 10^{-3}, 10^{-4}, 10^{-5}) used, and divided by 0.3, the volume used for plating. The result is given in PFU/mL.

4 Notes

1. Empirical experimentation to assess the virulence of the virus to be used must be done prior to other experiments to access the best inoculum for the desired phenotype.
2. All studies must be previously submitted to the Ethics Committee on Animal Experimentation. The entire procedure should be performed cautiously and within laminar biosafety flow hood, with the necessary PPEs (personal protective equipment) at the time of infection.

3. A flask of medium (75-cm^2) Vero with the confluence of 95% is enough for 3 plates of 24 wells.
4. Use DENV-2 or DENV-3 strains (10^1 PFU in 30 μL of PBS). For dilution, the virus stock must be thawed by vortexing, add PBS according to the dilution to be performed and performing a serial dilution. Keep samples on ice until starting the inoculation.
5. The maximum volume of inoculation should be 50 μL.
6. Clinical score and body weight loss are evaluated individually in each animal and should be evaluated every day until euthanasia on the third and/or fifth day post-DENV Infection. Since the behavioral analysis is subjective it is recommended to do it blindly. The evaluated parameters are described in Table 1.
7. The animals need to be previously anesthetized and rigorously evaluated for the absence perception of pain. After blood collection, we performed the cervical displacement and removal of the other organs. Euthanasia should be performed on the third or fifth day after infection or when there is a loss of more than 20% of body weight (humanitarian endpoint).
8. It is important to collect the blood without the rupture of the vena cava, since a platelet count must be properly performed.
9. The oxalate needs to be filtered (0.22 μm pore size) and freshly prepared on the day of euthanasia.
10. Each field at 100× magnification is equivalent to 0.01 mm^2.
    I. Therefore, multiply 0.01 by the total number of fields counted to obtain the total area of the analyzed field.
    II. Divide the total amount of mast cells by the area result (I).
    III. Divide the total degranulated mast cells by the area result (I).

    Example: Suppose you counted a total of 19 mast cells in 20 fields, 14 of which were degranulated. $20 \times 0.01 = 0.2$ mm^2

    $19/0.2 = 95$

    $14/0.2 = 70$

    Result: Result: 95 total mast cells per mm^2. Of the total mast cells, 70 are degranulated per mm^2.
11. When injecting a volume of any liquid into the tail of the animal, be aware: (i) if you feel resistance when tightening the needle inside the vein, remove the needle and try again; (ii) Do not inject the liquid quickly; (iii) Start attempts on the more distal portions of the veins, because if it starts at the base of the tail, the tissue will be perforated and will not allow further attempts.

12. The use of a Container for Mouse should be necessary for the inoculation of the virus into the tail vein, to ensure the maintenance of animal welfare.
13. Inject the inoculum with the maximum volume of 0.2 mL in the tail vein.
14. Indocyanine green (ICG) is a hydrophilic tricarbocyanine dye that exhibits affinity for plasma proteins such as albumin. After intravenous administration, ICG binds to albumin, circulates in the bloodstream, and then is depurated and eliminated exclusively by hepatocytes. Quantification of plasmatic concentration of ICG has emerged as a rapid, low-cost, and sensitive method for evaluating liver function. Impaired liver function is directly related to a decrease in ICG clearance.
15. The clearance of the water-soluble dye Indocyanine Green (IGC) is a quantitative indicator of liver function. After intravenous injection, IGC binds to albumin and disperses quickly in the blood. Like bilirubin and other toxins, IGC is exclusively removed from the bloodstream by the liver and then excreted uncharged into bile. The measurement of ICG clearance involves intravenous injection of ICG, followed by collection of venous blood after a 20–30 min interval. Thus, IGC tests express the concentration of dye retained in the bloodstream which is inversely proportional to liver function.

Funding

This study was funded by the Fundação de Amparo à Pesquisa do Estado de Minas Gerais (FAPEMIG), grant numbers RED-00202-22, APQ-02281-18, APQ-04650-23, APQ-02618-23, and APQ-04983-24. This work also received support from the National Institute of Science and Technology in Dengue and Host-Microorganism Interaction (INCT Dengue), funded by the Conselho Nacional de Desenvolvimento Científico e Tecnológico (CNPq, Brazil) [grant number 465425/2014-3], and by FAPEMIG [grant number 25036]. Additional support was provided by FINEP—Financiadora de Estudos e Projetos under MCTI/FINEP—MS/SCTIE/DGITIS/CGITS [grant number 6205283BBB28-4F9C-AA65-808FE4450542].

References

1. Adams MJ, Lefkowitz EJ, King AMQ et al (2017) Changes to taxonomy and the international code of virus classification and nomenclature ratified by the International Committee on Taxonomy of Viruses. Arch Viral 162: 2505–2538. https://doi.org/10.1007/s00705-017-3358-5
2. Burger-Calderon R, Gonzalez K, Ojeda S et al (2018) Zika virus infection in Nicaraguan households. PLoS Negl Trop Dis 12:

e0006518. https://doi.org/10.1371/journal.pntd.0006518

3. Rodenhuis-Zybert IA, Wilschut J, Smit JM (2010) Dengue virus life cycle: viral and host factors modulating infectivity. Cell Mol Life Sci 67:2773–2786. https://doi.org/10.1007/s00018-010-0357-z
4. Miner JJ, Diamond MS (2017) Zika virus pathogenesis and tissue tropism. Cell Host Microbe 21:134–142. https://doi.org/10.1016/j.chom.2017.01.004
5. Coyne CB, Lazear HM (2016) Zika virus—reigniting the TORCH. Nat Rev Microbiol 14:707–715. https://doi.org/10.1038/nrmicro.2016.125
6. Tang H, Hammack C, Ogden SC et al (2016) Zika virus infects human cortical neural progenitors and attenuates their growth. Cell Stem Cell 18:587–590. https://doi.org/10.1016/j.stem.2016.02.016
7. Vieira WT, Gayotto LC, Lima CP, Brito T (1983) Histopathology of the human liver in yellow fever with special emphasis on the diagnostic role of the councilman body. Histopathology 7:195–208. https://doi.org/10.1111/j.1365-2559.1983.tb02235.x
8. Bailey AL, Kang L-I, de Assis Barros D'Elia Zanella LGF et al (2020) Consumptive coagulopathy of severe yellow fever occurs independently of hepatocellular tropism and massive hepatic injury. Proc Natl Acad Sci 117:32648–32656. https://doi.org/10.1073/pnas.2014096117
9. Johnston LJ, King NJC, Halliday GM (2000) Langerhans cells migrate to local lymph nodes following cutaneous infection with an arbovirus. J Invest Dermatol 114:560–568. https://doi.org/10.1046/j.1523-1747.2000.00904.x
10. Marovitch M, Grouard-Vogel G, Eller M et al (2001) Human dendritic cells as targets of dengue virus infection. J Investig Dermatol Symp Proc 6:219–224. https://doi.org/10.1046/j.0022-202x.2001.00037.x
11. St. John AL (2013) Influence of mast cells on dengue protective immunity and immune pathology. PLoS Pathog 9:e1003783. https://doi.org/10.1371/journal.ppat.1003783
12. Fagundes CT, Costa VV, Cisalpino D et al (2011) Therapeutic opportunities in dengue infection. Drug Dev Res 72:480–500. https://doi.org/10.1002/ddr.20455
13. Raut CG, Deolankar RP, Kolhapure RM, Goverdhan MK (1996) Susceptibility of laboratory-bred rodents to the experimental infection with dengue virus type 2. Acta Virol 40:143–146. https://doi.org/8891094
14. Zompi S, Harris E (2012) Animal models of dengue virus infection. Viruses 4:62–82. https://doi.org/10.3390/v4010062
15. Costa VV, Fagundes CT, Souza DG, Teixeira MM (2013) Inflammatory and innate immune responses in dengue infection. Am J Pathol 182:1950–1961. https://doi.org/10.1016/j.ajpath.2013.02.027
16. Raut CG, Deolankar RP, Kolhapure RM, Goverdhan MK (1996) Susceptibility of laboratory-bred rodents to the experimental infection with dengue virus type 2. Acta Virol 40:143–146
17. Atrasheuskaya A, Petzelbauer P, Fredeking TM, Ignatyev G (2003) Anti-TNF antibody treatment reduces mortality in experimental dengue virus infection. FEMS Immunol Med Microbiol 35:33–42. https://doi.org/10.1111/j.1574-695X.2003.tb00646.x
18. Huang K-J, Li S-YJ, Chen S-C et al (2000) Manifestation of thrombocytopenia in dengue-2-virus-infected mice. J Gen Virol 81:2177–2182. https://doi.org/10.1099/0022-1317-81-9-2177

Chapter 15

Methods for Evaluating Neurotropic Viral Infections: Exploring Zika and Coronavirus Diseases

Jordane C. Pimenta, Jéssica A. B. Pereira, Danielle Cunha Teixeira, Felipe Rocha da Silva Santos, Luiz Pedro Souza-Costa, Manoela Couto, Victor R. M. Costa, Vinícius Amorim Beltrami, Victor Rodrigues Santos, Antônio Lúcio Teixeira, Mauro Martins Teixeira, Aline Silva de Miranda, Celso Martins Queiroz-Junior, and Vivian V. Costa

Abstract

Emerging infections represent a major challenge for public health due to the urgent need to develop efficient clinical protocols in a scenario with few available scientific studies. The experience with Zika virus (ZIKV) shares important points with the COVID-19 pandemic, since both viruses are associated with community transmission, limited diagnostics, significant perinatal neurological risks, prognostic uncertainty, and lack of effective therapies. Zika virus and SARS-CoV-2 infect cells from the central nervous system causing neuroinflammation and neurodegeneration. Moreover, these neurotropic infections have been associated with several behavioral manifestations including depression. Therefore, the development of murine models to better understand the short- and long-term effects of neurotropic infections is desirable. In this chapter, we describe several methods to study neuroinflammation and neurodegeneration in mice infected with Zika and betacoronavirus (murine hepatitis virus, MHV) in biosafety level 2 environments. The described strategies encompass techniques from viral inoculum preparation and infection, to textual and visual descriptions of tissue collection at euthanasia, as well as histopathological and behavioral analyses.

Key words Neuroinflammation, Neuroinfection, Behavior, Zika virus, Coronaviruses

---

The authors Jordane C. Pimenta, Jéssica A. B. Pereira, Danielle Cunha Teixeira have equally contributed to this chapter.

Supplementary Information The online version contains supplementary material available at https://doi.org/10.1007/978-1-0716-5019-6_15.

Lucas Secchim Ribeiro (ed.), *Models in Inflammation and Immunopharmacology*, Methods in Molecular Biology, vol. 2993, https://doi.org/10.1007/978-1-0716-5019-6_15,
© The Author(s), under exclusive license to Springer Science+Business Media, LLC, part of Springer Nature 2026

1 Introduction

Neuroinflammation brought on by viral illnesses has become a major global health issue. Several viruses enter the central nervous system (CNS) and infect neurons and glial cells, triggering the immune system antiviral reaction in order to clear the virus [1, 2]. In this process, however, viruses can lead to production of immune elements known to be detrimental in an imbalanced system, going beyond viral clearance and being even associated with neurodegeneration [3]. Reactive oxygen species (ROS), cytokines, chemokines, and secondary messengers produced by microglia, astrocytes, and endothelial cells are known to play a role in the neuroinflammatory reactions in the brain and spinal cord. Moreover, even the blood-brain barrier can be altered by inflammatory molecules like tumor necrosis factor (TNF), interleukin (IL) IL-6, IL-8, CCL5, and CXCL10, making it easier for immune cells to enter the central nervous system (CNS) and exacerbating the inflammatory deleterious effects.

Neurological diseases may develop even further as a result of the immunopathological processes initiated with the infection by members of the *Flaviviridae* family and, more recently, Betacoronaviruses. Out of the first group, Zika virus (ZIKV) infection is known to be initially systemic and can later reach the peripheral and central nervous system, causing severe neurological manifestations, including microcephaly in fetuses and Guillain-Barre syndrome (GBS) in adults [4, 5]. Once in the CNS, ZIKV infects endothelial cells, pericytes, astrocytes, and microglial cells, promoting the local production of pro-inflammatory cytokines and chemokines that will induce neurodegeneration and neuroinflammation [4]. Similarly, infection with SARS-CoV-2, a betacoronavirus, has been shown to increase critical inflammatory markers that are associated with encephalitis [3, 6, 7]. Encephalopathy and cerebrospinal fluid (CSF) abnormalities have been observed in COVID-19-positive patients, suggesting that the elevation of cytokines and chemokines in individuals infected with SARS-CoV-2 may lead to neuroinflammation and possibly to acute and long-term neurobehavioral consequences [6, 7].

To better comprehend how the modulation of the CNS-immune response can impair the cognitive neural processes of people who have been exposed to these viruses, murine Zika virus and Betacoronavirus infection models have been developed. For instance, type I interferon receptor (A129) knockout mice have been widely used as a suitable tool to study neuroinflammation and neurodegeneration induced by ZIKV infections. A murine model of coronavirus (MHV, a betacoronavirus/murine hepatitis virus) seems to emulate the clinical manifestations of SARS-CoV-2 and has been widely used in research, since this virus has tropism for several organs, including the CNS, and can cause neuroinflammation [8].

In this chapter, we discuss two infection models: (i) a ZIKV model in A129 mice and (ii) an MHV infection in C57BL/6J mice. We provide a detailed demonstration of viral inoculum preparation and subsequent in vivo inoculation; dissection and collection of specific brain areas, techniques to evaluate neuroinflammation and a wide range of behavioral tests.

2 Materials

For better understanding and study of neurological alterations caused by ZIKV infection, the most widely used experimental model is performed in immunocompromised A129 mice, which are deficient for IFN-α/β receptor (IFN-α/βR$^{-/-}$). 8-week-old mice of both sexes are infected with a clinical isolate of ZIKV (HS-2015-BA-01 strain isolated from a viremic and symptomatic patient from the state of Bahia, Brazil, in 2015). In addition to the use of the animal model, we also used Vero cell line (African green monkey kidney cells—*Cercopithecus aethiops*; ATCC #CCL-81) and C6/36 cells (*Aedes albopictus* clone C6/36), for virus propagation.

For the study of neurological alterations caused by coronavirus infection, here we demonstrate an infection model in C57BL/6J, 7 week-old mice of either sex infected by MHV-3 (GenBank accession no. MW620427.1) or with MHV-A59 (ATCC #VR-764), in 4-week-old, C57BL/6J mice. In this infectious model, murine fibroblast cells (cell line L929, ATCC #CCL1), are required.

2.1 General Cell Culture

1. 2-mL microcentrifuge cryotube.
2. Laminar flow hood.
3. Refrigerated microcentrifuge.
4. Sterile pipettes and pipette tips.
5. Incubator with humidified atmosphere of 5% CO_2 at 37 °C.
6. Inverted Microscope.
7. Sterile PBS: 137 mM NaCl, 2.7 mM KCl, 10 mM Na_2HPO_4, 2 mM KH_2PO_4 (pH 7.4).
8. Heated water bath at 37 ° C.
9. Centrifuge tubes (15 mL and 50 mL).
10. Trypsin/EDTA: 0.25% (w/v) Trypsin + 0.02% (w/v) EDTA, warmed up to 37 °C.
11. 175-cm^2 cell culture flasks.
12. Serological pipettes.
13. Electronic pipette.

2.2 Culture of Monolayer Cells and Preparation of ZIKV and MHV-3/MHV-A59 Stock

1. Frozen stock of ZIKV, MHV-3, or MHV-A59.
2. Roswell Park Memorial Institute's culture medium (RPMI) supplemented with fetal bovine serum (RPMI +2.5% FBS and RPMI +10% FBS), optionally supplemented with Gentamicin at 10 μg/mL.
3. Frozen stock of Vero cells (ATCC #CCL-81).

2.3 L929 Cell Culture

1. Frozen stock of MHV-3 or MHV-A59.
2. Dulbecco's modified Eagle medium (DMEM) supplemented with fetal bovine serum (DMEM +2.5% FBS and DMEM +7% FBS), optionally supplemented with Gentamicin at 10 μg/mL.
3. Frozen stock of L929 cells (ATCC #CCL-1).

2.4 ZIKV Study Model in Vivo

2.4.1 Preparation of the Viral Inoculum

1. Laminar flow hood.
2. Sterile pipettes and pipette tips.
3. Frozen aliquot of ZIKV.
4. Roswell Park Memorial Institute's culture medium (RPMI) or sterile saline solution (NaCl 0.9%) for virus dilution.
5. 1.5-mL or 2-mL microcentrifuge tubes.
6. A box with ice.
7. Tube shaker.
8. Microcentrifuge tube rack.

2.4.2 Intravenous Infection

1. A129 mice.
2. Microisolator.
3. Container for Mouse made in 4 mm Crystal Acrylic, with locking system by side handles on the lid; presence of side holes for air passage and non-slip feet; inside measurements in mm (H × W × L): 25 × 25 × 100.
4. 1-mL syringe.
5. 13 mm × 0.30 mm (30G) needle.
6. Sterile gauze.
7. Stock vial with ZIKV diluted at a concentration of 4×10^3 PFU in 100 μL of RPMI 1640 medium.
8. Infrared lamp.
9. Adhesive tape.
10. Sterile saline solution (0.9% NaCl).

2.4.3 Clinical Score

1. Precision scale.
2. Printed table for data registration.
3. Pen and Paper.

2.4.4 Intraocular Pressure (IOP) Evaluation

1. Tono-Pen ® Vet applanation tonometer.
2. Topical anesthetic (based on 1% tetracaine hydrochloride and 0.1% phenylephrine hydrochloride).
3. Tono-pen ® Tip Covers.
4. Parafilm plastic.

2.5 MHV Study Model

2.5.1 Preparation of the Viral Inoculum

1. Frozen aliquot of MHV-3 or MHV-A59.
2. Sterile pipettes and pipette tips.
3. Laminar flow hood.
4. Sterile saline solution (0.9% NaCl).
5. 1.5-mL or 2-mL microcentrifuge tubes.
6. A box with ice.
7. Tube shaker.

2.5.2 Intranasal Infection

1. Tubes with previously aliquoted viruses.
2. Microisolator.
3. Laminar flow hood.
4. Sterile pipette and pipette tip (100 μL or 200 μL).
5. Ketamine hydrochloride (100 mg/kg, final dose per animal).
6. Xylazine (5 mg/kg, final dose per animal).
7. Sterile saline solution (0,9% NaCl).
8. 1.5-mL or 2-mL microcentrifuge tubes.
9. 50-mL centrifuge tubes.

2.5.3 Clinical Score

1. Precision scale.
2. 30 × 30cm delimited space.
3. Printed table for annotation.
4. Pen.

2.6 Behavioral Analysis

All the tests detailed between the Subheadings 2.6.1 and 2.6.6 require an acoustically isolated room, stopwatch, pen, and paper for note-taking.

2.6.1 Open Field Test (OFT)

1. OFT apparatus (*see* **Notes 1** and **2**).
2. Matte black paper, 30 cm × 30 cm.
3. Phenotyper (Noldus, Information Technology, Leesburg, VA, EUA) coupled with computerized tracking software (Ex: EthoVision XT, Noldus Information Technology, Leesburg, VA, EUA) (*see* **Note 3**).
4. Infrared light (*see* **Note 4**) or white light (*see* **Note 5**).

5. Ethanol 70%.
6. Paper towels.

2.6.2 Elevated Plus Maze Test (EPMT) (See **Note 6**)

1. EPMT apparatus (*see* **Note 7**).
2. Experimental animals (*see* **Note 8**).
3. Phenotyper (Noldus, Information Technology, Leesburg, VA, EUA) coupled with computerized tracking software (Ex: Etho-Vision XT, Noldus Information Technology, Leesburg, VA, EUA) (*see* **Note 9**).
4. Light source (*see* **Note 10**).
5. Ethanol 70%.
6. Paper towels.

2.6.3 Sucrose Preference Test (SPT)

1. Animals (*see* **Note 11**).
2. Bottles for water or sucrose solution.
3. Cages that can hold simultaneously both bottles.
4. Sucrose 1%.
5. Filtered water.
6. Precision scale.

2.6.4 Marble Burying Test (MBT)

1. Animals (*see* **Note 12**).
2. Wood shavings.
3. Marbles/glass spheres, 25 units per cage (approximately 15 mm in diameter).
4. Cages with a grid (*see* **Note 13**).

2.6.5 Forced Swim Test (FST)

1. A beaker suitable for the animal's size.
2. Lukewarm water (*see* **Note 14**).
3. Camera/tracking software EthoVision XT (Noldus, Technology, Leesburg, VA, EUA).
4. Testing site with an adjacent room.
5. Paper towels.
6. Heating lamp.

2.6.6 Olfactory Discrimination Test (ODT)

1. Animals (*see* **Note 15**).
2. Separation boxes (*see* **Note 16**).
3. Clean wood shavings.
4. Wood shavings, unchanged from at least three previous days of regular use.
5. Paper towels.
6. Matte black paper.

7. Ethanol 70%.
8. Computer-coupled camera.

2.6.7 Y-Maze Test

1. 3-armed apparatus (Y-maze) (*see* **Note 17**).
2. Partition to block one of the arms of the labyrinth.
3. Acoustically isolated room with an adjacent site for the experimenter to be in (*see* **Note 18).**
4. Signs/clues around the maze (*see* **Note 19**).
5. Ethanol 70%.
6. Paper towels.
7. Pen.
8. Camera.
9. Phenotyper (Noldus, Information Technology, Leesburg, VA, EUA) coupled with computerized tracking software (Ex: Etho-Vision XT, Noldus Information Technology, Leesburg, VA, EUA) (*see* **Note 3**).
10. Stopwatch.

2.6.8 Grip Strength Test (GST)

1. Grip Strength test apparatus (*see* **Note 20**).
2. Ethanol 50%.
3. Paper towels.
4. Pen.
5. Precision scale.

2.7 Euthanasia

1. Laminar Flow Hood.
2. Surgical scissors and tweezers.
3. Anesthetic: Ketamine (100 mg/kg), Xylazine (10 mg/kg), diluted in sterile saline (0.9% sodium chloride).
4. Sterile PBS (see Subheading 2.1) or sterile saline solution (0.9% NaCl).
5. Ice box.
6. 1.5-mL tube for blood collection, treated with heparin (300 mM).
7. Sterile 2-mL microcentrifuge tubes.
8. 1-mL or 3-mL syringe.
9. 13 mm × 0.45 mm (26G) or 25 mm × 0.70 mm (22G) needles.
10. 70% ethanol.
11. Aluminum foil.

2.7.1 Dissection and Collection of Specific Brain Areas

1. C57BL/6J mice (MHV model) or A129 mice (ZIKV model).
2. Microisolator for mice.
3. Liquid nitrogen or ice box (*see* **Note 21**).
4. 2 stainless steel spatulas (6 mm wide × 18 cm long; weight 70 g) (*see* **Note 22**).
5. 1 stainless steel spatula (6 mm wide × 17.7 cm long; weight 75 g) (*see* **Note 23**).
6. Filter paper (80 g/m^2, 205 μm thick, 14-μm pores, 14 l/sm^2 air permeability) (*see* **Note 24**).
7. Sterile saline solution: 0.9% NaCl or sterile buffered PBS (*see* **Note 25**).
8. Styrofoam box (16 cm wide × 9.5 cm high) (*see* **Note 26**).
9. 2-mL microcentrifuge tubes previously identified.
10. Iris-type surgical scissors—Golgran (10.5-cm straight tip).
11. Laminar flow hood.
12. Petri dish.
13. 70% ethanol.
14. Aluminum foil.

*2.7.2 Brain Collection and Storage for 40 μm Sections (See **Note 27**)*

1. Stainless steel spatula (6 mm wide × 18 cm long; weight 70 g) (*see* **Note 22**).
2. Stainless steel spatula (6 mm wide × 17.7 cm long; weight 75 g) (*see* **Note 23**).
3. Sterile buffered PBS.
4. 4% PFA in PBS (Paraformaldehyde).
5. Solid Sodium Hydroxide (NaOH).
6. PFA solution: 4% PFA, diluted in PBS (*see* **Note 28**).
7. 15-mL centrifuge tubes filled with 7 mL of PFA solution (*see* **Note 29**).
8. Filter paper (80 g/m^2, 205 μm thick, 14-μm pores, 14 l/sm^2 air permeability).
9. Iris-type surgical scissors—Golgran 0.14-N (11.5 cm straight tip).
10. 70% ethanol.
11. Petri dish.

2.7.3 Brain Collection and Storage for 5 μm Sections

1. 1 stainless steel spatula (dimensions 6 mm wide × 18 cm long; weight 70 g); (*see* **Note 22**).
2. 1 stainless steel spatula (dimensions 6 mm wide × 17.7 cm long; weight 75 g) (*see* **Note 23**).

3. Filter paper (80 g/m^2, 205 μm thick, 14 μm most pores, 14 l/sm^2 air permeability).
4. 4.1 Iris-type surgical scissors—Golgran 0.14-N (11.5 cm straight tip).
5. 1 Plastic bottle with 500 mL of formalin.
6. Previously identified histological cassettes.
7. Surgical tweezers.

2.7.4 Dehydration of Brain Tissue for Freezing for 40 μm Sections

1. Brain tissues stored in centrifuge tubes with PFA.
2. Sterile PBS (see Subheading 2.1).
3. Sucrose solution at 30%, 20% and 10%, each diluted in PBS.
4. Refrigerated storage.
5. Magnetic stirrer.
6. Liquid disposal container for PFA.

2.7.5 Brain Tissue Freezing for Cryostat Sections

1. Brain tissues stored in 30% sucrose solution.
2. Isopentane.
3. Dry ice.
4. Ice box.
5. Tweezers.
6. Paper towels.
7. Aluminum foil in pieces of 8 cm × 8 cm.
8. Liquid disposal container for sucrose.
9. Identification tags for each animal.
10. Stapler.
11. Freezer at −80 °C.
12. Plastic beaker (50 mL).

*2.7.6 Gelatinization of Histological Slides for 40 μm Sections (See **Note 30**)*

1. Histological glass slides.
2. Glass container and slide supporter.
3. Tweezers.
4. Paper towels.
5. 0.01M HCl solution.
6. 400 mL Distilled water.
7. 2 g of gelatin powder.
8. 0.2 g of chrome alume.
9. Gelatin solution (*see* **Note 31**).

2.7.7 Obtaining and Storing 40 μm Brain Sections

1. Gelatinized histological slides.
2. Anti-freezing solution (PBS: Ethylene glycol: Glycerol at a 5: 3:2 proportion) (*see* **Note 32**).
3. Flat-bottomed 24-well cell culture plates (each well containing 700 μL of antifreeze solution).
4. Cryostat (*see* **Note 33**).
5. Ice box.
6. 2 soft bristle brushes (3 mm × 7 mm).
7. Pencil.
8. Beaker with 50 mL of tap water.
9. Tissue-Tek O.C.T.

2.8 Techniques for Assessing Neuroinflammation

For Subheadings 2.8.1 and 2.8.2, the following laboratory material and equipments are needed:

1. Plate shaker.
2. 24-well flat-bottom cell culture plate.
3. Aluminum foil.
4. BSA (Bovine serum albumin).
5. Triton X-100.
6. Sterile PBS (see Subheading 2.1).
7. Blocking solution (BSA 2%; Triton 0.2% Diluted in PBS).
8. 2 soft bristle brushes (3 mm × 7 mm).
9. Magnetic shaker.
10. Petri dish.

2.8.1 IBA-1 and GFAP Immunofluorescence for Cryostat Sections

1. Brain sections of 40 μm stored in a −20 °C freezer, previously placed in a 24-well plate, in antifreeze solution.
2. Blocking solution BSA 2%; Triton 0.2% diluted in PBS).
3. Primary antibodies anti-IBA-1 and anti-GFAP (*see* Table 5).
4. Fluorescent secondary antibody (*see* Table 5).
5. Pipettes and pipette tips.
6. Fluorescence microscope with blue wavelength 450-nm filter.
7. Gelatinized histological slides.
8. Glass cover slides 24 × 50 mm.
9. Fluoromont G.

2.8.2 Immuno-histochemistry Ki-67 for Cryostat Sections

1. Brain sections of 40 μm stored in a −20 °C freezer, previously placed in a 24-well plate, in antifreeze solution.
2. Primary antibody anti-rabbit anti-Ki-67, diluted in blocking solution 1:500.

3. Kit ABC (15 μL of A, 15 μL of B diluted in 1 mL of PBS) (*see* **Note 34**).
4. DAB (3.3′-Diaminobenzidine) Chromogen agent (VECTASTAIN® Elite® ABC Reagent and DAB substrate SK-4100): (2 volumes of buffer, 4 volumes of H_2O_2, 2 volumes of DAB).
5. Glass containers for staining slides.
6. 8 glass containers with Ethanol at 50%, 70%, 80%, 90%, 100%, ethanol + xylene, xylene I and xylene II.
7. Glass cover slides, 24 × 50 mm.
8. Tweezers.
9. Entellan (Merck).
10. Fume hood.
11. Pasteur pipette.

2.8.3 Immunohistochemistry IBA-1 and S100B for 5 μm Paraffinized Sections

1. Washing buffer TBS: 0.5 mM Tris (hydroxymethyl) aminomethane, 145 mM NaCl; Triton X-100 (pH 7.2–7.4).
2. PBS (see Subheading 2.1).

Buffered Wash Solutions

Antigen Recovery

1. Citrate buffer: 10 mM citric acid (pH 6.0).
2. EDTA solution: 1 mM $Na_2EDTA.2H_2O$ (pH 8.0).

Antibody Dilution Solution

1. PBS (see Subheading 2.1).
2. BSA 0.1% (w/v) in PBS.
3. TBS: 0.5 mM Tris-(hydroxymethyl)aminomethane, 145 mM NaCl (pH 7.2–7.4).

Antibodies

1. Mouse anti-Iba1 Antibody (concentration 6.6 ug/mL).
2. Mouse anti-S100B Antibody (concentration 93.3 ug/mL).
3. Goat Anti-Rabbit IgG (H + L) biotinylated, R.T.U (Ready To Use).

Blocking Solution

1. Protein block buffer (Abcam ab64226).
2. Prepare 200 mL of H_2O_2 3% (w/v) by mixing 20 mL of H_2O_2 30% (v/v) with 178 mL of TBS, followed by the addition of 2 mL of Triton ×-100.

Streptavidin-Peroxidase Solution

1. Vectastain Elite ABC-HRP Kit, Peroxidase (Standard) PK-6100 or similar.

DAB Chromogen Agent

1. DAB Chromogen System:1 mL DAB Substrate Buffer +30 μL DAB Chromogen.
2. Fume hood.
3. Fume masks.
4. Gloves.
5. Pipettes and pipette tips.
6. Proper disposal of DAB liquid and tips used in reagent pipetting.
7. Support for histological slides.
8. Container for solution containment and slide storage.

Reagents for Deparaffinization, Hydration, Dehydration, and Assembly

1. Xylene 100%.
2. Ethanol solutions at 70%, 80%, 90% and 100%.

Staining

1. Haematoxylin of Harris (Newprov).

Processing and Slide Assembly

1. Microscope slides for immunohistochemistry: 3 in × 1 in.
2. Incubator.
3. Vertical support for histological slides.
4. Containers for solution containment and slides support.
5. Containers for overnight slide incubation.
6. Funnel.
7. Filter paper (80 g/m2, 205 μm thick, 14-μm pores, 14 l/sm2 air permeability).
8. Hydrophobic pen: Super Pap Pen Liquid Blocker.
9. Pipettes and pipette tips.
10. Disposal for tips and liquids.
11. Glass cover slides 24 × 40mm.
12. Entellan (Merck).
13. Tweezers.
14. Histological slide storage box.

3 Methods

3.1 Vero Cell Culture and ZIKV Inoculum Production

1. Prepare a sterile environment in the laminar flow hood using UV light at least 15 min before the start of procedure.
2. Rapidly thaw a frozen vial of Vero cells in a 37 °C water bath. Remove the vial from 37 °C immediately after thawing.

3. Disinfect the external part of the cell vial with 70% ethanol, and in the laminar flow hood, transfer its content with a pipette into a 50-mL centrifuge tube.
4. Following, add 10 mL of warm RPMI +10% FBS. Centrifuge the tube for 5 min at 420 × *g*, at room temperature. Then, discard the supernatant.
5. Resuspend the obtained pellet in 1 mL of RPMI + 10% FBS, and transfer it to a 75-cm^2 cell culture flask containing 20 mL of RPMI + 10% FBS. Gently, move the flask to evenly distribute cells and place overnight in a 37 °C, 5% CO_2 humidified incubator and check daily for confluency.
6. When cells become a confluent monolayer (after 2 or 3 days), aspirate the culture medium. Wash the cells twice with PBS to remove any remaining serum and then pipette it off.
7. Overlay cells with warm trypsin + EDTA solution using enough volume to cover the monolayer (e.g., 2 mL for a 75-cm^2 flask). Incubate up to 5 min in a 37 °C, 5% CO_2 humidified incubator, and tap the side of the flask to completely detach the cells.
8. Add 4 mL of RPMI + 10% FBS to the cell culture flask with detached cells. Homogenize the cell suspension with a sterile pipette to disrupt any clumps.
9. Count cells using a Neubauer chamber and seed 2.5×10^4 viable cells/cm^2. The cells should be subcultured each 3–4 days post seeding.
10. For the infection, prepare 5 mL of inoculum for each 175-cm^2 tissue culture flask.
11. For the inoculum (ZIKV + culture medium), thaw the cryotube with the virus to be propagated in an ice bath. Upon thawing, transfer the contents of the vial into a 50-mL centrifuge tube, containing 4 mL of RPMI without FBS (final solution formed with 1 mL of viral stock + 4 mL of culture medium). Gently homogenize the solution using a micropipette.
12. Identify the flasks to be used before infection.
13. Aspirate the culture supernatant and rinse the flask with 15 mL of PBS, and with light movements, wash the cell monolayer. Decant or aspirate the PBS.
14. Repeat the previous step, **step 13**, twice.
15. Add the 5 mL of previously produced inoculum, ensuring that its distribution is homogeneous across the cell layer.
16. Place the flask(s) in a CO_2 incubator and gently homogenize the flasks for optimal viral adsorption every 15 min for 1 hour.
17. At the end of the viral adsorption time, add 45 mL of RPMI+ 2% FBS to the flask.

18. Place the flasks in an incubator intended for the cultivation of infected cells and monitor the cytopathic effects (CPE) daily, until the collection of the virus.
19. Once the CPE reaches 80%, perform viral collection as described below. It usually takes between 3 and 5 days after infection.
20. In a laminar flow hood intended for viral manipulation, remove all cells still adhered with a cell scraper.
21. Transfer the supernatant and the cells into a 50-mL centrifuge tube.
22. Centrifuge the contents at 2000 × *g* for 8 min, at 4 ° C.
23. Transfer the supernatant containing the viral copies to another 50-mL centrifuge tube.
24. Aliquot the collected stock and concentrated in cryovials identified with the date and viral isolate used.
25. Store the virus aliquots in a −80 °C freezer.

3.2 L929 Cell Culture and MHV Inoculum Production

The culture of L929 cells for production of the MHV inoculum can be executed following the procedure adopted for the Vero cell culture and ZIKV amplification, with the following differences:

- Overall, DMEM is used instead of RPMI.
- DMEM +7% FBS (instead of 10%, **step 5**).
- The virus collection point (~ 80% CPE) happens earlier, between 1 and 2 days after infection (instead of 3–5 days).

3.3 ZIKV Study Model

3.3.1 Preparation of the Viral Inoculum

1. Select the virus to use and thaw one vial of the stock inoculum on ice.
2. After thawing, the virus needs to be handled in completely sterile laminar flow with materials such as sterile pipettes and tips.
3. Inside the laminar flow, the vial of the virus must be homogenized in the vortex. According to the titer of the original ZIKV inoculum, dilute the aliquot of the virus with sterile saline solution (0.9% NaCl) to a final concentration of 4×10^4 PFU/mL. The final concentration and volume to be given is 4×10^3 PFU/mouse in 100 μL.
4. After preparing the inoculum to the final concentration and volume, the solution must be kept on ice until it is time to transfer it to a syringe that will carry out the immediate infection in the animals.

3.3.2 Intravenous Infection

1. On a clean experimental bench, take the microisolator with A129 mice, open it and expose them to the heating lamp for 2–3 min to dilate the tail veins (Fig. 1a) (*see* **Note 35**).

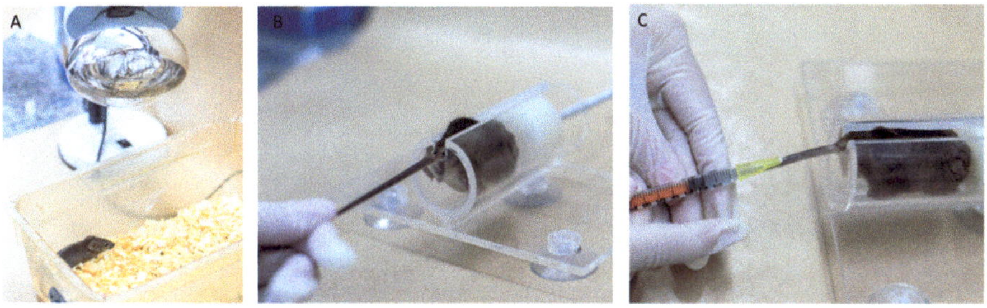

Fig. 1 Intravenous infection. (**a**) The box of mice from the experimental group to be infected will be exposed to a heat lamp for about 3 min for the tail vein to dilate. (**b**) After dilation, it is possible to see the tail vein located more laterally. (**c**) After locating the vein, place the needle parallel with the tail, carefully inserting the needle and lightly injecting the solution. A correct application should be fluid and there should be no resistance

2. Restrain each mouse on the mechanical container and expose the tail (Fig. 1b) (*see* **Note 36**).
3. Find the lateral tail vein by its purple color due to the dilation process (Fig. 1b). Anatomically, there are 2 thinner lateral veins that are more accessible because they are more superficial. There is also a deep dorsal vein, which is thicker but difficult to access.
4. Clean the animal's tail with a cotton pad soaked with 70% ethanol.
5. Position the needle with the bevel upwards.
6. Insert the needle parallel to the vein at the most distal position possible (Fig. 1c).
7. Slowly press the syringe plunger. If there is resistance, retry the insertion in a less distal position. A successful injection should offer no resistance.
8. Take out the needle, apply pressure with a gauze and proceed to the next mouse.

3.3.3 Intraocular Pressure (IOP) Evaluation

1. Separate the animals and calibrate the tonometer as specified (Fig. 2a).
2. Restrain the animal and add a drop of anesthetic over its eye; subsequently, wait for the anesthetic effect in accordance with the product's instructions.
3. Delicately press the tonometer at the mouse's eye, perpendicularly to the most apical portion of the cornea (Fig. 2b).
4. Each mouse should be measured 3 times. For better measurements the same eye should be evaluated on every animal.
5. For data analysis, plot the mean value expressing the mmHg unit of measurement.
6. The IOP evaluation should be performed at baseline, 3 and 5 days after infection, as well as prior to euthanasia.

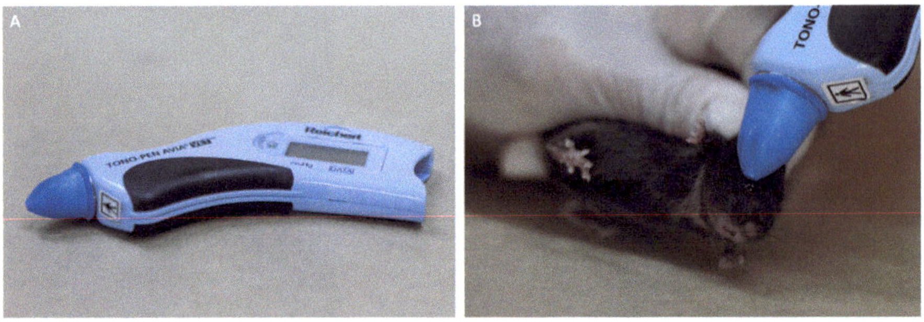

Fig. 2 Intraocular pressure. (**a**) Assemble the tonometer, place the cover and calibrate the tonometer according to the equipment's manual. (**b**) Gently restrain the mouse by holding the scruff of the neck and supporting the body, then place a drop of topical anesthetic in the animal's eye. Then, hold the tonometer perpendicular to the corneal surface, press the equipment button and carefully touch the animal's eye until the tonometer confirms the measurement. You must perform 3 measurements disregarding any percentage of error above 5%

3.4 MHV Study Model

3.4.1 Preparation of the Viral Inoculum

1. Select the virus aliquot to use and thaw one vial of the stock inoculum on ice.
2. After thawing, the virus needs to be handled in completely sterile laminar flow with materials such as sterile pipettes and tips.
3. Inside the laminar flow the vial of the virus must be homogenized in the vortex. According to the titer of the original MHV-3 or MHV-A59 inoculum, dilute the aliquot of the virus with sterile saline solution (0.9% NaCl). For MHV-3, an inoculum of 3×10^3 PFU/mouse are given to each mouse in 30 μL. For MHV-A59, an inoculum of 3×10^4 PFU/mouse are given to each mouse in 30 μL.
4. After preparing the inoculum to the final concentration and volume, the solution must be kept on ice until it is time to transfer it to a syringe that will carry out the immediate infection in the animals.

3.4.2 Intranasal Viral Inoculation

1. With the inoculum prepared, separate the animals and inject 60 μL of anesthetic solution containing ketamine (80 mg/kg) and xylazine (15 mg/kg) intraperitoneally (*see* **Note 37**).
2. After the mouse is anesthetized, choose the left or right nostril to initiate virus inoculation.
3. Using a pipette, apply 15 μL to the chosen nostril of all animals (Fig. 3a, b).
4. Inoculate all animals in the first nostril, waiting for completely aspiration of the droplet. Then, inoculate the remaining 15 μL into the other nostril for all animals.
5. After the inoculation is finished, wait for the animals to recover as the anesthetic wear off.

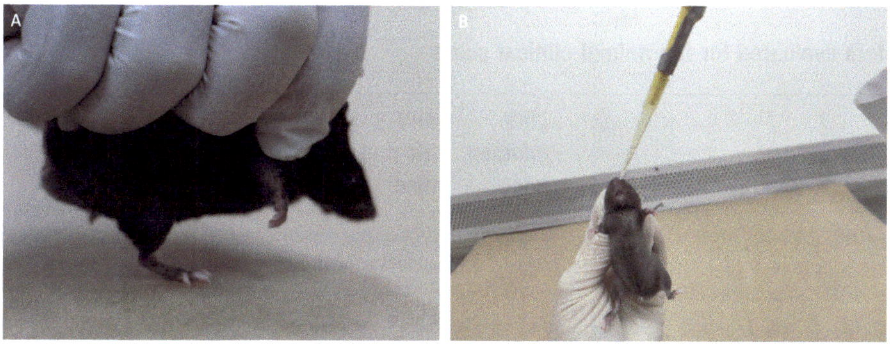

Fig. 3 Intranasal viral inoculation. (**a**) The mouse must be completely immobilized for the application of intraperitoneal anesthesia. (**b**) After the mouse is anesthetized, within the laminar flow with the aid of a pipette, 15 μL of the virus must be inoculated into the animal's nostril. After inoculation of all animals in the group, repeat the same process in the other nostril of the animal with 15 μL of the virus, totalizing 30 μL/mouse

3.4.3 Clinical Score of ZIKV and MHV Models

1. Weigh each animal on a precision scale, record the data, and return them to their respective microisolators (*see* **Note 38**).
2. Visually assess the clinical score of each animal, following the parameters shown in Table 1.

3.5 Behavioral Analysis

Behavior is the observable manifestation of neural activity [9]. Specific behaviors may inform about the neural activity of specific brain circuits and/or regions [Table 4]. Rodents are the most used animal model in behavioral neuroscience since they manifest behaviors that can be translated into cases of human diseases. There are already several behavioral tests for laboratory animals that address both locomotor and cognitive/emotional aspects. Here we will show the most commonly used tests in research, as well as the purpose of each one. However, before carrying out any test, some parameters need to be taken into account.

To assess whether the analysis of behavioral data in animal models is applicable to human conditions, it is generally classified based on three types of validity. **Construct validity** refers to the similarity between the origin or pathophysiological mechanisms of the condition in the animal model and in humans; **predictive validity** is observed when treatments produce similar effects in both the model and the human condition; and **apparent validity** occurs when the behaviors or symptoms presented by the animal reproduce the human disease [10].

Control groups should always be subjected to the same behavioral tests as experimental groups; At least ten animals per group are desirable to observe differences in the most common behaviors. Furthermore, significant results should be repeated in a second (or independent) cohort of animals [11]. Animals of different

Table 1
Parameters evaluated for the animal clinical score

Features	Healthy animal	ZIKV infected mice	MHV-3 infected mice	MHV-A59 infected mice	Procedure
Weight loss	Mice from 8 to 10 weeks, on average 20 g	Perform daily weight variation and score each parameter in: 0% (0 points), 1–5% (0.5 points), 6–10% (1 point), 20% perform euthanasia (2 points)			Weigh the animals from day 0 d.p.i to the day of euthanasia
Bristling	No bristling	May have bristly hair from the 2/3 day onward		Little to no effect	Daily assessment
Locomotion activity	Actively moves around the box, exploring the entire	Moves with less intensity, remaining in specific places in the box		Little to no effect	Daily assessment
Posture	Elevated posture, with lifting movements	More arched posture		Little to no effect	Daily assessment
Ocular secretion	No secretion	May present ocular secretion from the 2nd or 3rd day onward		Little to no effect	Daily assessment

Score for each parameter of each mouse: 0 for absent; 0.5 for mild; 1.0 for moderate and 2 for severe

sexes cannot be mixed to carry out the tests. Data obtained from both sexes should also not be combined, unless it has already been demonstrated that there is no difference between them in the behavior test performed. Behavioral analyses cannot be compared between different rodent strains [12]. The mice must be 2–-6 months old at the time of the test, except when the test performed has the objective of evaluating some specific parameter of young or elderly [12].

3.5.1 Open Field Test (OFT)

The Open Field Test (OFT) is a test used to evaluate locomotor activity, exploratory activity, and anxiety-like behavior in animals, mainly in rodents [13] (*see* **Notes 1** and **2**).

1. The arena must be cleaned with 70% ethanol and allowed to dry before use.
2. Direct the camera to the arena to record the test area and verify that the tracking system is ready for use (Fig. 4a).

3. Place the animals cages in the experimentation room to undergo acclimatization for at least 30 minutes before the test (*see* **Note 6**).
4. After the acclimatization period, before starting the recording, identify the animal/experimental group in the video that will be recorded.
5. Carefully place the animal in the arena and immediately return it to the center of the apparatus so that the test can be started. Record behavior for 30 minutes, the duration of the protocol may be adjusted depending on the study hypothesis or the condition being investigated.
6. At the end of the test, stop the recording and place the animal back in its cage and returned to its previous containment room.
7. Before starting the test with another animal or experimental group, the apparatus must be cleaned again with 70% ethanol to remove any type of odor, feces or urine.

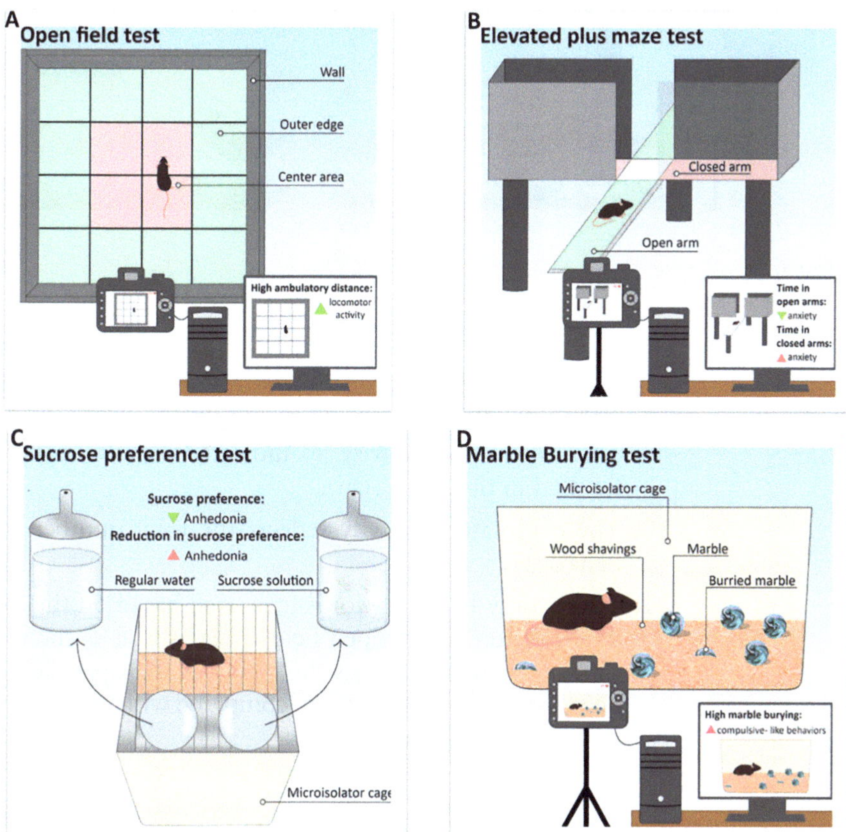

Fig. 4 Behavioral analysis. (**a**) Open field test. (**b**) Elevated plus maze test. (**c**) Sucrose preference test. (**d**) Marble Buried test. (**e**) Forced Swim test. (**f**) Olfactory Discrimination test. (**g**) Y-Maze test. (**h**) Grip strength test (Grip force test)

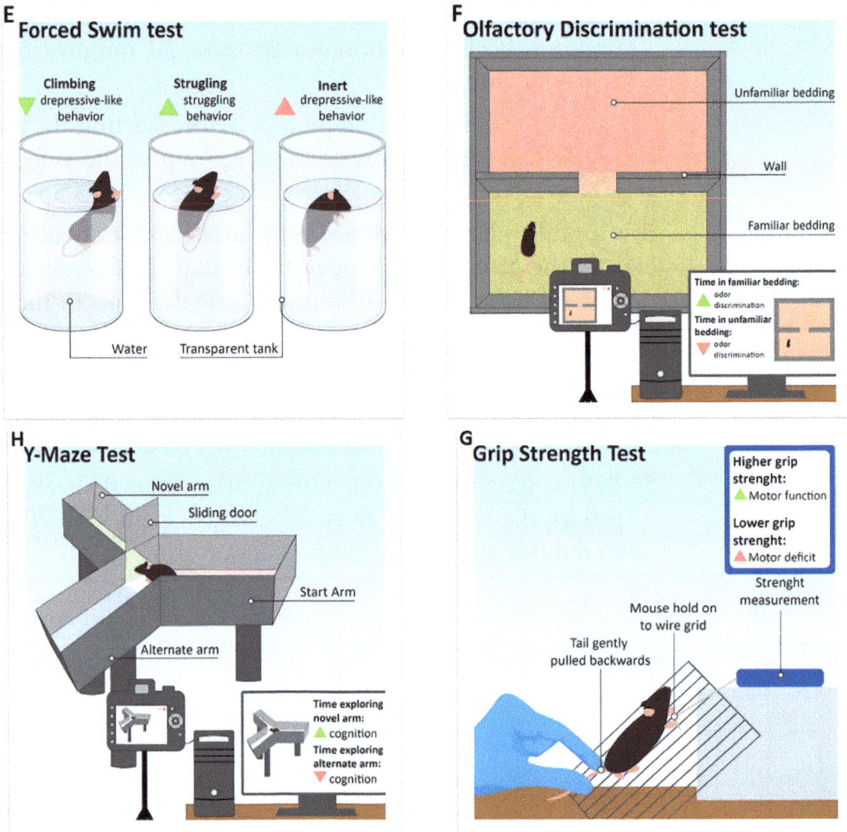

Fig. 4 (continued)

8. Verify if the test recording of the previous animal was saved correctly before identifying the video of the next animal or experimental group.
9. The locomotor activity is indicated by the total distance traveled in the arena (cm).
10. The exploratory activity reflects the total number of rearing behaviors.
11. The anxiety-like behavior is measured based on the time spent in the center of the open field arena as well as the number of times that the mice cross the center of the open field arena. A decrease in these parameters, compared with controls, indicates an anxiety-like behavior.

3.5.2 Elevated Plus Maze Test (EPMT)

The elevated plus maze test (EPMT) is a behavioral test widely used for studying anxiety-like behavior in rodents and the anxiolytic and anxiogenic effects of drugs. The test is based on the natural aversion of rodents to open and high places, as well as on the curious and spontaneous exploratory behavior that leads them to get to know new spaces [14]. The apparatus for carrying out the test has variations, but normally for mice, the EPM consists of four arms, which can be made of wood, acrylic or plastic, with an elevation between 30 and 50 cm in relation to the floor of the experimentation environment, with the open and closed arms measuring 25 to 35 cm in length by 5 to 10 cm in width, while the open arms have a small wall of 0.5 cm to prevent falls, the closed arms have structures of 15 cm in height [14, 15]. The open arms are placed on opposite sides, as well as the closed arms, and then the open and closed arms are positioned perpendicularly, crossing the center of the apparatus, where the test begins [15] (*see* **Note 39**).

1. Sanitize the EPM with 70% ethanol and let it dry.
2. Take the animals from their housing 30 min before the test for acclimatization in the experimentation room.
3. Verify if the video/camera tracking system is ready to use.
4. After the acclimatization period, before starting the recording, identify the animal/experimental group in the video that will be recorded.
5. Remove the rodent from its cage and place it in the center of the apparatus facing the open arm opposite to where the experimenter is standing (*see* **Note 11**).
6. Start recording the moment the animal is placed in the central area of the EPM (for 5 min).
7. At the end of the test, the recording must be stopped, the animal placed in its cage and returned to its previous containment room (Fig. 4b).
8. Before starting the test with another animal or experimental group, the apparatus must be cleaned again with 70% ethanol to remove any type of odor, feces or urine.
9. Verify if the test recording of the previous animal was saved correctly before identifying the video of the next animal or experimental group.
10. A list of possible issues in the EPMT and their solutions can be found in **Note 39.**
11. The anxiety-like behavior is generally indicated by a decrease in the number or percentage of time spent or arm entries in the open arms of the EPM. The absolute number of entries in the closed arms of the EPM is a measure of locomotor activity.

3.5.3 Sucrose Preference Test (SPT)

SPT is a reward-based test capable of assessing state of motivation and anhedonia (reduced ability to experience pleasure) in animal models [16]. The state of anhedonia leads to a reduced preference for the sugar solution, indicated by lower consumption [17].

1. Before starting the test, the animals must be acclimatized to the presence of two bottles of liquid, for at least 3 days, in the cage (*see* **Note 40**).
2. Animals may have a predilection for a particular bottle or drinking side, so to investigate possible preferences, water intake should be measured daily in both bottles for 3 days (*see* **Note 41**).
3. After the habituation period, the animals stay in pairs in a cage containing two bottles, one with a 1% sucrose solution and the other with filtered water (*see* **Note 42**).
4. The amount of water and sucrose solution consumed should be measured daily preferably at the same time.
5. At the end of the test, sucrose preference is calculated as a percentage of sucrose solution intake volume over total fluid intake volume and then averaged over the 3 test days (Fig. 4c).

$$\text{Sucrose preference} = \frac{V_{\text{Sucrose solution ingested}}}{V_{\text{Sucrose solution ingested}} + V_{\text{Water ingested}}} \times 100$$

3.5.4 Marble Burying Test (MBT)

The marble burying test (MBT) is based on the habitual burying and digging behavior of rodents, both in the wild and in the laboratory [18, 19]. Such behavior varies according to the species (it buries more or less) and has the objective of searching for food, burying objects whether they are harmful or not, as well as building nests and tunnels to protect against predators and to favor reproduction [20]. However, this habit can be characterized as pathological when the act of burying is incessant. The MBT is used to demonstrate compulsive-type behavior phenotypes [21]. According to Njung'e and Handley (1991), the MBT can also be used to detect anxiety-like behavior, however the act of burying is not encouraged by the presence of a novelty, in this case the marble, but reflects an obsessive-compulsive type behavior [21].

1. Fill the cages with wood shavings to a height of approximately 20 cm, forming a uniform platform.
2. Pick up the animals from their accommodations and take them to the experimentation room.
3. Place the marbles/glass spheres on the surface of the shavings in an equidistant way, about 4 cm apart from each other (for example: 25 spheres, arranged in 5 columns with 5 spheres in each column).
4. Place one animal in each cage and observe for 30 min (*see* **Note 43**).

5. At the end of the test, the animals are returned to their housing/housing cage and then the number of spheres buried by each animal is measured by two evaluators (only spheres that have had more than ⅔ of buried volume are considered).
6. The analysis is based on the number of spheres buried by the animals. When the experimental group buries/digs more spheres than the control group, there is evidence of obsessive-compulsive behavior (Fig. 4d).

3.5.5 Forced Swim Test (FST)

The forced swim test (FST) is the golden standard test for evaluating depressive-like behavior in rodents, being widely used in the screening of antidepressant drugs [22]. This test consists of placing rodents, individually, in cylinders filled with water to assess their behavior, that is, animals with symptoms of depression give up swimming earlier than healthy animals [22]. For mice, the animals remain in the water for 6 min, divided into a pre-test (the first 2 min) and the test itself (the remaining 4 min) (Tables 2, 3, and 4).

1. Mice must be moved from housing to the adjacent testing room for acclimatization (at least 30 min beforehand) (*see* **Note 44**).
2. Put water in beakers (~25 °C) so that the mouse does not touch its hind legs or tail on the bottom of the container.
3. Take note of the identification and group of the mouse submitted to the test.
4. Check if the camera is positioned correctly and the software is ready to use.
5. Start recording for 6 min (*See* **Note 45**) (Fig. 4e).
6. At the end of the test, interrupt the recording, remove the mice from the water, place them in a cage with a paper towel to dry with a warm light on over the box to provide heat (*see* **Note 46**).
7. Change the water in the beaker after each animal.

Table 2
Brain areas and the purpose of study

Brain area	Purpose of each area
Hippocampus	High capacity for neuronal plasticity, neurogenesis, cell proliferation, memory storage, mood and emotion
Pre-frontal cortex	Planning complex cognitive behavior, expressing personality, making decisions, and moderating social behavior.
Striatum	Regulation of muscle tone and strength and range and speed of movement
Cerebellum	Able to control motor activities by receiving muscle stimuli, maintaining muscle tone; keeper of body balance

Table 3
Inflammatory mediators and neurotrophic factor for quantification

Mediador	ZIKV model	MHV-A59 model	MHV-3 model
TNF	Hippocampus; Pre frontal cortex; Striatum; Cerebellum	Hippocampus; Pre frontal cortex;	Hippocampus; Pre frontal cortex; Striatum
IL-6	Hippocampus; Pre frontal cortex; Striatum; Cerebellum	Hippocampus; Pre frontal cortex;	Hippocampus; Pre frontal cortex; Striatum
IL-10	Hippocampus; Pre frontal cortex; Striatum; Cerebellum	Hippocampus; Pre frontal cortex;	Hippocampus; Pre frontal cortex; Striatum
IL-1β	Hippocampus; Pre frontal cortex; Striatum; Cerebellum		
CX3CL1	Hippocampus; Pre frontal cortex; Striatum; Cerebellum	Hippocampus; Pre frontal cortex;	Hippocampus; Pre frontal cortex; Striatum
CXCL1	Hippocampus; Pre frontal cortex; Striatum; Cerebellum		
BDNF	Hippocampus; Pre frontal cortex; Striatum; Cerebellum	Hippocampus; Pre frontal cortex;	Hippocampus; Pre frontal cortex; Striatum

8. Analyze animal behavioral data (*see* **Note 47**).
9. An increase in the immobility time (in seconds) is the main measure of depressive-like behavior.

3.5.6 Olfactory Discrimination Test (ODT)

The sense of smell plays a fundamental role in the lives of mammals, especially rodents, as it is involved in essential activities such as locating food, detecting predators, and regulating reproductive behaviors [23, 24]. Due to their high olfactory sensitivity, these animals tend to associate learning and memory processes with olfactory signals [24]. In addition, a relationship has been established between olfactory dysfunctions and several neurological diseases, and difficulty in discriminating odors may be related to future cognitive deficits. Therefore, olfactory tests in rodents have been

Table 4
Relating behavioral tests and the CNS

Behavioral test	Main meaning	Brain regions implicated
Open field	Sensorimotor test used to assess locomotor activity in rodent's models	Primary motor cortex Posterior parietal cortex Premotor cortex Supplementary motor area
Grip strength	Test used to assess muscle strength in rodents, for example, in a model of neuromuscular diseases	Primary motor regions Basal ganglia
Elevated plus maze	Test used to assess anxious-like behavior in rodent's models	Amygdala Insula Prefrontal cortex
Sucrose preference	Test that evaluates anhedonia behavior in rodents	Anterior cingulate cortex Medial prefrontal cortex Ventral prefrontal cortex Amygdala Striatum Hippocampus
Marble buried	Test employed as a model or measure to study anxiety and compulsive-like behavior	Anterior cingulate cortex Medial prefrontal cortex Ventral prefrontal cortex Amygdala Striatum
Forced swim	Test used to assess depressive-like behavior	Anterior cingulate cortex Medial prefrontal cortex Ventral prefrontal cortex Amygdala Striatum Hippocampus
Olfactory discrimination	Test used to assess the ability to detect and differentiate odors	Olfactory bulb Piriform cortex Entorhinal cortex Orbitofrontal cortex
Y-maze	Test used to evaluate spatial working memory and spatial reference memory	Prefrontal cortex Hippocampus Posterior parietal cortex

used as auxiliary tools in the diagnosis of conditions such as dementia and even in the identification of anosmia caused by COVID-19 [24]. Among the various methods available to investigate the functioning of the olfactory system, the protocol described in this study is based on the animal's ability to distinguish between a previously known environment—with used wood shavings—and a new environment, containing clean wood shavings.

Seclusion

1. First, separate the animals individually into cages with clean wood shavings (*see* **Note 48**).
2. Label each cage according to the group to which the animals belong.
3. Water and food should be added as normal, but wood shavings should not be changed until testing.

Acclimatization

1. Clean the ODT apparatus with 70% ethanol beforehand and dry with a paper towel.
2. Fill both apparatus compartments with clean wood shavings.
3. Start recording, place the animal in the center of the apparatus and let it explore the two compartments for 2 min (*see* **Note 49**).
4. After 2 min have elapsed, stop the recording, remove the animal from the apparatus and return it to the cage.
5. Remove the shavings from both compartments, clean the apparatus with 70% ethanol and dry with a paper towel.

Testing

1. After the ODT box is cleaned and dried, clean shavings must be placed in one compartment (unfamiliar odor) and wood shavings from the animal's isolation cage in the other compartment (familiar odor).
2. Start the recording and place the animal in the center of the apparatus so that it performs the olfactory exploration for 3 min (*see* **Note 50**).
3. At the end of this period, remove the animal from the test and return it to a clean cage.
4. Discard wood shavings from both compartments.
5. Clean the apparatus with a paper towel and 70% ethanol to perform a new test.
6. At the end of all tests, the videos must be analyzed (*see* **Notes 51** and **52**).
7. Preservation in the olfactory discrimination function is indicated by an increase in the time (in seconds) spent exploring the familiar compartment compared to the unfamiliar compartment (Fig. 4f).

3.5.7 Y-Maze Test

The Y-Maze is a behavioral test used to study short-term spatial learning, through the spontaneous alternation test, and memory, evaluated by the recognition memory test [25].

Spontaneous Alternation

1. Place the animals in the test room for a 30-minute habituation period prior to the experiment, allowing them to become familiar with the experimenter and the conditions of the test

room (light, sounds, temperature) (*see* **Note 53**). While the animals are habituating, clean the Y-maze with 70% ethanol and paper towel and ensure the apparatus is dry before starting the test.

2. Make sure the camera is reaching all the arms of the maze and that the software (if any) is ready to use.
3. Number the arms of the apparatus (1–3, and).
4. Take the animal from the adjacent room to the test room.
5. Start the recording.
6. Introduce the animal in the distal part of the arm identified as 1, directed toward the center of the maze, holding it by the tail.
7. Let the animal freely explore the Y-maze, undisturbed, for 5 min.
8. After that period, stop recording and make sure the video has been saved.
9. Take the animal out of the maze and place it in its home cage in the next room.
10. Clean the Y-maze with 70% ethanol and paper towel and ensure the apparatus is dry before starting another test.
11. At the end of the experiments, the videos can be analyzed by the tracking software or manually using a stopwatch. The number of entries in the arms and the alternation between them must be recorded (*see* **Note 54**) (Fig. 4g).
12. Calculate the alternation as percentage of alternation (%).

$$\text{Alternation} = \frac{\text{Number of alternations}}{(\text{Total number of arms entries-2})} \times 100$$

Short-Term Spatial/ Working Memory

1. Prepare the maze as specified in **steps 1** to **3** of the Spontaneous Alternation test (see Subheading 3.5.6.2).
2. Close one of the arms with a divider and note which arm was blocked (*see* **Note 55**).
3. Take the animal from the adjacent room to the test room and then, start recording.
4. Place the animal facing center on one of the arms and note which arm was chosen (*see* **Note 56**).
5. Let the animal explore the maze, undisturbed, for 15 min.
6. After that period, stop recording and make sure the video has been saved.
7. Take the animal out of the maze and place it in its home cage in the next room.

8. Remove the divider and wait for 1 hour, to avoid over-handling of the mice.
9. Close another arm and start recording again.
10. Place the same animal on the distal part of the same arm as used in the previous test, facing the center of the maze and let it explore for 5 min, undisturbed.
11. After that period, stop recording and make sure the video has been saved.
12. Take the animal out of the maze and place it in its home cage in the next room.
13. Analyze total distance covered during test on all arms using tracking software or stopwatch.
14. Analyze the total number of times the animal entered each arm and the time spent in each arm (*see* **Note 57**).

3.5.8 Grip Strength Test (GST)

The grip strength test (GST) is used to assess neuromuscular function in rodents. The grip strength task is performed by lifting the animal by the tail, having it grasp the gripping plate with the forelimbs or hindlimbs, and then it is pulled by the experimenter until it is released from the plate. The force, in Newtons (N), required to make the animal let go of the apparatus is a measure of grip strength [25, 26].

1. Clean the apparatus with 70% ethanol and dry with a paper towel.
2. Take the animals from the housing into the testing room.
3. Check that the gripping grid is securely connected to the sensor so that it does not move during the test.
4. Turn on the sensor and set it in peaking mode (*see* **Note 58**).
5. Restore the display sensor to zero.
6. Take the animal out of the cage, holding it by the base of the tail (Fig. 4h).

Fore Limb Strength

1. Place the animal horizontally on the grid, so that its front paws are caught in this grid.
2. After grabbing the grid, the animal must be pulled back by the experimenter through the tail.
3. Record the maximum force value displayed on the sensor screen when the animal is pulled.
4. Repeat this procedure three times to obtain three forelimb grip strength measurements

Fore and Hind Limb Strength

1. Hold the animal parallel to the grid until it grips the apparatus with all four paws.
2. Pull the rodent by the base of the tail and record the maximum value that appears on the sensor screen.
3. Repeat this procedure three times to obtain three measurements of forelimb and hindlimb grip strength.
4. Weight the animal on a precision scale (*see* **Note 59**).
5. Return the animal to its previous housing container.
6. Clean the grid with 70% ethanol and dry with a paper towel.

3.6 Euthanasia of ZIKV and MHV Models

For the euthanasia of the animals, the collection of the brain tissue must be performed carefully, using the surgical instruments shown in Fig. 5. To assist in that task, a video is available as Supplementary material (Video 1). Each of the steps in Subheading 3.6.1 is followed by a timestamp (*mm:ss*) from the video. To understand the functions of the specific brain areas to be dissected and the inflammatory mediators measured by ELISA, see Tables 2 and 3.

3.6.1 Dissection and Collection of Specific Brain Areas

1. With the animal anesthetized, decapitate the mouse with the surgical scissors (Fig. 5c) (***00:10)***.
2. From the most posterior dorsal part, cut the skin from posterior to anterior, until the olfactory region is reached, showing off the cranial vault ***(00:29)***.
3. With the scissors on a vertical position, find the suture called Bregma, perforate it with the tip of the scissor until you feel slight resistance and open the scissor (***00:58)***.
4. With the tip of the scissors remove the right and left parietal bones, exposing the cerebral hemispheres ***(01:26)***.

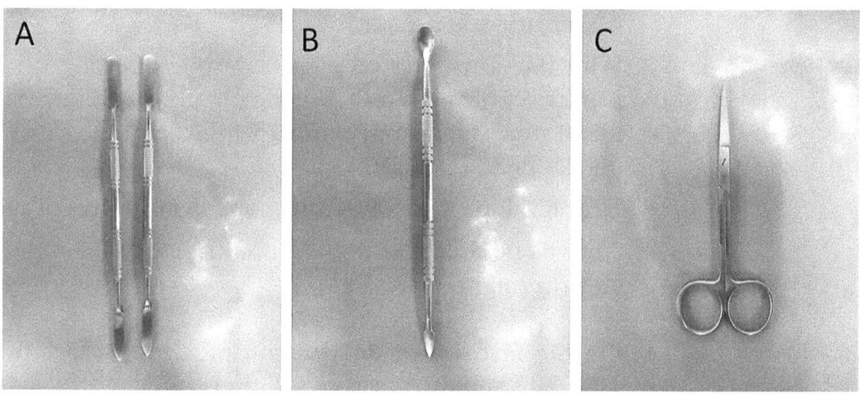

Fig. 5 Surgical instruments used in brain dissection. 18 cm long stainless steel spatula (**a**), 17.7 cm long stainless steel spatula (**b**), 11.5 cm long straight edge surgical scissors

5. With the tip of the scissors again, remove the occipital bone, exposing the cerebellum ***(01:56)***.
6. With the tip of the spatula (Fig. 5a), insert it slightly into the most posterior part of the cerebellum, gradually removing it. This will allow the total removal of the brain and deposit it on the moistened filter paper in the petri dish ***(02:38)***.
7. With the brain in the petri dish and filter paper, wash the tissue with saline ***(03:34)***.
8. With the spatula, separate the two hemispheres by making a cut on their separation line called sagittal suture ***(03:52)*** (*see* **Note 60**). At this point in the procedure the cerebellum is also split in half ***(03:57)***.
9. First, choose which cerebral hemisphere will remove the cerebral areas.
10. After that, use the spatula to remove the cerebellum and the prefrontal cortex (***04:29)*** and immediately place in microcentrifuge tube and transfer to nitrogen (for snap freezing) or ice.
11. With the two spatulas in hand, expose the hippocampus by finding it by its characteristic structure (*see* **Note 61**) and, with light movements towards the outside of the brain, push it with the side of the spatula ***(05:27–06:13)***.
12. With the removal of the hippocampus, the striatum becomes visible. Find it by its feature of presenting white striations/stripes ***(06:36)*** (Fig. 6) (*see* **Note 62**).
13. With light movements, use the tip of the spatulas to separate and remove the striatum from the hemisphere where it is located ***(06:50 until 07:51)***.

3.6.2 Brain Collection and Storage for 40 μm Sections

1. By using the materials described in the previous section, the remaining cerebral hemisphere will be collected as a whole, as the collection of specific cerebral areas in the other hemisphere has already taken place.
2. With the remaining cerebral hemisphere resting on the filter paper over the petri dish, use the spatula from Fig. 5b to collect it and store it in its respective 15-mL centrifuge tubes containing 4% PFA.
3. Repeat this process for all remaining hemispheres of all animals.
4. After tissue collection, store all centrifuge tubes in the refrigerator overnight.

3.6.3 Dehydration of Brain Tissue for Freezing for 40 μm Sections

1. On the first day of dehydration process, prepare tubes with 10%, 20% and 30% sucrose solutions (w/v).
2. Remove all centrifuge tubes with brain tissue in PFA from the refrigerator. Discard all sucrose from the centrifuge tube and

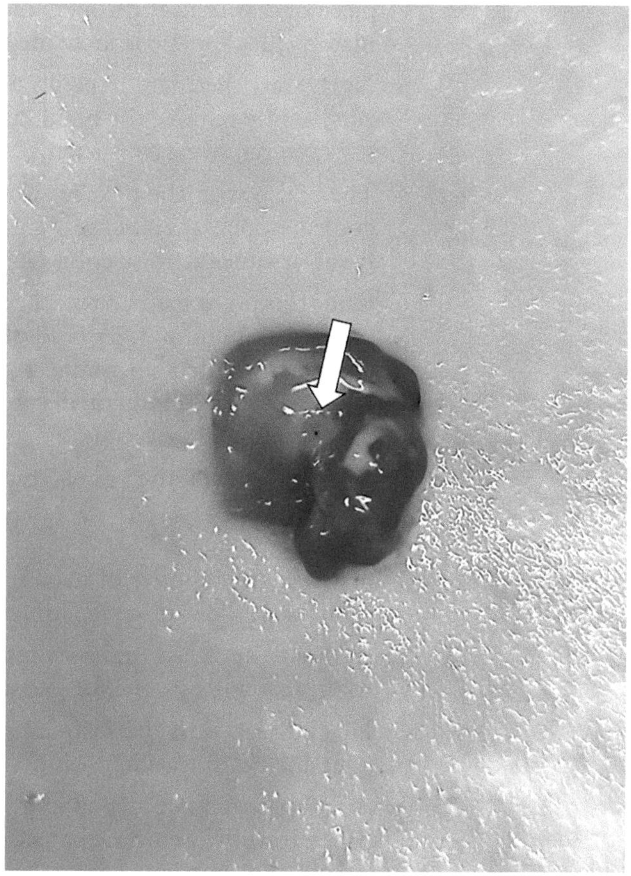

Fig. 6 Cerebral striatum dissected from the C57BL/6J mouse hemisphere. Note the white striations characteristic of the area (white arrow)

fill the same tube with 7 mL of 10% sucrose solution (*see* **Note 63**).

3. Repeat this process for all the tubes, and once they are all filled with the new solution, store them in the refrigerator overnight.
4. On the second day of the dehydration process, discard the 10% sucrose solution, change it to the 20% sucrose solution and repeat **item 3**.
5. On the third day of the dehydration process, discard the 20% sucrose solution and add the 30% sucrose solution, repeating **item 3**.

3.6.4 Brain Tissue Freezing for 40 μm Sections

1. Previously label animals in experimentation before carrying out the freezing process.
2. Start the process by cooling the isopentane with dry ice. To do this, place dry ice cubes filling half of a styrofoam box. Fill a 500 mL beaker halfway with isopentane. After that, place a

piece of dry ice in the isopentane and keep it there for approximately 20 s, for the isopentane to cool.

3. Next, start the brain freezing process. With the help of tweezers, gently remove the brain samples and place them on paper towel to remove excess sucrose.
4. Then, also with the help of tweezers, transfer the brain to the cooled isopentane solution and wait a few seconds until the tissue assumes a white color (*see* **Note 64**).
5. Once this point has been reached, remove the brain and place it on the aluminum foil. Fold the sheet so that the brain is completely rolled up, staple the corresponding identification tag and store the brain on dry ice until the process is completed for all samples (*see* **Note 65**).
6. After freezing all the brains, transfer and store them in a −80 ° C.

3.6.5 Gelatinization of Histological Slides for 40 μm Sections

1. Separate the number of glass staining dishes according to the number of slides to be gelatinized, based on your experiment.
2. At least one glass slide staining dish for hydrochloric acid solution wash and one for gelatin solution.
3. Fill the rack with the histological slides and insert it into the container with washing solution, leaving it for 5 min.
4. After that, remove the support with the slides, dab it on the paper towel to remove the excess solution. Place the support now in the container with gelatin solution and leave for 5 min.
5. After the elapsed time, dab the support on the paper towel again and place it in the incubator for the slides to dry overnight.

3.6.6 Obtaining and Storing 40 μm Brain Slides (See ***Note 66****)*

1. With the previously prepared materials, turn on the cryostat and check the temperature of the device and the angle of the support table.
2. Separate and identify the gelatinized slides that will be filled with brain sections of the first animal. Also identify the number of wells of interest for each biological replicate, to also store sections in the 24-well plate with antifreeze solution.
3. Unpack the frozen brain, place it on the removable cryostat holder and apply the Tissue TEK layer.
4. Once the Tissue TEK is completely frozen, fit the support with the brain into the cryostat, adjust and level it.
5. With the brain properly adjusted to the machine, use the brush to capture each cut that will be relocated in a slide or plate well.
6. Repeat this process for all brains and, at the end, store the 24-well plate and slides in a −20 °C freezer until use.

3.7 Neuro-inflammation Assessment Techniques

Two models of immunohistochemistry technique will be described: (I) immunofluorescence and immunohistochemistry in 40 μm brain sections made in cryostat; (II) immunohistochemistry in brain sections of 5 μm size, performed using the paraffin embedding method.

3.7.1 IBA-1 and GFAP Immunofluorescence for Cryostat Sections

1. Prepare the necessary materials and solutions in advance (*see* Subheading 2.8).
2. Remove the brain slices stored in antifreeze solution from the −20 °C freezer.
3. In a new 24-well plate, fill the wells that will be used for washing with 500 μL of PBS
4. Using the brush, separate the sections of each animal of interest and transfer them to a Petri dish with PBS. This makes the cut section completely visible, making it possible to determine whether the region of interest is present for the purpose of using the technique moving forward.
5. Also with the help of the brush, transfer the cuts from the Petri dish to the respective well of the 24-well plate.
6. Wash the sections 5 times, 10 min each in moderate agitation. Change the PBS solution after each wash.
7. At the end of the 5 washing rounds, transfer the sections to new wells containing 500 μL of blocking solution and keep stirring for 1 hour.
8. After the time in blocking solution, transfer the cuts to wells containing 500 μL of the primary antibody diluted in blocking solution (1:500 dilution), and keep stirring overnight (*see* **Note 67**).
9. In the next day, remove the sections from the primary antibody well and perform 5 washing rounds of 10 minutes in PBS, transferring the well sections as before.
10. After the 5 washing rounds, dilute the secondary antibody in blocking solution (dilution 1:1500).
11. Transfer the sections to wells containing 500 μL of the diluted secondary antibody solution. Keep the plate covered with aluminum foil, stirring for 1 hour.
12. After the incubation time, remove the sections from the antibody well and transfer to wells with PBS and wash 3 times, 10 min each. Change the PBS with each wash.
13. At the end of the 3 washing rounds, transfer the cuts to the Petri dish containing PBS for mounting the slides (*see* **Note 68**).

14. Identify the first gelatinized slide according to your experimental group in question. Repeat this process for subsequent experimental groups on different slides.
15. With the help of one of the brushes, remove a slice from the Petri dish and transfer it to the gelatinized slide. With the help of the two brushes, open the entire cut so that it exposes the entire fabric. Use drops of PBS to facilitate the handling of the cut (*see* **Note 69**).
16. Repeat this process for all cuts.
17. After all sections have dried on the slide, mount them with a coverslip as follows. Pipette 80 μL of Fluoromont onto the sections, dividing this amount into three parts of the slide. Next, place one end of the cover slip against the back of the slide and release the cover slip. Squeeze lightly with your fingers to remove any possible bubbles on the tissue.
18. Wait for the coverslip to dry well and seal the slide with nail polish (*see* **Note 70**).
19. Store these slides in boxes, in a refrigerator at 4 °C.
20. An example of the stained sample can be seen in Fig. 7. Positive cells for IBA-1 (Fig. 7a) and GFAP (Fig. 7b) are highlighted by white arrows.

3.7.2 Immunohistochemistry Ki-67 for Cryostat 40 μm Sections

1. Execute **steps 1** to **11** from Subheading 3.7.1.
2. After the incubation time, remove the sections from the well with antibody and transfer to new wells containing PBS. Perform 7 washing rounds of 4 min each, changing wells and PBS after each wash.
3. Prepare the ABC Kit 30 min before use. With this solution ready, pipette 500 μL into new wells and transfer the slices, keeping them in agitation for 1 hour.

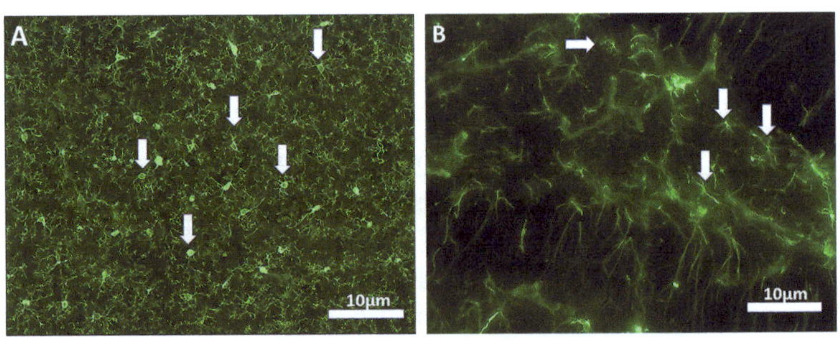

Fig. 7 Immunofluorescence IBA-1 and GFAP for 40 μm cryostat sections. (**a**) IBA-1 labeling in mouse brain at 40× magnification. (**b**) GFAP labeling in mouse brain at 40× magnification

4. Remove the slice and transfer to a new well with PBS, performing 2 washing rounds, 10 min each. Change wells and PBS after each wash.
5. Prepare the DAB solution at the time of use and transfer only the sections from a single well to time the staining time of the sections.
6. After defining the staining time, repeat the process of the previous topic for the remaining sections, using the same standardized time first, for all. All sections from each well should be placed together in the DAB solution.
7. After the sections are stained, remove them from the well containing DAB and transfer to another well containing PBS and wash 4 times, 10 min each.
8. After finishing the washing rounds, transfer the slices to the respective gelatinized slides and wait for them to dry.
9. With the histological section dry on the slides, proceed to make the xylene-ethanol battery for fixation. In this process, transfer the slides to a glass container with 50%, 70%, 80%, 90% ethanol, ethanol + xylene, xylene I and, finally, xylene II. Let the slides rest in the solution for 3 min at each solution. This process should be done using a fume hood.
10. After 3 min in xylene II, pipette two drops of Entellan over the histological sections and place the cover slip on it and wait for it to dry.
11. Store the histological slides in a box at room temperature.
12. Figure 8 shows a tissue with a positive staining for Ki-67, pointed by black arrows.

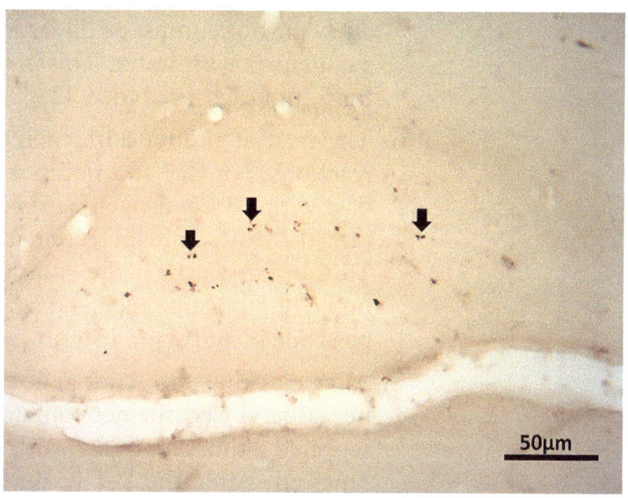

Fig. 8 Immunohistochemistry for Ki-67 in 40-μm cryostat sections. The black arrows indicate cells with a positive staining

3.7.3 Immunohistochemistry for IBA-1 and S100B for 5 μm Paraffinized Sections

Day 1

1. After the standard processing of the histological tissue sections on immunohistochemistry slides, leave the slides in a suitable oven at 60 °C for 30 min to fix the section on the slide. In the meantime, prepare all necessary solutions for immunohistochemistry (*see* Subheading 2.8).
2. *Deparaffinization step*: inside a fume hood, place the slides in a histological slide holder, use a suitable container to place enough volume of xylene (xylene I), carefully immerse the slide holder and leave it for 30 min. After 30 min, the support containing the slides must be transferred to another container containing xylene (xylene II) for 10 min and consecutively to a new container containing xylene (xylene III) for 5 min.
3. *Hydration step*: after removing the support containing the slides from the container immersed in xylene III, it is necessary to wait for the slides to dry until all the xylene evaporates. Soon after, immerse the support with the slides in a container containing 100% ethanol (ethanol I), for 5 min, after that time has elapsed, the slides should consecutively go through two more rounds of ethanol 100% (ethanol II and III), then 90% ethanol, 80% ethanol and finally 70% ethanol, for 5 min each. After the incubation with the different ethanol solutions, the slides must be kept in a container with water for 5 min.
4. After the slides are hydrated, the antigen retrieval is executed (*see* **Note 71**). The slides must be immersed in the recovery solution (citrate or EDTA buffer depending on the standardization of the primary antibody) inside a container, kept in a water bath at 96 °C for 20 min.
5. After 20 min, the bowl must be removed from the water bath and placed at room temperature, the slides must be kept in the buffer until both are cooled down (wait for ca. 20 min). Soon after, the buffer must be discarded, the slides must be washed in a container with water. This process must be repeated 3 times for 5 min each (*see* **Note 72**).
6. The next step after antigen recovery is blocking nonspecific reactions. For this, it is necessary to block endogenous peroxidase with a 3% hydrogen peroxide solution. The slides must be placed in the container containing the blocking solution for 15 min, after which the solution must be discarded and the procedure must be repeated again with the same 3% hydrogen peroxide solution for 15 min (*see* **Note 73**).
7. After blocking with endogenous peroxidase, the slides should be washed in a container containing TBS. This process must be repeated 3 times for 5 min, with each wash the solution must be renewed. After the last wash, the slides should be removed from the container, one by one, to use the hydrophobic pen, as it will keep the blockage and the primary antibody on top of

Table 5
Antibodies and recommended antigenic retrieval for 40 μm and 5 μm slices

Slice	Antibody	Concentration	Recovery antigenic/solution
40 μm	IBA-1/GFAP rabbit or mouse	1:500 both	No need
5 μm	IBA-1/S100B rabbit	1:150/1:75	Citrate buffer/EDTA buffer

the tissue on the slide. The slide should be dried around the cut with the aid of a piece of paper, to avoid drying out the histological sections, the procedure should be done quickly.

8. Once the slides are drawn with the hydrophobic pen, immediately add 20 μL per section. Incubate for 5 min to prevent nonspecific bindings. After 5 min, just discard the excess solution to add the primary antibody.
9. Dilute the primary antibody in 0.1% BSA at predetermined concentration in standardization assays. Pipette 20 microliters of the solution per section (do not wash after blocking, just remove the excess). Incubate overnight in a humid chamber in the refrigerator (*see* **Note 74**) (*see* Table 5).

Day Two

1. After overnight incubation, the slides should be placed in the slide holder to be washed in TBS for 5 min. This process must be repeated 3 times, and in each round the washing solution must be changed. After washing, the slides must be dried around the section for the addition of the biotinylated secondary antibody (*see* Table 5; **Note 75**).
2. Add 20 μL per cut of biotinylated secondary antibody solution on the previously dried sections, delimited with a hydrophobic pen. Incubate in a humid chamber for 30 min at room temperature (*see* **Note 76**).
3. After incubation, the slides should be washed with TBS, 3 rounds of 5 min each. Afterwards, the sections must be previously dried to be incubated with streptavidin-peroxidase. The preparation of the streptavidin-peroxidase solution must be done 30 min before use.
4. After adding 20 μL of streptavidin-peroxidase solution per section, incubate in a humid chamber for 30 min at room temperature. Afterwards, the slides must be washed with TBS, 3 rounds of 5 min each, changing the solution after each wash.
5. After washing the slides and preparing the DAB (*see* **Note 7**), remove excess liquid from the slides around the cuts without letting them dry out. Add 20 μL of DAB over the sections and observe the occurrence of the reaction: the sections that

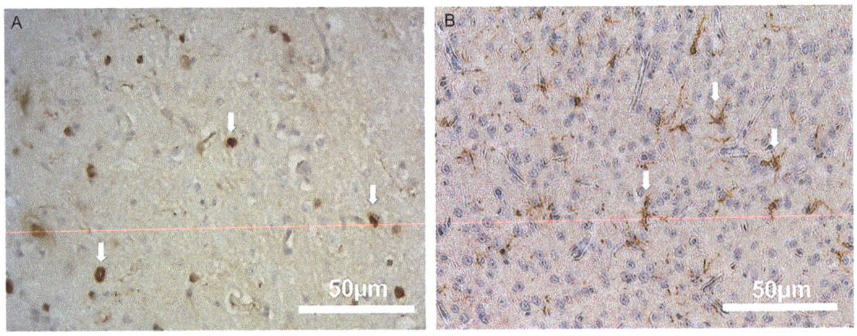

Fig. 9 Immunohistochemistry IBA-1 and S100B for 5 µm paraffinized sections. (**a**) S100B labeling in mouse brain at 40× magnification. (**b**) IBA-1 labeling in mouse brain at 40× magnification

received the primary antibody should turn slightly brown. The reaction is usually fast and should not exceed 1 min. Stop the reaction by placing the slides in a container with distilled water.

6. After reaction with DAB, slides should be washed with water for 10 min. After washing, the slides must be quickly immersed in hematoxylin, which will serve as a counterstain. After immersion in hematoxylin, the slides must be washed with water, 3 times for 5 min. Always remembering to change the water in each wash.
7. After staining and washing the slides, dehydrate the sections through ethanol in different concentrations. First, the support containing the slides will be placed in a container containing 70% ethanol for 3 min, then the slides will be transferred to another container containing 80% ethanol consecutively for 90% ethanol, 100% ethanol I, 100% ethanol II, 100 ethanol % III for 3 more minutes each.
8. Next, inside a fume hood, the slides must be transferred to a container containing xylene I for 5 min. Soon after, they will be transferred to xylene II and xylene III for 5 min each. After xylene III, each slide must be removed individually to mount the coverslip. With the aid of tweezers, the slide will be placed on a flat, dry surface. Then, immediately add the Entellan glue onto the cover slip, which is placed on top of the histological section. After placing the coverslip, press lightly with tweezers until all air pockets are removed from the section. Allow the slides to dry, analyze, and count the antibody stains under the microscope at 40× magnification (Fig. 9).

4 Notes

1. The apparatus for carrying out this test was developed by Hall [27] and consists of a quadrangular arena made of transparent acrylic, measuring 45 cm × 45 cm, with walls approximately 22 cm high, to prevent the animal escapes during the test [28]. One study evaluated the effect of open field apparatus size on rodent behavior and observed increased activity in larger arenas. However, this study showed that despite the large variations in the size of the arenas (40 × 60 cm to 260 × 400 cm), behavioral changes were practically insignificant [29]. The bottom of the arena must be subdivided into quadrants of the same size to facilitate the analysis and the walls must be covered with dark paper to prevent the animals from viewing the external environment and thus interfering with the test [28]. Most of the time, the rodents spend more time in the peripheral zone of the apparatus when compared to the time they stay in the central area, as locomotion close to the wall allows the rodents to perform thigmotaxis, a practice in which the animals bump their vibrissae on surfaces vertical and feel safer and less exposed to predators [30]. Testing time depends on the researcher's objective. The literature shows a wide variation in this time, but the most cited protocol used a total time of 15 min, with 5 min of acclimatization, followed by 10 min of testing [31]. Several parameters such as exploration, walking, rearing, stretching, cleaning, sitting, turning, jumping, sleeping, defecation is measured and evaluated in OPT [26], however normally in our models we analyze the total distance covered (evaluates locomotion), time spent in the center and distance covered in the center and/or number of entrances to the center (evaluate anxiety-like behavior).
2. OFT is widely used to assess the effects of drugs and the motivation of animals to perform activities. Therefore, the number of times the animal is submitted to the test influences its behavior and motivation, that is, after repeated tests the animals are habituated and tend to reduce their locomotion throughout the sessions [32]. Therefore, the objective of the test must first be evaluated, and then decide whether a single test will be performed or whether it will be repeated.
3. Phenotyper/tracking software streamlines behavioral testing analysis. However, the videos can be made using a camera that is capable of clearly recording the movements of the animals and the analysis can be performed manually, based on the observation of the videos and stopwatches. When using an aerial video tracking system, opt for a matte floor in the arena to avoid reflections that could interfere with the analyses [33].

4. The use of infrared light is optional. The purpose of its use consists in the fact that it is an imperceptible light to the eyes of rodents, but it is captured by the camera and the recording can be performed. Mice in the dark tend to perform activities spontaneously, as they are nocturnal animals.
5. The lighting used during the OFT test is very variable, and can be infrared or white. Dim white lighting of about 30 lux is generally used, as brighter light may increase thygmotaxis and/or decrease locomotion [32].
6. The use of computerized tracking systems allows animals to be tested individually (one animal per arena) and simultaneously. The arenas are isolated with dark paper so that the presence of one animal does not interfere with the behavior of the other. However, if there are not several arenas, the test must be performed with one animal at a time.

   The parameters usually evaluated in the EPMT are [33]:

   (a) Length of stay in open arms.
   (b) Latency time in the center of the apparatus.
   (c) Time spent in closed arms.
   (d) First site chosen after the beginning of the test (open arm, closed arm, or if it remained in the center of the apparatus).
   (e) Percentage of time spent in open arms relative to total test time.
   (f) Percentage of time spent in closed arms in relation to total test time.
   (g) Total number of entries into open and closed arms.
   (h) Total number of open arms entries.
   (i) Total number of entries into closed arms.
   (j) Percentage of entries into open arms relative to the total number of entries.
   (k) Percentage of entries into closed arms relative to the total number of entries.
   (l) Total distance covered in open and closed arms.
   (m) Total distance covered in open arms.
   (n) Total distance covered in closed arms..
   (o) Percentage of the distance covered in closed arms in relation to the total distance covered.
   (p) Percentage of distance covered in open arms in relation to total distance covered.
7. The EPMT platform is placed inside an empty circular tank to protect animals that fall or try to escape during the test.

8. The animals must be transferred to the experimentation room at least 30 min before the test to allow them to get used to the environmental conditions.
9. Phenotyper and tracking software make behavioral testing analysis easier and faster. However, the videos can be made using a camera that is capable of clearly recording the movements of the animals and the analysis can be performed manually, from the observation of the videos and the use of stopwatches.
10. The light in the experimentation room must illuminate the four arms of the apparatus evenly, that is, one arm cannot be brighter than the other. With regard to light intensity, a study evaluated different light intensities (0, 1, 3, 10, 30, 100, and 300 lux) and demonstrated that there is an increase in the exploration of open arms at levels of 0 and 1 lux, but there is no change from 3 lux to 300 lux [34].
11. All animals must be handled in the same way and subjected to the same procedures prior to testing. Furthermore, the rodents must be placed in the maze in the same position, as there are already differences when they are placed facing an open arm or if they are positioned facing a closed arm [35].
12. The animals must be housed in groups of 3 to 5 per cage, kept in a facility with controlled temperature (~23 °C), under a 12-hour light/dark cycle, free access to food and water.
13. The cages used are rectangular and measure 30 cm × 30 cm × 50 cm.
14. The animals must be monitored during the entire test through the video transmitted to a computer external to the experimentation room, as in situations of extreme fatigue they may start to drown. If this occurs, the animals must be immediately removed from the water and excluded from the experiment [36].
15. The animals that will be submitted to ODT are placed in individual cages and are isolated for 48 hours. The shavings used in the family compartment during the test come from the animal's cage in isolation, so each animal will have its familiar odor [37, 38].
16. Acclimatization must be recorded to record the time the animal spent in each compartment and thus assess whether there was any preference for the side of the apparatus that could compromise the true result of the test [38].
17. The Y-maze has three arms oriented at 120° angles to each other, with dimensions of 21 cm × 7 cm × 15.5 cm, the arms being equal in length.
18. There is a need to have a room adjacent to the test room to minimize noise and prevent the experimenter from interfering

with the test, as noise is associated with increased stress and cognitive impairment in animals. The camera must be attached to a computer so that the researcher can view the test remotely (in the adjacent room).

19. It is considered an entry when the animal places all four paws on one arm of the maze (1–3, or). An alternation is recorded when the animal enters the three arms consecutively.
20. The apparatus that measures force in rodents consists of a stainless steel trapezoidal grid, a base plate and a sensor that measures force. Handle the apparatus carefully, as it can be irreversibly damaged if subjected to loads greater than the rated capacity of the force sensor withstands [40].
21. Prefer the use of liquid nitrogen over ice. The more gelatinous morphology and appearance of the brain tissue means that it needs to be frozen very quickly.
22. Spatula brand suggestion: Yaxun. The use of this spatula from another brand refers to the availability of the material in the market. It is a spatula belonging to a dental kit or maintenance kit for notebooks and tablets. Pay attention only to the measurements and model provided. Spatula with flattened ends and rectangular shape.
23. Spatula brand suggestion: Glonxd. The use of this spatula from another brand refers to the availability of the material on the market. It is a spatula belonging to a dental kit or maintenance kit for notebooks and tablets. Pay attention only to the measurements and model provided. Spatula with one end in a triangular shape and the other in a concave shape.
24. The use of filter paper helps to prevent the brain tissue from slipping during the dissection of areas and the total brain.
25. Saline solution or PBS is necessary for washing the tissue while the dissection is performed, so as not to dry it out.
26. The use of the Styrofoam box with these specific measurements provides a better visualization of details of brain areas within the laminar flow hood.
27. Both collection and dissection of brain areas and collection of the entire cerebral hemisphere should be performed as quickly as possible to avoid rapid tissue disintegration.
28. Dilute the 4% PFA to the final solution volume of interest. The dilution of the PFA in PBS must be carried out in a magnetic stirrer at 57 °C, with this temperature being maintained and controlled; The weighing and dilution of the PFA must be carried out in a fume hood. At the end of PFA dilution, add approximately 0.4 g of NaOH.

29. Preparation of centrifuge tubes with PFA must be carried out prior to the day of euthanasia.
30. Gelatinized slides are necessary for better adhesion of the 40 μm sections. Sections of this thickness are difficult to handle on commercially acquired marked blades. The researcher can try to use the gelatinized slide optionally and compare the two types of slides with a 40 μm cut.
31. The dilution of gelatin powder must be done in water heated to a temperature of 60 °C, in a magnetic stirrer. After complete dilution add chrome alum, no bubbles can occur during gelatin preparation. Keep the gelatin solution stored in a refrigerator, but at the moment of use it must be at room temperature. It cannot be cold, as the slide will dry out with the formation of bubbles.
32. Prepare the antifreeze solution by diluting all reagents in PBS one at a time while stirring on a magnetic stirrer. Keep the solution stored in the refrigerator.
33. For the tissue cutting procedure, the researcher must have technical training to use the cryostat that will be used. Pay attention to the ideal temperature of the cryostat chamber (17 °C) and the angle of the table, as the angle must always be the same to avoid tissue variation.
34. The ABC kit must be prepared at the time of use. Reagent C must be stirred in a magnetic stirrer for 30 min and, at the end of that time, used immediately.
35. The researcher must stick to the observation of animals that are exposed to the heating lamp, as exposure to heat beyond the necessary time can cause animal death.
36. Secure the animal container on the experimental table with the aid of adhesive tape, as to avoid movement of the container, the animal and the consequent loss of access to the caudal vein.
37. The concentration of anesthetic to be injected per animal may depend on its weight. This volume of 60 μL corresponds to an animal weighing around 20 g. For animals weighing less than this, apply part of the anesthetic volume and observe the animal's reaction to anesthesia. The ideal anesthetic point is when the animal is relaxed but is able to perform aspiration through the nose.
38. Weighing and clinical scoring of animals should be performed at the same time each day to avoid bias.
39. Problems/solutions in EPMT.
    (a) Rodents fall into open arms: this is a rare event, but when it occurs, the experimenter must pick up the animal and return it to the EPM. The behavior of that animal must be

recorded and evaluated, but will be excluded from the analysis. The researcher continues testing that animal to ensure that exposure in the maze was consistent across animals, particularly when subjected to other tests [41].

(b) Immobility of rodents in open arms: It can be considered an atypical situation and when it occurs, the experimenter must record this event. This immobility may be due to some noise or movement while the animal is in the open arms of the maze, which are considered criteria for excluding data from this test. The animal continues to be filmed so that its exposure to the maze is compatible with that of the other test animals. Standing still for a long time, without the occurrence of noise or interruptions, would be included in the analysis if the time spent in the open arms was greater than 30% of the total test time [41].

40. The bottles used in the habituation of the animals contain water. Previous exposure to sucrose can increase intake and preference to peak values, which reduces test sensitivity [42]. If it is necessary to use sucrose solution to avoid effects such as neophobia, one should choose to use sucrose solutions with descending concentrations [43].

41. Mice that show lateral preference, even after inverting the bottles over the days, should be excluded from the test analysis [17].

42. Bottles should be rotated daily to minimize the effect of side preference. In addition, it is important that the sucrose solution is made up every day and replaced in the bottles, because the exposure of this solution to room temperature for 3 days can lead to the growth of microorganisms, which may impair the effectiveness of the test [17].

43. Standardize the place in the box where the animals started the test, for example, always place them in the center. The test can be performed with several animals simultaneously, as long as they are in separate boxes and far from each other.

44. Five days before starting the FST protocol, the mice should be handled for 2 min daily [37].

45. To speed up the process, several animals can be submitted to the FST concomitantly, however the beakers must be separated/isolated so that one animal cannot see the other during the test. Also, be careful not to mix animals from different cages in the drying box [37].

46. Animals should be watched while drying to ensure they are recovering. A thermometer should be placed in the cage to monitor the temperature and keep it at 37 °C [37].

47. For mice, generally, the last 4 min of the video are analyzed. The parameters (immobility time, climbing, and swimming) must be standardized so that the analysis is homogeneous [37]:
    (a) Consider it as motionless time when the animals are floating and only moving their heads to breathe.
    (b) Consider it as climb/escape time when the animals make quick movements with the front paws trying to get out of the water.
    (c) Consider it as swimming when animals move their hind and/or front legs to stay on the surface.

    The immobility parameter is the most frequently analyzed behavior in FST experiments, which is indicative of depressive-like behavior in animals. However, it is important to evaluate mobility behaviors during the FST, for example, antidepressants that increase serotonergic neurotransmission cause an increase in the swimming time of the animals [44].
48. The animals that will be submitted to ODT are placed in individual cages and are isolated for 48 hours. The wood shavings used in the familiar (previously known) compartment during the test come from the animal's cage in isolation, so each animal will have its familiar odor [38].
49. Acclimatization must be recorded to record the time the animal spent in each compartment and thus assess whether there was any preference for the side of the apparatus that could compromise the true result of the test [38].
50. When starting the recording, show the familiar compartment on the camera, as the analysis of the videos will be blind. In addition, it is important to identify the number and group of the animal being tested in each recording [38].
51. The analysis of the videos must be carried out blindly, that is, the analyst must only know which compartment is the familiar one and this is indicated at the beginning of each recording. With the aid of a stopwatch, the time (in seconds) that each animal remains in the familiar compartment and in the non-familiar compartment is recorded. To calculate the olfactory discrimination index, in percentage, the following calculation must be performed [38]:

$$\text{Olfactory discrimination index} = \frac{\text{Time in familiar compartment (TCF)-Time in non-familiar compartment (TCNF)}}{\text{Total exploration time (TCF + TCNF)}} \times 100$$

52. Animals with an intact sense of smell tend to spend more time in the familiar compartment, while animals with olfactory impairment, as they cannot distinguish odors, will spend similar times in both compartments [38]. If animals need to undergo multiple behavioral tests, choose to perform the

olfactory discrimination test first so that other procedures/odors do not interfere with the ODT result [40].

53. Euthanasia or blood collection should be prohibited in the behavior room, as well as in the adjacent areas at least 24 hours before the test, as this interferes with the behavior of the animals in any test that is performed.
54. It is considered an entry when the animal places all four paws on one arm of the maze (1–3, or). An alternation is recorded when the animal enters the three arms consecutively [22].
55. The closed arm with the divider must be chosen at random and the lock must be alternated between animals [22].
56. The arm chosen to place the animal must also be chosen randomly and alternated between animals [22].
57. Animals with intact spatial memory should enter and spend more time in the arm where the divider was removed, the "new arm". However, it has already been observed that the animals tend to stay in the initial arm in the initial moments, and therefore, this arm can be disregarded in the comparison. Animals with compromised spatial memory will not be able to distinguish between the new arm and the others [22].
58. The sensor's peak mode allows the maximum force exerted by the rodent to be measured. Force is given in grams [25].
59. Animals should be weighed, as studies have shown that a reduction in body weight leads to a decrease in grip strength [45].
60. The use of the spatula with the concave end supports the cerebellum at the moment of separation of the two hemispheres. Without this support, the cerebellum can separate from the rest of the organ.
61. To locate the hippocampus visually look for its appearance reminiscent of the shape of a banana. Consider studying neuroanatomy for easier location of this region, in addition to its shape.
62. Not every striatum presents white striae clearly, therefore neuroanatomical knowledge for the location of this region is also necessary.
63. The brain tissue has to remain in PFA for 24 hours before being exchanged for sucrose solution.
64. It is necessary for the researcher to standardize the time for freezing the tissue, evaluating the whitish color and the freezing time. The brain cannot remain too long in the cooled isopentane and not less than the minimum time. Isopentane, as the various tissues freeze, starts to heat up, therefore, cool the isopentane several times with a piece of dry ice when it is

noticed that the brain is taking a longer time to whiten. This is significant from heating the isopentane.

65. The process of removing the tissue from isopentane and storing it in aluminum foil must be carried out as quickly as possible. Care must be taken not to staple the brain when stapling the identification tag. Always staple it to the edges of the aluminum foil after wrapping the brain in it.
66. The brains stored in −80 °C freezer should be transferred to −20 °C freezer, 24 hours before using the cryostat.
67. Dilute the antibody only at the time of use. Overnight incubation time is required to be at least 16 hours.
68. Use the petri dish placed diagonally and on a dark background. This makes the cuts easier to see.
69. Make sure that the entire section is open and that no part bends. This bending impairs the visualization of the cell marking and consequently the cell count.
70. It is necessary to use enamel to seal the slide. This provides greater durability of the fluorescence labeling.
71. After preparing the solutions, immediately turn on the water bath at 96 °C and place a vat with the antigenic recovery solution (the choice of recovery solution depends on each primary antibody) to heat.
72. While washing the slides, the water must be changed every 5 min.
73. 3% Peroxidase Blocking Solution must be prepared immediately prior to use. The blocking process must be carried out 2 times consecutively, that is, it will be necessary to prepare 2 blocking solutions.
74. The primary antibody must be diluted before blocking with protein block. Remember to put only 0.1% BSA without primary antibody in some sections to be the negative control for immunohistochemistry. If primary antibody concentration is not standardized, start at 0.5 microgram/mL. It is essential to know the species of origin of the antibody in order to choose the corresponding secondary antibody.
75. Excess solution should be removed one at a time to avoid drying out the tissue.
76. The secondary antibody-containing solution to be used depends on the source of the primary antibody, and the chosen antibody must react against the species in which the primary antibody was made.

Acknowledgments

We would like to thank Dr. Clarice Arns and Dr. Ricardo Durães-Carvalho from Universidade Estadual de Campinas (UNICAMP, São Paulo, Brazil) for kindly providing us with the MHV-3 strain. This study was funded by the Fundação de Amparo à Pesquisa do Estado de Minas Gerais (FAPEMIG), grant numbers RED-00202–22, APQ-02281–18, APQ-04650-23, APQ-02618-23, and APQ-04983-24; and by the Coordenação de Aperfeiçoamento de Pessoal de Nível Superior (CAPES/Brazil), through the project: CAPES—Programa 9951—Programa Estratégico Emergencial de Prevenção e Combate a Surtos, Endemias, Epidemias e Pandemias, AUX 0641/2020—Process 88881.507175/2020–01. This work also received support from the National Institute of Science and Technology in Dengue and Host-Microorganism Interaction (INCT Dengue), funded by the Conselho Nacional de Desenvolvimento Científico e Tecnológico (CNPq, Brazil) [grant number 465425/2014-3], and by FAPEMIG [grant number APQ 03606-17 and 25036]. Additional support was provided by FINEP—Financiadora de Estudos e Projetos under MCTI/FINEP—MS/SCTIE/DGITIS/CGITS [grant number 6205283BBB28-4F9C-AA65-808FE4450542], and by the International Society of Neurochemistry—Committee for Aid and Education in Neurochemistry (ISN-CAEN) [project number 29344, ISID/ICB/DMORF/Zika Virus].

References

1. Ismail FS, Faustmann TJ, Faustmann PM, Corvace F (2024) Microglia as potential key regulators in viral-induced neuroinflammation. Front Cell Neurosci 18:1426079. https://doi.org/10.3389/fncel.2024.1426079
2. Ginhoux F, Greter M, Leboeuf M, Nandi S, See P, Gokhan S, Mehler MF, Conway SJ, Ng LG, Stanley ER, Samokhvalov IM (2010) Fate mapping analysis reveals that adult microglia derive from primitive macrophages. Science 330(6005):841–845
3. Blackhurst BM, Funk KE (2023) Viral pathogens increase risk of neurodegenerative disease. Nat Rev Neurol 19(5):259–260
4. Komarasamy TV, Adnan NAA, James W, & Balasubramaniam VRMT (2022) Zika Virus Neuropathogenesis: The Different Brain Cells, Host Factors and Mechanisms Involved. Frontiers in immunology, 13, 773191. https://doi.org/10.3389/fimmu.2022.773191
5. Tan ST, Kwan AT, Rodríguez-Barraquer I, Singer BJ, Park HJ, Lewnard JA, Sears D, Lo NC (2023) Infectiousness of SARS-CoV-2 breakthrough infections and reinfections during the omicron wave. Nat Med 29 (February). https://doi.org/10.1038/s41591-022-02138-x
6. Edén A, Kanberg N, Gostner J, Fuchs D, Hagberg L, Andersson LM, Lindh M, Price RW, Zetterberg H, Gisslén M (2021) CSF biomarkers in patients with COVID-19 and neurologic symptoms: a case series. Neurology 96(2):e294–e300. https://doi.org/10.1212/WNL.0000000000010977
7. Farhadian S, Farhadian S, Glick LR, Vogels CBF, Thomas J, Chiarella J, Casanovas-Massana A, Zhou J, Odio C, Vijayakumar P, Geng B, Fournier J, Bermejo S, Fauver JR, Alpert T, Wyllie AL, Turcotte C, Steinle M, Paczkowski P, Barakat LA (2020) Acute encephalopathy with elevated CSF inflammatory markers as the initial presentation of COVID-19. BMC Neurol 20(1):1–5. https://doi.org/10.1186/s12883-020-01812-2

8. Körner RW, Majjouti M, Alejandre Alcazar MA, Mahabir E (2020) Of mice and men: the coronavirus mhv and mouse models as a translational approach to understand sars-cov-2. Viruses 12(8):1–26. https://doi.org/10.3390/v12080880
9. Gomez-Marin A, Paton JJ, Kampff AR, Costa RM, Mainen ZF (2014) Big behavioral data: psychology, ethology and the foundations of neuroscience. Nat Neurosci 17(11): 1455–1462. https://doi.org/10.1038/nn.3812
10. Takao K, Yamasaki N, Miyakawa T (2007) Impact of brain-behavior phenotypying of genetically-engineered mice on research of neuropsychiatric disorders. Neurosci Res 58(2):124–132. https://doi.org/10.1016/j.neures.2007.02.009
11. Holmes A, Yang RJ, Lesch KP, Crawley JN, Murphy DL (2003) Mice lacking the serotonin transporter exhibit 5-HT1A receptor-mediated abnormalities in tests for anxiety-like behavior. Neuropsychopharmacology 28(12): 2077–2088. https://doi.org/10.1038/sj.npp.1300266
12. Crawley JN (2007) Mouse behavioral assays relevant to the symptoms of autism. Brain Pathol 17(4):448–459. https://doi.org/10.1111/j.1750-3639.2007.00096.x
13. Podhorna J, Brown RE (2002) Strain differences in activity and emotionality do not account for differences in learning and memory performance between C57BL/6 and DBA/2 mice. Genes Brain Behav 1(2):96–110. https://doi.org/10.1034/j.1601-183x.2002.10205.x
14. Komada M, Takao K, Miyakawa T (2008) Elevated plus maze for mice. J Vis Exp 22:1088. https://doi.org/10.3791/1088
15. Pinto WB, De Rezende V et al (2012) Elevated plus maze test: applications and contributions in the study of neuropsychiatric diseases in animal models. Biol Models Res Technol 1(1): 102–120
16. Overstreet DH (2011) Modeling depression in animal models. Psychiatric Disorders:125–144
17. Eagle DE, Hybels CF, Proeschold-Bell RJ (2018) Perceived social support, received social support, and depression among clergy. J Soc Pers Relat 36(7):2055–2073. https://doi.org/10.1177/0265407518776134
18. Deacon RMJ (2006) Digging and marble burying in mice: simple methods for in vivo identification of biological impacts. Nat Protoc 1(1):122–124. https://doi.org/10.1038/nprot.2006.20
19. Smithers RHN (1983) XXIII. Families CRICETIDAE and MURIDAE, rats and mice. In: The mammals of the southern-African subregion, 1st edn. University of Pretoria, Pretoria, pp 220–296
20. Wolmarans DW, Stein DJ, Harvey BH (2016) Of mice and marbles: novel perspectives on burying behavior as a screening test for psychiatric illness. Cogn Affect Behav Neurosci 16: 551–560. https://doi.org/10.3758/s13415-016-0413-8
21. Thomas A, Burant A, Bui N, Graham D, Yuva-Paylor LA, Paylor R (2009) Marble burying reflects a repetitive and perseverative behavior more than novelty-induced anxiety. Psychopharmacology 204(2):361–373. https://doi.org/10.1007/s00213-009-1466-y
22. Kraeuter AK, Guest PC, Sarnyai Z (2019) The forced swim test for depression-like behavior in rodents. In: Guest P (ed) Pre-clinical models. Methods in molecular biology, vol 1916. Humana Press, New York. https://doi.org/10.1007/978-1-4939-8994-2_5
23. Schellinck HM, Price SR, Wong MJ (2007) Using ethologically relevant tasks to study olfactory discrimination in rodents. Chem Signals Vertebrates 11:71–80. https://doi.org/10.1007/978-0-387-73945-8_6
24. Godoy MD, Voegels RL et al (2015) Olfaction in neurologic and neurodegenerative diseases: a literature review. Int Arch Otorhinolaryngol 19(2):176–179. https://doi.org/10.1055/s-0034-1390136
25. Ribeiro, R., Silva, E.G., Moreira, F.C. et al. Chronic hyperpalatable diet induces impairment of hippocampal-dependent memories and alters glutamatergic and fractalkine axis signaling. Sci Rep 13, 16358 (2023). https://doi.org/10.1038/s41598-023-42955-9
26. Rossi, Leonardo et al. "Neuromuscular defects after infection with a beta coronavirus in mice." Neurochemistry international vol. 169 (2023): 105567. https://doi.org/10.1016/j.neuint.2023.105567
27. Hall CS. Emotional behavior in the rat. I. Defecation and urination as measures of individual differences in emotionality. J Comp Psychol. 1934;18(3):385. https://doi.org/10.1037/h0071444
28. Eilam D. Open-field behavior withstands drastic changes in arena size. Behav Brain Res 2003;142(1–2):53–62
29. Valle, F. P. Effects of strain, sex, and illumination on open-field behavior of rats. The American journal of psychology, v. 83, n. 1, p. 103–111, 1970.

30. Sadakata T, Shinoda Y, Oka M, Sekine Y, Furuichi T (2013) Autistic-like behavioral phenotypes in a mouse model with copy number variation of the CAPS2/CADPS2 gene. FEBS Lett 587(1):54–59. https://doi.org/10.1016/j.febslet.2012.10.047
31. Walsh RN, Cummins RA (1976) The open-field test: a critical review. Psychol Bull 83(3): 482–504. https://doi.org/10.1037/0033-2909.83.3.482
32. Gould TD, Dao DT, Kovacsics CE (2009) The open field test. In: Gould TD (ed) Mood and anxiety related phenotypes in mice: characterization using behavioral tests. Humana Press/Springer Nature, pp 1–20. https://doi.org/10.1007/978-1-60761-303-9_1
33. Lister RG (1987) The use of a plus-maze to measure anxiety in the mouse. Psychopharmacology 92(2):180–185. https://doi.org/10.1007/BF00177912
34. Pellow S, Chopin P, File SE, Briley M (1985) Validation of open:closed arm entries in an elevated plus-maze as a measure of anxiety in the rat. J Neurosci Methods 14(3):149–167. https://doi.org/10.1016/0165-0270(85)90031-7
35. Eagle L, Hamann M, Low DR (2016) The role of social marketing, marine turtles and sustainable tourism in reducing plastic pollution. Mar Pollut Bull 107(1):324–332. https://doi.org/10.1016/j.marpolbul.2016.03.040
36. Yankelevitch-Yahav R, Franko M, Huly A, Doron R (2015) The forced swim test as a model of depressive-like behavior. J Vis Exp 97:52587
37. Noseda ACD, Lima MMS (2022) Olfaction and melatonin: the use of the olfactory discrimination test. In: Jockers R, Cecon E (eds) Melatonin. Methods in molecular biology, vol 2550. Humana, New York. https://doi.org/10.1007/978-1-0716-2593-4_41
38. Wang S, Yu Y, Feng Y, Zou F, Zhang X, Huang J, Zhang Y, Zheng X, Huang XF, Zhu Y, Liu Y (2016) Protective effect of the orientin on noise-induced cognitive impairments in mice. Behav Brain Res 296:290–300. https://doi.org/10.1016/j.bbr.2015.09.024
39. Girgis CM, Cha KM, Houweling PJ, Rao R, Mokbel N, Lin M, Clifton-Bligh RJ, Gunton JE (2015) Vitamin D receptor ablation and Vitamin D deficiency result in reduced grip strength, altered muscle fibers, and increased myostatin in mice. Calcif Tissue Int 97(6): 602–610. https://doi.org/10.1007/s00223-015-0054-x
40. Maurissen J, Marable B, Andrus A, Stebbins K (2003) Factors affecting grip strength testing. Neurotoxicol Teratol 25:543–553. https://doi.org/10.1016/S0892-0362(03)00073-4
41. Strekalova T, Steinbusch HW (2010) Measuring behavior in mice with chronic stress depression paradigm. Prog Neuro-Psychopharmacol Biol Psychiatry 34(2):348–361
42. Strekalova, T., Gorenkova, N., Schunk, E., Dolgov, O. and Bartsch, D. (2006). Selective effects of citalopram in a mouse model of stress-induced anhedonia with a control for chronic stress. Behav Pharmacol 17(3): 271–287
43. Cryan JF, Valentino RJ, Lucki I (2005) Assessing substrates underlying the behavioral effects of antidepressants using the modified rat forced swimming test. Neurosci Biobehav Rev 29(4–5):547–569. https://doi.org/10.1016/j.neubiorev.2005.03.008
44. Andreatini R, Bacellar LFS (1999) The relationship between anxiety and depression in animal models: a study using the forced swimming test and elevated plus-maze. Braz J Med Biol Res 32(9):1121–1126. https://doi.org/10.1590/S0100-879X1999000900011
45. McBroom K (2021) A comparison of Zika virus and COVID-19: clinical overview and public health messaging. J Midwifery Womens Health 66(3):334–342. https://doi.org/10.1111/jmwh.13230
46. Garcia AMB, Cardenas FP, Morato S (2005) Effect of different illumination levels on rat behavior in the elevated plus-maze. Physiol Behav 85:265–270
47. Walf A, Frye C (2007) The use of the elevated plus maze as an assay of anxiety-related behavior in rodents. Nat Protoc 2:322–328. https://doi.org/10.1038/nprot.2007.44

Chapter 16

In Vivo Preclinical Platforms to Study Coronavirus Infection

Larisse de Souza Barbosa Lacerda, Rafaela das Dores Pereira, Natália Nóbrega, Jordane C. Pimenta, Vinícius Amorim Beltrami, Manoela Gonzaga Gontijo do Couto, Luiz Pedro de Souza-Costa, Filipe Resende, Celso Martins Queiroz-Junior, Mauro Martins Teixeira, and Vivian V. Costa

Abstract

The COVID-19 pandemic caused by SARS-CoV-2 has increased the attention to viral diseases due to the global rapid dissemination, high transmissibility, and potential lethality. In these situations, the development of pre-clinical models is crucial to better understand the molecular aspects of the pathogenesis of such condition. Thus, animal models of *Betacoronavirus* infection that mimic disease features observed during COVID-19 in humans are fundamental tools to uncover disease mechanisms. In this chapter, we describe in detail the animal models and the pathogenesis of the disease caused by different Betacoronaviruses in rodents. The main differences between the murine models used for coronavirus studies are discussed, such as the mouse model for mouse hepatitis virus (MHV) infection as well as humanized model (K18-hACE2), as well as hamster models for infection with SARS-CoV-2. Methods concerning virus manipulation and animal infection are also described in detail, along with important notes on different techniques to assess clinical outcomes and protocols of sample collection for further biochemical, immune, microbiological and histopathological examination.

Key words Animal model, Viral infection, MHV, SARS-CoV-2, COVID-19

1 Introduction

The World Health Organization declared in March 2020 the coronavirus disease 19 (COVID-19) as a pandemic [1]. The outbreak originated in Hubei province, in China, spread throughout the world, and promoted health and socioeconomic impacts, as it exposed vulnerable populations, overloaded health systems, and caused a high rate of deaths [1, 2]. The disease consists of a severe

The authors Larisse de Souza Barbosa Lacerda and Rafaela das Dores Pereira have been equally contributed equally to this work.

Lucas Secchim Ribeiro (ed.), *Models in Inflammation and Immunopharmacology*, Methods in Molecular Biology, vol. 2993, https://doi.org/10.1007/978-1-0716-5019-6_16,
© The Author(s), under exclusive license to Springer Science+Business Media, LLC, part of Springer Nature 2026

acute respiratory syndrome, followed by systemic commitment caused by Severe Acute Respiratory Syndrome Coronavirus 2 virus (SARS-CoV-2) [3, 4]. SARS-CoV-2 infection and its pathogenicity in humans were initially reported in the lungs [5], but later studies identified vulnerability to this coronavirus infection in other organs, including the liver, brain, kidneys, and intestine [6]. Thus, COVID-19 is now considered an acute and chronic multisystemic syndrome [7].

Overall, the infected individual may remain asymptomatic or experience mild to moderate symptoms in ~75–80% of cases, including fever, dry cough, dyspnea, chest pain, fatigue, myalgia, headache, dizziness, abdominal pain, diarrhea, nausea, and vomiting [8–10]. A portion of those infected (~20–25%) progress to severe conditions with complications such as pneumonia, drop in oxygen saturation, hypoxemia, acute respiratory syndrome, arrhythmia, shock, acute heart injury, acute kidney injury, and thrombolytic events [3, 4, 10]. The public health interventions for COVID-19 consisted of vaccination, testing, and also the use of personal protective equipment (PPE), personal hygiene practices, municipal lockdowns, travel restrictions, and physical distancing [11]. The challenge of containing the pandemic was compounded by the emergence of new variants of SARS-CoV-2 that are more infectious and resistant to available therapies [7]. Additionally, the distribution of vaccines, medications, and tests to nations is not equanimeous, exposing vulnerable populations and compromising the control and erradication of the disease [12].

The study of COVID-19 in experimental models is challenging, given that SARS-CoV-2 does not infect wild-type (WT) mice. Thus, strategies such as the use of transgenic mice expressing the human receptor for SARS-CoV-2 (hACE2) and the use of hamsters, which are susceptible to the infection, have been employed. Due to SARS-CoV-2, these animal models must be performed in Biological Safety Level 3 (BSL3) facilities, thereby, other coronaviruses were proposed to mimic COVID-19 experimentally in Biological Safety Level 2 (BSL2) facilities, which are more viable and affordable.

The Mouse Hepatitis Virus (MHV) infection is used as a mouse model for diseases such as multiple sclerosis and hepatitis [12, 13]. At least in part, this is due to the different tropisms that each strain of MHV display. For instance, the strains MHV-1 to MHV-3 and MHV-A59 exhibit a pulmonary tropism, which initiates with the replication of the virus in nasal respiratory and olfactory epithelium, causing mild necrosis of these tissues and consequently disseminating to different organs, such as lung, liver, spleen, and bone marrow [14–16]. The tropism for the lungs that some strains can exhibit makes MHV infection a suitable model for mirroring some aspects observed during COVID-19 in humans. As showed by Yang and colleagues, intranasal infection of C57/BL6 mice with MHV-A59 caused acute pneumonia with typical markers of severe inflammation, such as production of TNF and IL-1β, hemorrhages, and fibrosis of alveolar walls [17]. In addition, Andrade and co-workers showed that intranasal infection with MHV-3 causes acute lung lesion and production of pro-inflammatory mediators in the lungs, with alterations of mechanical and ventilatory parameters and coagulopathy. Still, mice also exhibit viremia and organ viral spread, which lead to death at approximately 5 days post infection [18]. The MHV infection model presents some differences in coagulopathy, cellular infiltrate profile in the lungs and receptor for viral adsorption compared with SARS-CoV-2 infection. Therefore, caution must be taken concerning the use of the MHV model as a platform to mirror all aspects of COVID-19 manifestations. Despite these limitations, the use of MHV as a platform to evaluate the potential of anti-inflammatory drugs may be interesting, as these drugs act on a common axis of the inflammatory response elicited in both humans and mice.

2 Materials

2.1 Viral Infection in Mice (MHV or SARS-CoV-2) or Hamster (SARS-CoV-2)

1. Male or female mice (C57BL/6J, SV129, BALB/C, K18h-ACE2), weighing 18–20 g, from 5 to 8 weeks old (*see* **Note 1**).
2. Male or female Syrian hamsters, weighing 100–150 g, from 5 to 8 weeks old (*see* **Note 2**).
3. Tubes containing aliquots with a known titer of MHV-3 (GenBank MW620427), MHV-A59 (ATTC-VR 764) (*see* **Note 3**) or SARS-CoV-2 virus (GenBank MT710714) (*see* **Note 4**).
4. Sterile aqueous solution of 0.9% NaCl (w/v).
5. Sterile 1.5- or 2.0-mL tubes.
6. Pipettes and sterile and tips (1000, 200, 20 μL).
7. Vortex.
8. 1-mL Syringe and 30G needle (mice), 26G needle (hamster).
9. Anesthetic solution for (a) *mice*: 1.8 mL of 10% ketamine hydrochloride (w/v) + 1.5 mL of 2% xylazine (w/v), q.s. p 10 mL of sterile saline, so that, when intraperitoneal injecting 100 μL of the solution, a mouse weighing 20 g receives 90 mg/kg of ketamine hydrochloride and 15 mg/kg of xylazine; and (b) *hamsters*: 10 mL of 10% ketamine hydrochloride (w/v) + 2.5 mL of 2% xylazine (w/v), q.s.p 10 mL of sterile saline, so that, when intraperitoneal injecting 250 μL of the solution, a hamster weighing 100 g receives 200 mg/kg of ketamine hydrochloride and 10 mg/kg of xylazine.

10. Animal facilities with certification for BSL-2 (MHV) (*see* **Note 5**) or BSL-3 (SARS-CoV-2) (*see* **Note 6**).
11. Clean microisolator cages.
12. Weight scale.
13. 70% Ethanol.
14. Pen and Paper.
15. Hemocytometer, or Neubauer chamber and microscope slide for blood smear.
16. Anticoagulant for blood collection (700 mM EDTA).
17. Microtubes pre-treated with EDTA for blood collection.
18. Cooler/box with ice.

2.2 Clinical Score

1. Score table.
2. Animals, identified by the tail or ear.
3. Clean microisolator cages.

2.3 Euthanasia

1. Anesthetic solution: as stated in Subheading 2.1.
2. Microtubes pre-treated with EDTA for blood collection.
3. 2.0-mL sterile microtubes for tissue collection.
4. A box suitable for ice or liquid nitrogen.
5. Sterile scissors and tweezers.
6. 1.0-mL syringe.
7. 22G and 30G needles.
8. Laminar flow hood.
9. Histological cassette for organs.
10. 70% Ethanol.

2.4 Hematological Alterations

1. Anesthetic solution: as stated in Subheading 2.1.
2. Sterile scissors and tweezers.
3. 1-mL Syringe.
4. 22G, 26G, and 30G needles.
5. Microtubes pre-treated with EDTA for blood collection.
6. EDTA (700 mM).
7. Pipette and tips (100 or 200 μL).
8. Petri dish.
9. Celltac MEK-6500 K hemocytometer (if available).

2.5 Viral Titration

1. Cell lines in culture: L929 (ATCC-CCL1) for MHV model or Vero (ATCC-CCL81) for SARS-CoV-2 model.
2. 24-well plate.

3. Trypsin/EDTA: 0.25% (w/v) trypsin and 0.02% (w/v) EDTA, at 37 °C.
4. Sterile saline (0.9% NaCl, w/v) or sterile PBS.
5. Incubator with humidified atmosphere of 5% CO_2, at 37 °C.
6. Racks for tubes.
7. 50-mL tubes.
8. Pipettes 1000, 200, 100 μL and tips.
9. Fetal Bovine Serum (FBS).
10. Dulbecco's Modified Eagle's Medium (DMEM), supplemented with 2 or 10% FBS (for L929 cells).
11. Roswell Park Memorial Institute (RPMI), supplemented with 2 or10% FBS (for Vero cells).
12. Solution of 0.4% (w/v) of carboxymethylcellulose (CMC) in DMEM, supplemented with 2% of FBS (v/v) (*see* **Note 7**).
13. Solution of 0.6% (w/v) of carboxymethylcellulose (CMC) in DMEM, supplemented with 2% of FBS (v/v) (*see* **Note 7**).
14. Automatic pipetting device and serological pipettes.
15. Containers for disposal of solids and liquids.
16. Vortex.
17. Water bath.
18. Tissue samples homogenizer.
19. Ice box.
20. Calculator.
21. Timer.
22. 1.5- and 2.0-mL tubes.
23. Inverted microscope.
24. Neubauer chamber.
25. 10% buffered formaldehyde (v/v): dilute formaldehyde in PBS, adjust the pH to 7.2.
26. Violet crystal 0.1% (w/v), diluted in 10% buffered formaldehyde solution.
27. Pen.

3 Methods

3.1 Viral Infection in Mice or Hamster (MHV or SARS-CoV-2)

1. In the BSL-2 or BSL-3 facility, prepare the class II type A1 biological safety cabinet creating a sterile environment using UV light for 15 min before starting the experiment and placing all the needed material inside it.

2. Take a microisolator cage containing the mice or hamsters and weigh the animals inside the laminar flow hood.
3. Take the viral stock solution from the −80 °C freezer, transport it into a box of ice to the laminar flow hood, allow it to thaw (~ 5 min) and then vortex the vial for 30 s.
4. Perform the adequate dilution of the virus (in accordance with the required inoculum—*see* **Notes 8** and **9**) in sterile saline.
5. Vortex the diluted inoculum and keep it in a cooler/ice with box until the infection of the animals.
6. Take a microisolator cage containing the animals into the laminar flow hood.
7. Attach a 1-mL syringe with a needle (30G for mice or 26G for hamster) and fill it with the appropriate anesthetic volume (*see* **Note 10**).
8. Take an animal and immobilize it (*see* **Note 11,** Fig. 1/*see* **Note 12**, Fig. 2). With your dominant hand and the syringe at an angle of 45 degrees insert the needle through the animal's skin with the bevel facing up and intraperitoneally inject the anesthetic (Fig. 3).
9. Put the animal back to the microisolator and wait the anesthetic effect (around 5 min).

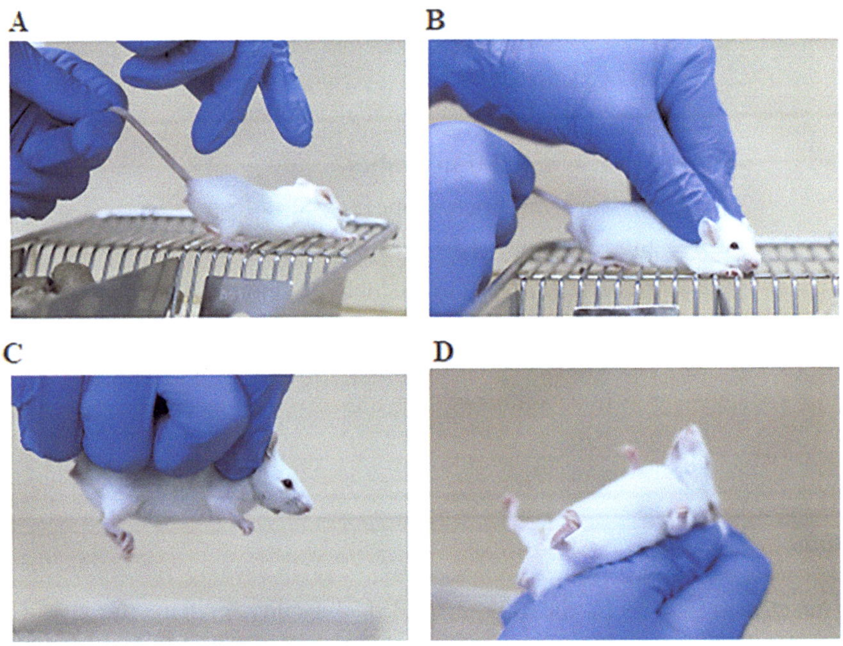

Fig. 1 Immobilization procedure in mice. Restraining mice is important to perform the procedure of the treatment and anesthesia. Adequate immobilization reduces animal stress and promotes the experiment safer. (**A-D**) Steps of mice immobilization

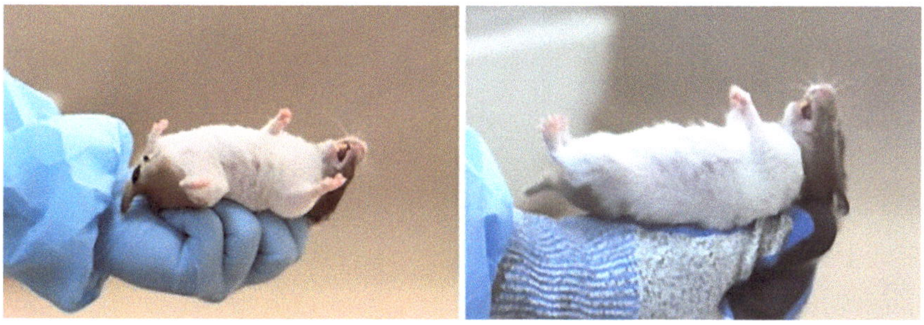

Fig. 2 Full hand restraint of hamsters. Restraining hamsters can be challenging for inexperienced researchers and this is a necessary procedure to anesthetize and treat the animals. The photos exemplify a way of restraining the hamsters using surgical gloves (**a**) or protection gloves (**b**)

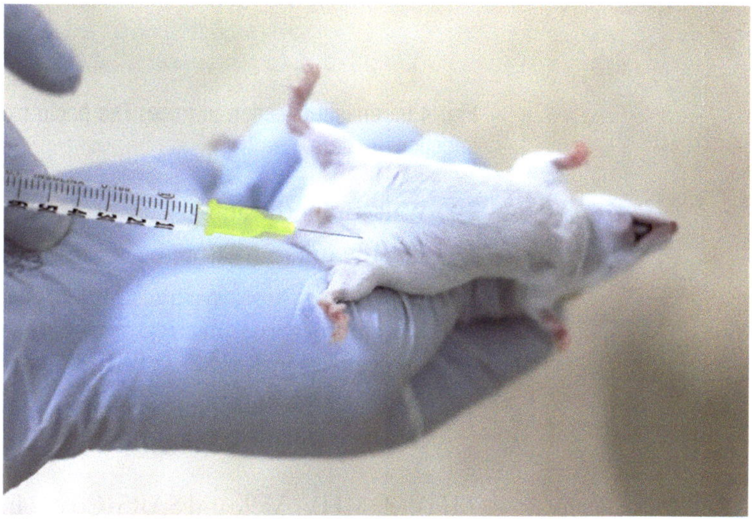

Fig. 3 Anesthesia in mice. After restraining mice, position the mouse for the realization of the anesthesia, according to the figure. To optimize the procedure, incline down the mouse. Thereby, the intestine and viscera dislocate and reduce the risk of damage to these tissues

10. Take the anesthetized animal in a vertical position with one hand and, slowly pipette 15 μL (mice)/50 μL (hamster) of the virus solution into the animal's nostril in order to allow it to aspirate each drop of the solution (*see* **Note 13,** Fig. 4).
11. Wait 2 minutes before repeating the previous step to avoid drowning. Each animal should be inoculated with a total of 30 μL (mice)/100 μL (hamsters).
12. Put the animals back to the cage. Disinfect the external surfaces of the microisolator.
13. Repeat the steps from 6 to 12 until all animals are infected. Control animals should receive sterile saline instead of the virus (*see* **Note 14**).

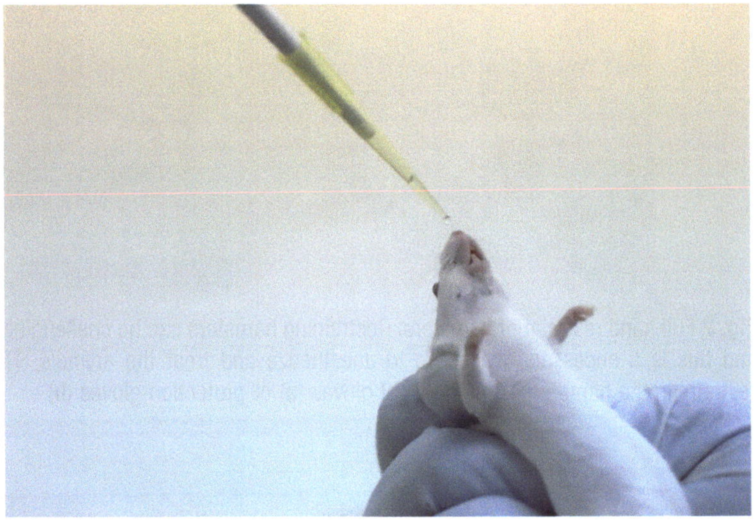

Fig. 4 Intranasal infection in mice. The photo exemplifies the procedure of the infection

14. Wait for all the animals are awake from the anesthesia.
15. Put the microisolators back into the ventilated racks.
16. Discard the tips, syringes and tubes in appropriate recipients containing virus inactivators, such as sodium hypochlorite, as preconized in BSL-2 or BSL-3 facility standards.
17. Disinfect the biological safety cabinet with 70% ethanol and irradiate it with UV light for 15 min.

3.2 Clinical Score (MHV or SARS-CoV-2)

MHV-3, MHV-A59, and SARS-CoV-2 infection results in systemic clinical manifestations (*see* **Note 15**). Follow the steps below to investigate a score based in different evaluation parameters.

1. Put each animal in a clean microisolator cage and evaluate each of them visually regarding the parameters in Table 1. The experimental endpoint is reached when the weight loss exceeds 20% of the initial value.
2. Return the animal to the original cage and repeat the process to all the other animals.

3.3 Euthanasia

1. In the BSL-2 or BSL-3 facility, prepare biological safety cabinet by placing all the needed materials inside it and creating a sterile environment using UV light for 15 min before starting the experiment.
2. Fill a syringe with anesthetic solution in the volume of 150 μL (for mice) or 750 μL (for hamsters).
3. With your non-dominant hand, grasp the animal skin from the back in order to immobilize the animals' head and body (*see*

Table 1
Clinical score

Clinical parameter	Score (value)			
	Absent (0)	Mild (1)	Moderate (2)	Severe (3)
Lethargy	Normal exploratory behavior. Rapidly moves in response to auditory or touch stimuli	Limited exploratory behavior to only one or two microisolator corners. Slowly moves in response to auditory or touch stimuli	Stationary behavior. No response to auditory or touch stimuli	–
Ruffled fur	Smooth fur	Ruffled facial fur	Ruffled dorsal body fur	–
Rushed posture	No dorsal curvature	Mild to moderate dorsal curvature. Hind paws are placed closer to the front paws	Severe dorsal curvature. Hind paws almost touches front paws	–
Eye appearance	Open eyes without secretion	Open eyes with little secretion	Half-closed eyes with secretion	Fully closed eyes with secretion
Cachexia	No evident loss of muscle. Dorsal bones cannot be felt when you handle the mice	No evident loss of muscle. Dorsal bones can be slightly felt when you handle mice	Evident loss of muscle. Dorsal bones are palpable but not prominent	Evident loss of muscle. Dorsal bones are prominent
Breathing	Normal breathing with coordinated cycles and no arrhythmia	Sporadic faster breathing periods followed by normal breathing cycles	Shortness of breath with cycles faster than normal. Usually, rustles can be heard.	–

Notes 10 and **13**). With your dominant hand, inject the anesthetic intraperitoneally in the animal.

4. Wait for the animal to become completely anesthetized. For that, moderately press the paw and check any movement or reflex.
5. Expose the abdominal cavity through a median incision starting from the lower abdominal region and ending at the inferior portion of the animal's xiphoid process of the sternum.
6. Find the inferior vena cava in the posterior part of the peritoneal cavity and, using a 22G needle coupled to a syringe pre-treated with EDTA, start to draw the blood of the inferior vena cava. Transfer the collected blood to a tube pre-treated with EDTA.

Table 2
Techniques and assays usually employed

	Tissue specimens							
Assay methods	**Brain**	**Spleen**	**Liver**	**Blood**	**Plasma**	**Testicles**	**Heart**	**Optic nerve**
H&E staining	X	X	X			X	X	
qPCR	X	X	X	X	X	X	X	X
ELISA	X	X	X		X	X	X	
IHC	X	X	X			X	X	
Myeloperoxidase activity	X	X	X					
Transaminase activity	X				X			
Indocyanine green clearance			X					
Hemocytometer				X				
Viral titration	X	X	X	X	X	X	X	X

7. If the removal of other tissues is desired, prioritize the brain and then collect other organs. A list of suggested tissue specimens to be collected and the techniques that can be applied to them can be found in Table 2. For those cases, transfer each tissue to a pre-identified 2-mL microcentrifuge tube and place the tube with the organ on ice or liquid nitrogen until all the collection is done.
8. Although the exsanguination via vena cava often guarantees the euthanasia, make sure the animal is dead. Then, properly discard the carcasses using the approved regulations of the experimental facility.
9. After the experiment is finished, sanitize the cages, clean the workbench and the laminar flow with alcohol 70%, then use UV light for 15 minutes. Clean and sterilize the surgical materials.

3.4 Hematological Alterations

The infection by MHV-3, MHV-A59, and SARS-CoV-2 results in strong hematological alterations. Of note, mice develop progressive thrombocytopenia and leukopenia, the last as a result of lymphopenia. Therefore, changes in total and differential leukocyte counts help us understand the severity of the infection. For this analysis, we collect blood from the mice's portal vein and use the automated hemocytometer (*see* **Note 16**) to subsequent analyses:

1. After euthanasia, gently move the intestines and liver aside to better visualize the posterior vena cava. After moving the organs, visualize the widest portion of the posterior vena cava, located between the kidneys.

2. Flush a 1-mL syringe with EDTA multiple time to avoid clotting during collection. Insert a 22G-needle attached to the syringe into the vein, with the bevel facing up, at an angle of approximately 15 degrees.
3. Draw blood very slowly and once done, slowly dispense the blood in a microtube flushed with EDTA.
4. With the automated hemocytometer, aspirate 30 μL of the blood and perform total and differential leukocyte count. Manual counting is also a valid alternative.

3.5 Viral Titration

3.5.1 Preparation of the Cell Monolayer

1. The viral titration requires a 24-well plate containing L929 (MHV) or Vero (SARS-CoV-2) cells at a density of 1×10^5 cells/well, forming a homogeneous cell monolayer. Here are the details for this procedure.
2. Prepare biological safety cabinet by placing all the needed materials inside it and creating a sterile environment with UV light for 15 min before starting the experiment.
3. Take the flasks containing the L929/Vero cells currently in culture.
4. In an inverted microscope, verify the viability of the cells and their confluence, ideally at 80%.
5. Remove the supernatant from the flasks and add 5 mL of saline sterile, close and wash the cell monolayer. Discard the liquid and repeat the process.
6. Then, add 2 mL of trypsin/EDTA solution in the flask and distribute the reagent for all over the cell monolayer to release the cells. Incubate for 5 minutes at 37 °C.
7. Next, add 10 mL of DMEM (for L929) or RPMI (for Vero) supplemented with 10% FBS and wash the internal surface of the flask to detach and homogenize the cell suspension.
8. Collect the cell supernatant and transfer it to a 50-mL tube.
9. Count the cells using a hemocytometer/Neubauer chamber.
10. Adjust the cell density with supplemented medium in order to achieve 1×10^5 cells/mL in a 24-well plate and add 1 mL of the cell suspension per well in a 24-well plate.
11. Incubate the 24-well plate at 37 °C, with 5% CO_2 for 24 h.

3.5.2 Viral Titration

1. Prepare biological safety cabinet by placing all the needed materials inside it and creating a sterile environment with UV light for 15 min before starting the experiment.
2. Make sure the entire surface of each well in the plate prepared on the previous day is covered by a cell monolayer (100% confluence).

3. Fluid samples like plasma or whole blood can be used directly or diluted. Save at least 50 uL of the sample for the titration.
4. Solid samples, like tissues, must undergo homogenization to ensure that the virus present in the cells is in suspension. For that, weigh an empty, sterile tube, transfer the tissue to it, and weigh the tube again to obtain the weight of the tissue. Next, add 9 times (mg to uL) the weight of the tissue of medium supplemented with 2% FBS (e.g., 100 mg of tissue + 900 uL of supplemented medium) and mechanically dissociate the tissue until a homogeneous suspension is achieved, without clumps;
5. For samples of both solid and fluid origin, make 5 serial dilutions per sample in a 1:10 proportion using supplemented medium as diluent, vortexing the source tube and changing the tips for each dilution. Keep the original samples and their dilutions on ice or in the fridge.
6. Remove the plates with the cells from the incubator and, in a sterile hood, remove all the supernatant, leaving only a thin layer of medium in each well so as not to dry out the monolayers.
7. Next, gently add 100 uL of every sample (stock +5 serial dilutions) to its respective well, from the most diluted to the most concentrated. Keep the rest of the samples in the −80 °C freezer until the conclusion of the viral titration.
8. Take note of the plate identification and which well was used for each dilution.
9. Then, gently tilt the plate around to make sure the inoculum covers the entire well.
10. Incubate the plate at 37 °C for 1 hour to promote viral adsorption. Carefully move the plate in circular motion to promote adequate viral adsorption throughout the well.
11. After that, carefully add 1 mL of DMEM medium containing 0.4% of CMC (for L929 cells) or 0.6% CMC (for Vero cells).
12. Keep the plate in the incubator for 48 hours.
13. At the end of the incubation period, add 2 mL of 10% buffered formaldehyde solution to each well and wait for a minimum of 1 hour at room temperature for a complete fixation.
14. Carefully, completely discard the fluid on top of the fixed cells and add 1 mL of 1% violet crystal (w/v), incubating at room temperature for 30 min.
15. Next, remove the dye and gently wash the plates with water.
16. Allow the wells to completely dry at room temperature before proceeding to counting.
17. Finally, count the number of plaques in each well for each sample. Multiply the number of plaques by the inverse of the

sample dilution factor (e.g., 10^1, 10^2, 10^3, 10^4) and again by 10 to account for the initial dilution from the original sample and convert it to PFU/mL.

4 Notes

1. Importantly, the severity of the disease caused by murine beta-coronaviruses might differ depending on the animal lineage. C57/BL6J and SV-129 mice are more resistant to the viruses when compared to BALB/C animals, so one can use such differences to mimic susceptible and/or resistant individuals. The age of the animals is also another relevant point, since younger mice (4–5 weeks) are more susceptible when compared to mice that are considered adults (8–12 weeks). Therefore, the virus strain, the animal lineage, and its age must be taken into account to mimic the study model that is of interest to you. Since the expression of angiotensin-converting enzyme 2 receptors (ACE2) is limited in mice, experimental strategies were designed to promote suitable murine models for the development of SARS-CoV-2 infection. For instance, the expression of human ACE2 in genetically modified mice (K18-hACE2 mice) is driven by K18 promoter that has transgenic efficiency to induce hACE2 expression in epithelial cells including in the liver, kidney, and gastrointestinal tract. A comparative table of the different models of coronavirus infection can be found in Table 3.
2. Syrian hamsters (*Mesocricetus auratus*) naturally express ACE2 with high affinity to the SARS-CoV-2 spike glycoprotein receptor-binding domain [19, 20]. Therefore, these animals are naturally susceptible to SARS-CoV-2 infection. When infected, the hamsters develop clinical symptoms closely related to the symptoms of COVID-19 patients, such as rapid breathing and weight loss, besides the histopathological signs of lung inflammation associated with significant local viral titers [21]. Although dependent on the viral inoculum, the infection is self-limited and does not cause severe disease and death of the animals.
3. Animals infected intranasally with MHV-3 display significant weight loss, thrombocytopenia, leukopenia, and lung damage at the third day post infection of an inoculum of 10^3 PFU. Lung injury is characterized by an increase in specific leukocyte populations in the peribronchiolar region, such as polymorphonuclear and mononuclear cells, alveolar hyperplasia, and hemorrhagic foci. Such pulmonary changes are accompanied by an increase in cytokines (TNF, IL-1-beta, IL-6, IL-10, IFN-γ) and chemokines (CXCL-1, CCL2, CCL3, CCL4, and CCL5) [18]. After 3 days after infection, it is observed that the

Table 3
Comparative table of the 3 models of coronavirus infection

	Viral models		
Features	**MHV-3**	**MHV-A59**	**SARS-CoV-2**
Virus genus	Betacoronavirus	Betacoronavirus	Betacoronavirus
Receptor	CEACAM-1	CEACAM-1	ACE-2
Animal models	Mice (C57BL/6, BALB/C, Sv129)	Mice C57BL/6	Mice K18-hACE-2 and hamsters
Biosafety level	BSL-2	BSL-2	BSL-3
Route of viral inoculation	Intranasal (15 μL/nostril)	Intranasal (15 μL/nostril)	Intranasal (15 μL/nostril in mice and 50 μL/nostril in hamster)
Higher tropism	Liver	Lung	Lung
Extrapulmonary damage	Hepatitis and liver necrosis. Brain, small intestine, and colon had minimal to mild leukocyte infiltration. Testicular damage	Mild infiltration of leukocytes in the liver. Mild blood vessel activation and gliosis in the brain cortex	Gliosis and necrosis in the brain in K18-hACE2 mice. Hamsters present only mild SNC alterations
Clinical score	Accentuated	Mild	Accentuated in mice, mild in hamsters
Lethality	High (~ 100%) Inoculum 10^3 PFU/mL	Absent or low (~10%) Inoculum 10^4 PFU/mL	High (~100%) in mice, absent or low in hamsters Inoculum $10^4/10^5$ PFU/mL
Peak of lung disease	3 dpi Inoculum: 10^3 PFU/mL	5 dpi Inoculum: 10^4 PFU/mL	5 to 7 dpi Inoculum: $10^4/10^5$ PFU/mL
Type of COVID-19 mimics	Severe	Mild	Severe/moderate
Cell types- viral propagation	L929 (ATCC CCL-1)	L929 (ATCC CCL-1)	Vero (ATCC CCL81)

damage caused by the virus is not restricted to the lung, causing systemic damage, such as liver damage, increased plasma cytokines, and dissemination of the virus to various tissues, leading to the death of the animals by day 6 post infection [18]. MHV-A59 infection mimics a milder aspect of COVID-19. It has a main tropism for the lung and brain, with increased leukocyte infiltration, hemorrhagic regions, and fibrosis in the alveolar walls observed 2 to 11 days after intranasal infection. In addition to these clinical manifestations, the infection induced by MHV-A59 provides an increase in important markers of inflammation such as TNF, IL-6, IP-10, IL-1β, IFN-β, and IFN-γ, despite no death being observed [17]. MHV-A59 is shown to be a model for study associated with a cell death mechanism known as panoptosis, contributing to the synergistic activation of several pathways of cell death such as apoptosis, pyroptosis, and necrosis [22].

4. SARS-CoV-2 is a betacoronavirus with positive-sense genome, unsegmented, enveloped RNA, containing single RNA strand [20, 23]. During the COVID-19 pandemic, the SARS-CoV-2 virus presented a plethora of mutations, which resulted in different variants. Among them, it is worth mentioning: B.1.1.7 (Alpha), B.1.351 (Beta), P1 (Gamma), B.1.617.2 (Delta), and B.1.1.529 (Omicron) [1, 24].
5. Procedures involving MHVs are performed in BSL-2 laboratories, as murine betacoronaviruses are considered class 2 risk pathogens. It means that the individual risk is moderate and the risk for the community is considered low. Microorganisms of this class might cause infections, but there are effective therapeutic and prophylactic measures that limit the risk of its propagation.
6. All the procedures involving SARS-CoV-2 must be performed in BSL-3 facilities. SARS-CoV-2 is classified as a class 3 risk (high individual risk and moderate risk for the community) pathogen. They represent a risk if disseminated in the community and the environment and can spread from person to person.
7. To prepare the carboxymethylcellulose-DMEM solutions, prepare equal volumes of 2× CMC and 2X DMEM (already supplemented with FBS and Penicillin-Streptomycin/ 10,000 U/mL). For example, to prepare the CMC 0.4% with-DMEM: add 4 g of CMC in 500 mL of distilled water, mix, autoclave and reserve; meanwhile, prepare a 2× supplemented DMEM: if a concentrated medium is not available, dilute 10.4 g of DMEM (powder with glutamine), 2.5 g of $NaHCO_3$, and 10 mL of Penicillin-Streptomycin (10,000 U/mL) in 500 mL of Milli-Q water. After complete dissolution, filter the solution through a 0.22 um filter and add 20 mL of FBS. Finally, mix the medium and the CMC solution until it is homogeneous.
8. To induce mild infection in mice by MHV-A59, use an inoculum of 3.0×10^4 PFU/mL and for a moderate/severe infection in mice by MHV-3, use an inoculum of 3.0×10^3 PFU/mL. It is recommended to calculate the final volume in accordance with the number of mice to be infected plus 2, to account for possible losses.

9. To induce a moderate/severe infection in hamsters we used the inoculum of 10^4 PFU per hamster. It is recommended to calculate the final volume in accordance with the number of mice to be infected plus 2, to account for possible losses.
10. Proper anesthesia of the animal is fundamental, as it allows the entire virus solution to be inoculated efficiently into the respiratory tract of the mouse, ensuring the correct inoculum of the virus. The process for sedation of the animal must be carried out carefully, generally with a volume of 100 μL (mice) or 200 μL (hamster) of anesthetic solution. It is important to mention that the younger and thinner the mice are, the greater care must be taken with anesthesia.
11. To immobilize the mouse, support the animal on the cage rail, then hold the tail by the base, gently pulling it backwards. Thus, the animal tends to move forward and hold the grid with its forepaws. At that time, the researcher will position the index finger and thumb involving the skin of the neck, near the ears, hold and carefully pull the fold of the skin in the opposite direction of the animal. The rest of the skin from the animal's back will be grabbed with the other fingers of the researchers. Hence, the mouse will present its ventral face facing upwards, and its tail will be trapped between the ring finger and the tip of the little finger. The containment should not be too tight, as it can make it difficult for the animal to breathe. Conversely, the containment cannot either be too loose, as accidents can occur (the animal can turn and bite or scratch the operator). Nevertheless, well-performed immobilization generates less animal stress (Fig. 1).
12. For the restraint of hamsters, the whole hand should be used to grasp the animal skin from the back in order to immobilize the animals' head and body. Protection gloves can be used to avoid accidents.
13. When the mouse/hamster is fully anesthetized, hold the animal as mentioned in **Notes 11** and **12** and position it vertically. Place the pipette tip containing 15 μL (mice)/50 μL (hamsters) of virus inoculum into the nostril and slowly eject the solution. Ensure that the mouse/hamsters are inhaling the preparation by observing the drop of inoculum entering the nostril of the animal (Fig. 4).
14. To prevent any possibility of contamination of the material used, such as pipettes and surfaces, it is advisable to start the experiment with the control group, which receives sterile saline. This will increase the accuracy and reliability of the experiment.
15. For greater reliability, the clinical score must be conducted by a blinded pathologist. The clinical score must be done in a quiet place, since sounds and movements close to the microisolator

cage can alter the animal's behavior and, therefore, the scoring. The endpoint criteria must be approved by the Animal Care and Use Committee (ACUC) and strictly followed during the evaluation of the clinical score.

16. While using an automated hemocytometer, aspirating directly from the EDTA tube might result in blood clot getting stuck and clogging the equipment. In order to avoid it, pipette 30 μL of unclotted blood in a clean tube and aspirate the blood. In the absence of a hemocytometer, the total leukocyte count in the blood can be performed using a Neubauer chamber with Turk's solution as counterstaining. The differential leukocyte count can be performed using the Panoptic kit for staining the whole blood smear performed on a glass slide.

Acknowledgments

We thank the contribution of Dr. Felipe Rocha da Silva Santos and Dra. Rayane Aparecida Nonato Rabelo for images.

Funding This study was funded by the Fundação de Amparo à Pesquisa do Estado de Minas Gerais (FAPEMIG), grant numbers RED-00202–22, APQ-02281–18, APQ-04650-23, APQ-02618-23, and APQ-04983-24; and by the Coordenação de Aperfeiçoamento de Pessoal de Nível Superior (CAPES/Brazil), through the project: CAPES—Programa 9951—Programa Estratégico Emergencial de Prevenção e Combate a Surtos, Endemias, Epidemias e Pandemias, AUX 0641/2020—Process 88881.507175/2020–01. This work also received support from the National Institute of Science and Technology in Dengue and Host-Microorganism Interaction (INCT Dengue), funded by the Conselho Nacional de Desenvolvimento Científico e Tecnológico (CNPq, Brazil) [grant number 465425/2014-3], and by FAPEMIG [grant number 25036]. Additional support was provided by FINEP—Financiadora de Estudos e Projetos under MCTI/FINEP—MS/SCTIE/DGITIS/CGITS [grant number 6205283BBB28-4F9C-AA65-808FE4450542].

References

1. World Health Organization (WHO) (2020) Origin of SARS-CoV-2. Available at: https://iris.who.int/server/api/core/bitstreams/17648610-ce1a-492e-afa1-625caee1e84a/content. Accessed 11 Oct 2025
2. Nicola M, Alsafi Z, Sohrabi C et al (2020) The socio-economic implications of the coronavirus pandemic (COVID-19): a review. Int J Surg (London, England) 78:185–193. https://doi.org/10.1016/j.ijsu.2020.04.018
3. Harapan H, Itoh N, Yufika A et al (2020) Coronavirus disease 2019 (COVID-19): a literature review. J Infect Public Health 13(5): 667–673. https://doi.org/10.1016/j.jiph.2020.03.019
4. Zhai P, Ding Y, Wu X et al (2020) The epidemiology, diagnosis and treatment of

COVID-19. Int J Antimicrob Agents 55(5): 105955. https://doi.org/10.1016/j.ijantimicag.2020.105955

5. Tay MZ, Poh CM, Rénia L et al (2020) The trinity of COVID-19: immunity, inflammation and intervention. Nat Rev Immunol 20(6): 363–374. https://doi.org/10.1038/s41577-020-0311-8
6. Puelles VG, Lütgehetmann M, Lindenmeyer MT et al (2020) Multiorgan and renal tropism of SARS-CoV-2. N Engl J Med 383(6): 590–592. https://doi.org/10.1056/NEJMc2011400
7. Raman R, Patel KJ, Ranjan K (2021) COVID-19: unmasking emerging SARS-CoV-2 variants, vaccines and therapeutic strategies. Biomolecules 11(7):993. https://doi.org/10.3390/biom11070993
8. Huang C, Wang Y, Li X et al (2020) Clinical features of patients infected with 2019 novel coronavirus in Wuhan, China. Lancet (London, England) 395(10223):497–506. https://doi.org/10.1016/S0140-6736(20)30183-5
9. Wang D, Hu B, Hu C et al (2020) Clinical characteristics of 138 hospitalized patients with 2019 novel coronavirus-infected pneumonia in Wuhan, China. JAMA 323(11): 1061–1069. https://doi.org/10.1001/jama.2020.1585
10. Zhu N, Zhang D, Wang W et al (2020) A novel coronavirus from patients with pneumonia in China, 2019. N Engl J Med 382(8):727–733. https://doi.org/10.1056/NEJMoa2001017
11. Phua J, Weng L, Ling L et al (2020) Intensive care management of coronavirus disease 2019 (COVID-19): challenges and recommendations. Lancet Respir Med 8(5):506–517. https://doi.org/10.1016/S2213-2600(20)30161-2
12. Rabaan AA, Mutair AA, Hajissa K et al (2022) A comprehensive review on the current vaccines and their efficacies to combat SARS-CoV-2 variants. Vaccine 10(10):1655. https://doi.org/10.3390/vaccines10101655
13. Singh M, Khan RS, Dine K et al (2018) Intracranial inoculation is more potent than intranasal inoculation for inducing optic neuritis in the mouse hepatitis virus-induced model of multiple sclerosis. Front Cell Infect Microbiol 8: 311. https://doi.org/10.3389/fcimb.2018.00311
14. Homberger FR, Zhang L, Barthold SW (1998) Prevalence of enterotropic and polytropic mouse hepatitis virus in enzootically infected mouse colonies. Lab Anim Sci 48(1):50–54
15. Barthold SW, Smith AL (1984) Mouse hepatitis virus strain – related patterns of tissue tropism in suckling mice. Arch Virol 81(1–2): 103–112. https://doi.org/10.1007/BF01309300
16. Körner RW, Majjouti M, Alcazar MAA, Mahabir E (2020) Of mice and men: the coronavirus MHV and mouse models as a translational approach to understand SARS-CoV-2. Viruses 12(8):880. https://doi.org/10.3390/v12080880
17. Yang Z, Du J, Chen G et al (2014) Coronavirus MHV-A59 infects the lung and causes severe pneumonia in C57BL/6 mice. Virol Sin 29(6):393–402. https://doi.org/10.1007/s12250-014-3530-y
18. Andrade ACDSP, Campolina-Silva GH, Queiroz-Junior CM et al (2021) A biosafety level 2 mouse model for studying Betacoronavirus-induced acute lung damage and systemic manifestations. J Virol 95(22): e0127621. https://doi.org/10.1128/JVI.01276-21
19. Chan JF, Zhang AJ, Yuan S et al (2020) Simulation of the clinical and pathological manifestations of coronavirus disease 2019 (COVID-19) in a Golden Syrian hamster model: implications for disease pathogenesis and transmissibility. Clin Infect Dis 71(9):2428–2446. https://doi.org/10.1093/cid/ciaa325
20. Imai M, Iwatsuki-Horimoto K, Hatta M et al (2020) Syrian hamsters as a small animal model for SARS-CoV-2 infection and countermeasure development. Proc Natl Acad Sci USA 117(28):16587–16595. https://doi.org/10.1073/pnas.2009799117
21. Lee AC, Zhang AJ, Chan JF et al (2020) Oral SARS-CoV-2 inoculation establishes subclinical respiratory infection with virus shedding in Golden Syrian hamsters. Cell Rep Med 1(7): 100121. https://doi.org/10.1016/j.xcrm.2020.100121
22. Karki R, Lee S, Mall R et al (2022) ZBP1-dependent inflammatory cell death, PANoptosis, and cytokine storm disrupt IFN therapeutic efficacy during coronavirus infection. Sci Immunol 7(74):eabo6294. https://doi.org/10.1126/sciimmunol.abo6294
23. Astuti I (2020) Severe acute respiratory syndrome coronavirus 2 (SARS-CoV-2): an overview of viral structure and host response. Diabet Metabol Syndrome 14(4):407–412. https://doi.org/10.1016/j.dsx.2020.04.020
24. Centers for Disease Control (CDC) (2021) COVID-19: SARS-CoV-2 variant classifications and definitions. https://www.cdc.gov/coronavirus/2019-ncov/variants/variant-classifications.html. Assessed 21 Feb 2023

Index

Lucas Secchim Ribeiro (ed.), *Models in Inflammation and Immunopharmacology*, Methods in Molecular Biology, vol. 2993, https://doi.org/10.1007/978-1-0716-5019-6,
© The Editor(s) (if applicable) and The Author(s), under exclusive license to Springer Science+Business Media, LLC, part of Springer Nature 2026

MIX
Papier aus verantwortungsvollen Quellen
Paper from responsible sources
FSC® C105338

If you have any concerns about our products,
you can contact us on
ProductSafety@springernature.com

In case Publisher is established outside the EU,
the EU authorized representative is:
Springer Nature Customer Service Center GmbH
Europaplatz 3, 69115 Heidelberg, Germany

Printed by Libri Plureos GmbH
in Hamburg, Germany